CARDIOVASCULAR SOFT TISSUE MECHANICS

Cardiovascular Soft Tissue Mechanics

Edited by

STEPHEN C. COWIN
Department of Mechanical Engineering,
New York Center for
Biomedical Engineering,
The City College,
New York, U.S.A.

JAY D. HUMPHREY
Department of Biomedical Engineering,
Texas A&M University,
College Station, TX, U.S.A.

Reprinted from *Journal of Elasticity*, Vol. 61, Nos. 1–3 (2000)

KLUWER ACADEMIC PUBLISHERS
Dordrecht / Boston / London

Library of Congress Cataloging-in-Publication Data

A C.I.P. Catalogue record for this book is available from the Library of Congress.

ISBN 1-4020-0220-3

Published by Kluwer Academic Publishers,
P.O. Box 17, 3300 AA Dordrecht, The Netherlands.

Sold and distributed in North, Central and South America
by Kluwer Academic Publishers,
101 Philip Drive, Norwell, MA 02061, U.S.A.

In all other countries, sold and distributed
by Kluwer Academic Publishers
P.O. Box 322, 3300 AH Dordrecht, The Netherlands.

Printed on acid-free paper

Printed in the Netherlands

TABLE OF CONTENTS

Foreword

This special volume of the *Journal of Elasticity* represents the first in a new program dedicated to the occasional publication of collections of invited, reviewed papers of topical interest. The purpose of this program is to spotlight the developments and applications in the mechanics of materials within specific areas that can enhance growth and provide insight for the advancement of the field as well as promote fundamental understanding and basic discovery.

Soft Tissue Mechanics is an area of biomechanics that draws heavily upon fundamental ideas and material models from nonlinear elasticity and viscoelasticity. A major goal of this research is to understand those mechanics properties of heart, artery, collagen and skeletal muscle tissue that can be used for the diagnosis of health problems and the improvement of human life. This volume illustrates how experiment, modeling and computation is currently employed in this emerging field.

May 2001

ROGER FOSDICK
Editor-in-Chief

Journal of Elasticity **61:** ix–xii, 2000.

Preface

There are two primary areas for the application of elasticity in the biomechanics of tissues: hard tissue mechanics (e.g., bone, teeth, horns, etc.) and soft tissue mechanics (e.g., skin, tendons, arteries, etc.). The distinguishing feature between these tissue types is the amount of physiological "normal" deformation they experience. While "hard" tissues only experience small deformations, soft tissues typically experience large deformations. From a biomechanics viewpoint soft tissues fall within the realm of finite elasticity.

The seven papers of this volume [1–7] present a glimpse into current research on soft tissue mechanics as well as some future directions. The seven papers concern tissues within the cardiovascular system: three focus on arteries, three on the heart, and one on biaxial testing of planar tissues such as heart valves. Given that cardiovascular disease continues to be the leading cause of death in the developed world, the importance of such research is clear.

Table I. Topics contained in the seven papers. A single X indications that a topic is discussed and a double X indicates that a topic is emphasized

TOPIC /Paper	Holzapfel, Gasser & Ogden [1]	Humphrey & Canham [2]	Rachev [3]	Nash & Hunter [4]	Usyk, Mazhari & McCulloch [5]	Taber & Perucchio [6]	Sacks [7]
Arteries	XX	XX	XX				
Heart				XX	XX	XX	
Experimental methods	X	X	X	X	X		XX
Theoretical developments	X	X	X	X	X	X	X
Hyperelastic constitutive models	X	X	X	X	X	X	X
Orthotropic symmetry	X	X	X	X	XX	Trans Iso	X
Composite/micro-structural models	XX	X	XX	X	X	X	XX
Collagen	X	X	X	X	X		X
Structure/function	X	X	X	X	X	X	
Morphogenesis						XX	
Growth		XX		X		XX	
Remodeling		XX	XX			XX	
Residual stress	X			X			
Visco- & poro-elasticity				X			X
Constitutive models for active stresses			X		X	X	

Table I reveals specific topics that are covered in each of the seven papers and many of the common assumptions. Note, in particular, that most of the proposed constitutive relations are motivated directly by data on the underlying microstructure, and especially the orientations of a structurally important protein (collagen) that forms as undulated cross-linked fibers. For this reason, orthotropy and, to a lesser extent, transverse isotropy are often assumed. Note, too, that each paper either reports or relies on multiaxial mechanical data. Hence, consistent with the known importance of structure-function relations in physiology, these papers reveal that it is essential to combine microstructural information with mechanical data in soft tissue mechanics. Whereas it has been known since the early 1960s that rigorous experimental determination of constitutive (response) functions and values of the associated material parameters is extremely challenging in nonlinear elasticity and viscoelasticity [8], the structurally-motivated approach has become the method of choice – it was championed early on by Lanir [9] but now appears in many forms.

Although many aspects of soft tissue mechanics can be regarded as simply applied finite elasticity, there are numerous characteristics that require significant extensions of the traditional theory. For example, given that both arteries and the heart contain muscle, there is a need to quantify the so-called active (contractile) response in addition to the passive (non-contractile) response. Such relations must not only be structurally motivated, they must ultimately include the kinetics of calcium transport in the muscle. Constitutive relations for active behavior are discussed in [3–6]. Despite good progress, this area remains a significant challenge. Over the last twenty years, separate advances in biochemistry, cell biology, genetic engineering, and biomechanics have focused attention on the ubiquitous role of growth and remodeling of tissues. This general topic is discussed in [2–4, 6]; see, too, the review by Taber [10]. Although Skalak [11] emphasized the importance of growth mechanics nearly twenty years ago, and noted that incompatible growth would result in residual stresses, only recently has this topic found heightened interest. Not only does a living tissue continually "turnover" via a delicate balance between the production of new material and removal of old (i.e., the process of maintenance), it is often subject to changing loads – mechanical, chemical, thermal, etc. In response to these changes, there may be changes in the rates of turnover of the various constituents. This can lead to atrophy (loss of material), fibrosis (gain of excessive fibrous material), or hypertrophy (overall increase in mass of the material). This general area remains a significant challenge, and there is a pressing need for more experimental data and new theoretical ideas.

In order for the microstructural constitutive theories to provide accurate predictions of overall tissue response they should incorporate microstructural morphological data at the fiber and bulk tissue level. There are demonstrations of the failure to do this in the biomechanics literature in that one soft tissue constitutive model that matches the data of one experimental investigation does not match the data of a second experimental investigation. We suggest that the failure of the modeling match in the two situations is likely due to the non-incorporation of microstructural morphological data at the fiber and bulk tissue level. Such data are likely to be unique for each specimen because of the adaptive capacity of living tissues.

One of the most obvious consequences of growth and remodeling is that of micro-evolutionary adaptations – that is, refinements in the structure and mechanical properties of a particular tissue in response to a particular need. For example, whereas a person may develop a callus on the hand after sustained manual labor (a short term adaptation), the skin forming the soles of the feet is already thickened in the human fetus "in preparation" for its intended function (a long term adaptation). In both cases, of course, it is the genes that allow such adaptations to occur in response to, in this case, mechanical loads. Understanding better how such mechanical information is transduced by the cells – that is, processed and programmed – is one of the most exciting and important challenges in soft tissue mechanics.

There have been many advances in biomechanics over the last 35 years, and much has been learned. Nonetheless, it appears that we are at a unique time in the history of this field, one that promises major advances in our ability to understand and thus control tissue mechanics. The implications of this are far-reaching – from better prevention of injury, to better diagnosis of disease, to improved treatment of disease via both traditional interventions as well as the engineering of a new class of replacement tissues. Fundamental to the needed advances, however, are extensions of the nonlinear theories of mechanics to include muscle activation, growth and remodeling via mechanotransduction mechanisms, solid-fluid interactions, etc. We hope that this volume serves, in part, to heighten interest of the elasticity community in open problems in the elasticity of living tissues and their role in advancing the quality of life.

STEVE COWIN

Department of Mechanical Engineering
New York Center for Mechanical Engineering
Convent Avenue at 140 Street
New York, NY 10031
U.S.A.

JAY HUMPHREY

Biomedical Engineering
233 Zachry Engineering Center
Texas A&M University
College Station, TX 77843-3120
U.S.A.

References

1. G.A. Holzapfel, T.C. Gasser and R.W. Ogden, A new constitutive framework for arterial wall mechanics and a comparative study of material models. *J. Elasticity* **61** (2000) 1–48.
2. J.D. Humphrey and P.B. Canham, Structure, mechanical properties, and mechanics of intracranial saccular aneurysms. *J. Elasticity* **61** (2000) 49–81.
3. A. Rachev, A model of arterial adaptation to alterations in blood flow. *J. Elasticity* **61** (2000) 83–111.
4. M.P. Nash and P.J. Hunter, Computational mechanics of the heart. *J. Elasticity* **61** (2000) 113–141.

5. T.P. Usyk, R. Mazhari and A.D. McCulloch, Effect of laminar orthotropic myofiber architecture on regional stress and strain in the canine left ventricle. *J. Elasticity* **61** (2000) 143–164.
6. L.A. Taber and R. Perucchio, Modeling heart development. *J. Elasticity* **61** (2000) 165–197.
7. M.S. Sacks, Biaxial mechanical evaluation of planar biological materials. *J. Elasticity* **61** (2000) 199–246.
8. F.J. Lockett, *Nonlinear Viscoelastic Solids*. Academic Press (1972), Chapter 6.
9. Y. Lanir, A structural theory for the homogeneous biaxial stress–strain relationships in flat collageneous tissues. *J. Biomechanics* **12** (1979) 423–436.
10. L. Taber, Biomechanics of growth, remodeling and morphogenesis. *Applied Mechanics Reviews* **48** (1995) 487–545.
11. R. Skalak, G. Dasgupta, M. Moss, E. Otten, P. Dullemeijer and H. Vilmann, Analytical description of growth. *Journal of Theoretical Biology* **194** (1982) 555–577.

Journal of Elasticity **61:** 1–48, 2000.

A New Constitutive Framework for Arterial Wall Mechanics and a Comparative Study of Material Models

GERHARD A. HOLZAPFEL and THOMAS C. GASSER
Institute for Structural Analysis – Computational Biomechanics, Graz University of Technology, 8010 Graz, Schiesstattgasse 14-B, Austria

RAY W. OGDEN
Department of Mathematics, University of Glasgow, University Gardens, Glasgow G12 8QW, U.K.

Received 20 April 2000; in revised form 8 December 2000

Abstract. In this paper we develop a new constitutive law for the description of the (passive) mechanical response of arterial tissue. The artery is modeled as a thick-walled nonlinearly elastic circular cylindrical tube consisting of two layers corresponding to the media and adventitia (the solid mechanically relevant layers in healthy tissue). Each layer is treated as a fiber-reinforced material with the fibers corresponding to the collagenous component of the material and symmetrically disposed with respect to the cylinder axis. The resulting constitutive law is orthotropic in each layer. Fiber orientations obtained from a statistical analysis of histological sections from each arterial layer are used. A specific form of the law, which requires only *three* material parameters for each layer, is used to study the response of an artery under combined axial extension, inflation and torsion. The characteristic and very important residual stress in an artery *in vitro* is accounted for by assuming that the natural (unstressed and unstrained) configuration of the material corresponds to an open sector of a tube, which is then closed by an initial bending to form a load-free, but stressed, circular cylindrical configuration prior to application of the extension, inflation and torsion. The effect of residual stress on the stress distribution through the deformed arterial wall in the physiological state is examined.

The model is fitted to available data on arteries and its predictions are assessed for the considered combined loadings. It is explained how the new model is designed to avoid certain mechanical, mathematical and computational deficiencies evident in currently available phenomenological models. A critical review of these models is provided by way of background to the development of the new model.

Mathematics Subject Classifications (2000): 74B20, 74E10, 74L15, 92C10.

Key words: biomechanics, arteries, artery wall, material models, constitutive laws, finite deformations, nonlinear elasticity.

1. Introduction

In the last few years there has been a significant growth in interest in the mechanical properties of biological soft tissue treated from the continuum mechanical perspective, and, in particular, in the mechanics of arterial tissue. An excellent starting point for the study of the mechanics of arteries is the wide-ranging review by Humphrey [32], who, more recently [33], has conducted a comparative study of a small number of constitutive models used in the literature to describe the mechanical response of arteries. One important motivation for such studies is the belief that mechanical factors may be important in triggering the onset of atherosclerosis, the major cause of human mortality in the western world. In order to fully understand these mechanical influences it is necessary to have reliable constitutive models for the artery. Moreover, several clinical treatments, such as percutaneous transluminal angioplasty [3] can only be studied in detail if a reliable constitutive model of the arterial wall is available.

In vivo the artery is a pre-stretched material under an internal pressure load and it is essential to use a theory which takes account of the resulting finite deformation. The aim of this paper is therefore to provide a systematic study of the mechanical properties of arteries based on the continuum theory of large deformation elasticity. We begin, in Section 2, by giving a brief description of the histological structure of arterial walls, a summary of the main deformation geometries used in the experimental determination of the mechanical properties of arteries and an outline of the general characteristics of the mechanical response of arteries. It is emphasized that the vast majority of constitutive models used in the literature are phenomenological in nature and do not take account directly of the histological structure.

In Section 3 we summarize the theoretical framework to be used as the background for the description of the arterial mechanics. This consists of the general equations governing the elastic response of an *anisotropic* material based on the use of an elastic free-energy function. The equations are then specialized in order to consider the circular cylindrical geometry appropriate for the analysis of extension, inflation and torsion of an artery, which is treated as a thick-walled circular cylinder. In the absence of the applied loading, it should be emphasized, an artery is not stress free since if cut along a radius it will spring open to form an open sector. In this paper, for simplicity and in order to produce a distribution of residual stress in the unloaded configuration, we assume that the opened-up configuration is unstressed, although it is known that in general such a configuration is not unstressed (see, for example, Vossoughi et al. [64]). In general, there may also be residual stresses in the axial direction, but we do not allow for these in the present work. The assumed stress-free configuration is taken to correspond to an open sector of a circular cylindrical tube and is designated as the reference configuration of the material. The stressed but unloaded circular cylindrical shape is recovered by application of an initial bending deformation. Thus, the overall deformation from the reference configuration consists of bending, axial extension, inflation and torsion. This provides a composite deformation of sufficient generality to allow a comparative judgement of the predictions of different material models to be made.

In Section 4 a range of both two- and three-dimensional phenomenological models adopted in the literature for the study of *elastic* arteries is examined on the basis of the theory in Section 3 from a comparative point of view. Their performance is assessed critically against a number of criteria, and certain deficiencies of the models are highlighted. Some anisotropic models are able to provide a full three-dimensional description of the state of stress in an artery, but at the expense of incorporation of a large number of material constants, which may lead to parameter identification problems. On the other hand, oversimplification through use of isotropy, as in Delfino et al. [10], is also evident. Several models, including that in Fung et al. [18], are based on formulations which may be associated with geometrical simplifications (for example, the membrane approximation) and are not suitable for analysis of the through-thickness stress distribution in an artery or for the treatment of shearing deformations. They can, however, be used to *simulate* the deformation in special cases, such as that corresponding to axial extension and inflation of an artery regarded as a thin-walled (or thick-walled) circular cylindrical tube. An approach which uses the incremental elastic moduli is also found in the literature (for some examples see the data book [1] edited by Abé et al.) but is not discussed here since it is inappropriate for the finite deformation analysis with which we are concerned.

One problem which arises in making comparisons is that each different model in the literature is based on data from a different artery (and generally from different animals). Nevertheless, we carry out a systematic and detailed evaluation of several of the most commonly used models in respect of combined extension, inflation and torsion of a thick-walled tube and with residual stresses incorporated. Moreover, a certain convexity property is introduced and checked for each model to provide an indication of its mechanical, mathematical and computational efficacy.

This provides the background for the introduction, in Section 5, of a new model which aims to circumvent the difficulties encountered with some other models. Specifically, the new model takes account of the architecture of the arterial wall. The artery is treated as a two-layer thick-walled tube, the two layers representing the media (the middle layer of the artery) and the adventitia (the outer layer). These are the main (solid) mechanically relevant components in healthy arteries. Thus, a third layer (the intima) is disregarded in this work, although it is not difficult to account for this on a similar basis as for the other layers if the need arises, which would be the case for arteries subject to pathological intimal change.

Each layer is composed of a non-collagenous matrix, which is treated as an isotropic material, and two families of collagen fibers helically wound along the arterial axis and symmetrically disposed with respect to the axis (but with different orientations in the two layers). These fibers induce the anisotropy in the mechanical response such that the overall response of each layer is orthotropic and is accounted for by the constitutive theory of fiber-reinforced solids. Their contribution to the strain energy is modeled using a pair of preferred directions identified in the reference configuration, and from which structure tensors characterizing the anisotropy are formed. The model is *structural* in the sense that it involves two layers and within each layer information about the orientations of the collagen fibers, obtained from a statistical analysis of histological sections of each arterial layer (see Holzapfel et al. [27]), is incorporated. The material parameters included in the model, although phenomenologically based, reflect the structural characteristics within each arterial layer. The properties of the matrix material are described in terms of a single material constant and those of the fiber bundles by two additional constants in each layer.

With a specific form of the model used for illustrative purposes, the predictions of the model are examined in detail and compared with those from some phenomenological models. The predictions of the model agree well with the typical mechanical response of arteries observed in experiments. The Cauchy stress distributions through the deformed arterial wall in the physiological state are also determined in order to illustrate the significant difference made by incorporation of the residual stresses. Moreover, the three-dimensional model introduced here is consistent with the convexity requirements that ensure mechanically and mathematically reliable behavior. It also admits an efficient numerical implementation within the finite element method, an aspect which is discussed in detail in [26], in which there is an extension to viscoelasticity (suitable for the modeling of muscular arteries). Extension to elastoplasticity is discussed in [20] and [21]. Thus, more complex boundary-value problems, possibly of clinical relevance, can be solved on the basis of the proposed model.

Section 6 contains a summary of the results and concludes with an assessment of the advantages of the new model.

2. Histology and Typical Mechanical Behavior of Arterial Walls

Efficient constitutive descriptions of arterial walls require a fundamental knowledge and understanding of the entire arterial histology, i.e. the morphological structure, and an extensive investigation of the particular arterial wall of interest. Additionally, this is of crucial importance for the understanding of the general mechanical characteristics of arterial walls and the components that provide the main contributions to the deformation process.

This brief overview is included only for the purpose of clarifying the macroscopic and microscopic structure of arterial walls and to provide essential information for scientists without a background in biology or physiology. For a more detailed exposition of the different mechanical characteristics of the interrelated arterial components and the overall functioning of the blood vessel (which constitutive models aim to characterize) see, for example, the reviews by Rhodin [48] and Silver et al. [54].

2.1. ARTERIAL HISTOLOGY

This paper is concerned with the *in vitro* passive behavior of arteries. Hence, *in vivo* effects such as the vasa vasorum, nerve control, humoral control, perivascular connective tissue, etc. and neighboring organs such as the pulsating heart are not relevant and are not therefore discussed here.

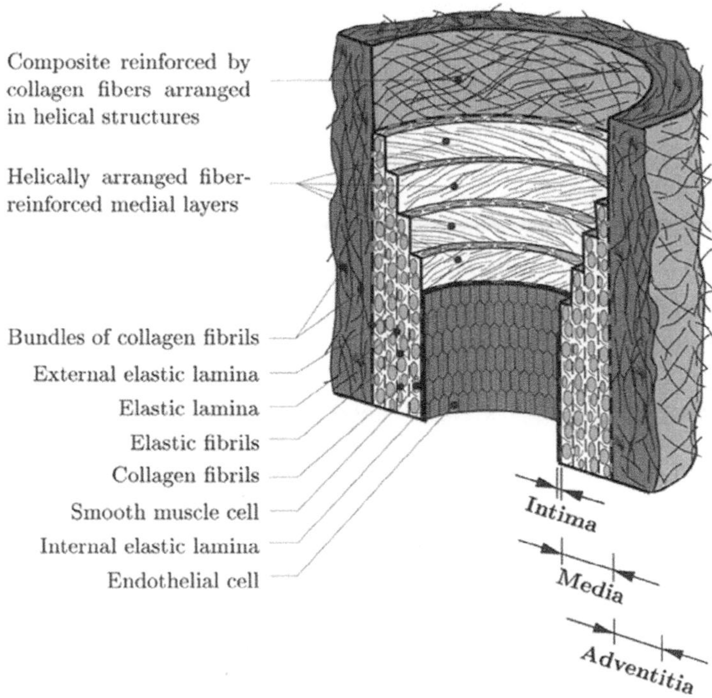

Figure 1. Diagrammatic model of the major components of a healthy *elastic* artery composed of three layers: intima (I), media (M), adventitia (A). I is the innermost layer consisting of a single layer of endothelial cells that rests on a thin basal membrane and a subendothelial layer whose thickness varies with topography, age and disease. M is composed of smooth muscle cells, a network of elastic and collagen fibrils and elastic laminae which separate M into a number of fiber-reinforced layers. The primary constituents of A are thick bundles of collagen fibrils arranged in helical structures; A is the outermost layer surrounded by loose connective tissue.

In general, arteries are roughly subdivided into two types: *elastic* and *muscular*. Elastic arteries have relatively large diameters and are located close to the heart (for example, the aorta and the carotid and iliac arteries), while muscular arteries are located at the periphery (for example, femoral, celiac, cerebral arteries). However, some arteries exhibit morphological structures of both types. Here we focus attention on the microscopic structure of arterial walls composed of three distinct layers, the *intima* (*tunica intima*), the *media* (*tunica media*) and the *adventitia* (*tunica externa*). We discuss the constituents of arterial walls from the mechanical perspective and emphasize those aspects which are important to researchers interested in constitutive issues. Figure 1 shows a model of a healthy elastic artery.

2.1.1. *Intima*

The intima is the innermost layer of the artery. It consists of a single layer of endothelial cells lining the arterial wall and resting on a thin basal membrane (basal lamina). There is also a subendothelial layer whose thickness varies with topography, age and disease. In healthy young muscular arteries, however, the subendothelial layer is almost non-existent. In healthy young individuals the intima is very thin and makes an insignificant contribution to the *solid* mechanical properties of the arterial wall. However, it should be noted that the intima thickens and stiffens with age (*arteriosclerosis*) so that the mechanical contribution may become significant.

It is known that pathological changes of the intimal components may be associated with *atherosclerosis*, the most common disease of arterial walls. It involves deposition of fatty substances, calcium, collagen fibers, cellular waste products and fibrin (a clotting material in the blood). The resulting build-up is called *atherosclerotic* plaque. It may be very complex in geometry and biochemical composition. In later stages the media is also affected. These pathological changes are associated with significant alterations in the mechanical properties of the arterial wall. Hence, the mechanical behavior of atherosclerotic arteries differs significantly from that of healthy arteries.

2.1.2. *Media*

The media is the middle layer of the artery and consists of a complex three-dimensional network of smooth muscle cells, and elastin and collagen fibrils. According to [48] the fenestrated elastic laminae separate the media into a varying number of well-defined concentrically fiber-reinforced medial layers. The number of elastic laminae decreases toward the periphery (as the size of the vessels decreases) so that elastic laminae are hardly present in muscular arteries.

The media is separated from the intima and adventitia by the so-called internal elastic lamina and external elastic lamina (absent in cerebral blood vessels), respectively. In muscular arteries these laminae appear as prominent structures, whereas in elastic arteries they are hardly distinguishable from the regular elastic laminae. The orientation of and close interconnection between the elastic and collagen fibrils, elastic laminae, and smooth muscle cells together constitute a continuous fibrous helix (Faserschraube) [52, 58]. The helix has a small pitch so that the fibrils in the media are almost circumferentially oriented. This structured arrangement gives the media high strength, resilience and the ability to resist loads in both the longitudinal and circumferential directions. From the mechanical perspective, the media is the most significant layer in a healthy artery.

2.1.3. *Adventitia*

The adventitia is the outermost layer of the artery and consists mainly of fibroblasts and fibrocytes (cells that produce collagen and elastin), histological ground substance and thick bundles of collagen fibrils forming a fibrous tissue. The adventitia is surrounded continuously by loose connective tissue. The thickness of the adventitia depends strongly on the type (elastic or muscular) and the physiological function of the blood vessel and its topographical site. For example, in cerebral blood vessels there is virtually no adventitia.

The wavy collagen fibrils are arranged in helical structures and serve to reinforce the wall. They contribute significantly to the stability and strength of the arterial wall. The adventitia is much less stiff in the load-free configuration and at low pressures than the media. However, at higher levels of pressure the collagen fibers reach their straightened lengths and the adventitia changes to a stiff 'jacket-like' tube which prevents the artery from overstretch and rupture.

2.2. TYPICAL MECHANICAL BEHAVIOR OF ARTERIAL WALLS

Each constitutive framework and its associated set of material parameters requires detailed studies of the particular material of interest. Its reliability is strongly related to the quality and completeness of available experimental data, which may come from appropriate *in vivo* tests or from *in vitro* tests that mimic real loading conditions in a physiological environment.

In vivo tests seem to be preferable because the vessel is observed under real life conditions. However, *in vivo* tests have major limitations because of, for example, the influence of hormones and nerval control. Moreover, data sets from the complex material response of arterial walls subject to simultaneous cyclic *inflation*, axial *extension* and *twist* can only be measured in an *in vitro* experiment. Only with such data sets can the anisotropic mechanical behavior of arterial walls be described completely. In addition, in *in vitro* experiments the contraction state (active or passive) of the muscular media has to be determined. This can be done with appropriate chemical agents.

For pure inflation tests of straight artery tubes, which is the most common two-dimensional test, see the early work [2] (in which shear deformations are not considered). Since arteries do not change their volume within the physiological range of deformation [4], they can be regarded as incompressible materials. Hence, by means of the incompressibility constraint we may determine the mechanical properties of three-dimensional specimens from two-dimensional tests [36]. It is important to note that uniaxial extension tests on arterial patches (strips) provide basic information about the material [24] but they are certainly not sufficient to quantify completely the anisotropic behavior of arterial walls. Other uniaxial extension tests on small arterial rings (so-called ring tests) are also insufficient [9]. In general, a segment of vessel shortens on removal from the body, as was first reported in [15]. The *in vivo* pre-stretch in the longitudinal direction must therefore be reproduced within *in vitro* tests [37].

Each non-axisymmetric arterial segment (such as a bifurcation or a segment with sclerotic changes) under combined inflation and axial extension develops significant shear stresses in the wall. Hence, in order to characterize the shear properties of arterial walls shear tests are required. In shear tests *either* the angle of twist of an arterial tube subjected to transmural pressure, longitudinal force and torque [12] *or* the shear deformation of a rectangular arterial wall specimen subjected to shear forces [65] is measured. Additionally, one can classify mechanical tests according to the strain rates used (quasi-static or dynamic) and to whether the loading is performed cyclically or discontinuously (creep and relaxation tests). It has been known for many years that the load-free configuration of an artery is not a stress-free state [61]. In general, a load-free arterial ring contains residual stresses. It is of crucial importance to identify these in order to predict reliably the state of stress in an arterial wall, and this has been the aim of many experimental investigations (see, for example, the bending tests on blood vessel walls in [68]).

The mechanical behavior of arteries depends on physical and chemical environmental factors, such as temperature, osmotic pressure, pH, partial pressure of carbon dioxide and oxygen, ionic concentrations and monosaccharide concentration. In *ex vivo* conditions the mechanical properties are altered due to biological degradation. Therefore, arteries should be tested in appropriate oxygenated, temperature controlled salt solutions as fresh as possible. For an overview of experimental test methods used to verify material parameters, see [32] and the references contained therein.

As indicated in Section 2.1 the composition of arterial walls varies along the arterial tree. Hence, there seems to be a systematic dependence of the shape of the stress–strain curve for a blood vessel on its anatomical site. This fact has been demonstrated several times experimentally; see, for example, the early works [51, 37] and [8]. Although the mechanical properties of arterial walls vary along the arterial tree, the general mechanical characteristics exhibited by arterial walls are the same. In order to explain the typical stress–strain response of an arterial wall of smooth muscles in the passive state (governed mainly by elastin and collagen fibers), we refer to Figure 2. Note that the curves in Figure 2 are schematic, but based on experimental tension tests performed in the authors' laboratory (some of which is described in a recent paper [28]).

As can be seen, a circumferential strip of the media subjected to uniaxial cyclic loading and unloading typically displays pronounced stress softening, which occurs during the first few load cycles. The stress softening effect diminishes with the number of load cycles until the material exhibits a *nearly repeatable* cyclic behavior, and hence the biological material is said to be 'pre-conditioned' (compare with, for example, the characteristic passive behavior of a bovine coronary artery in [32], p. 33). Thus, depending on the type of artery considered, the material behavior may be regarded as (perfectly) elastic for proximal arteries of the elastic type, or viscoelastic for distal arteries of the muscular type, often modeled as pseudoelastic (see, for example, [18]). For a definition of the term *pseudoelasticity* in the context of biomechanics the reader is referred to [18].

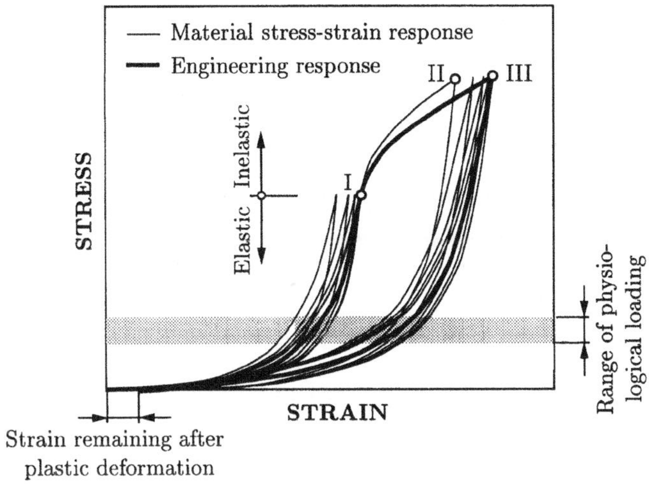

Figure 2. Schematic diagram of typical uniaxial stress–strain curves for circumferential arterial strips (from the media) in passive condition (based on tension tests performed in the authors' laboratory): cyclic loading and unloading, associated with stress softening effects, lead to a pre-conditioned material which behaves (perfectly) elastically or viscoelastically (nearly repeatable cyclic behavior) – point I. Loading beyond the (visco)elastic domain up to point II leads to inelastic deformations. Additional loading and unloading cycles display stress softening again until point III is reached. Then the material exhibits (perfectly) elastic or viscoelastic response. The thick solid line indicates the (approximate) engineering response of the material.

Healthy arteries are highly deformable composite structures and show a nonlinear stress–strain response with a typical (exponential) stiffening effect at higher pressures, as illustrated in Figure 2. This stiffening effect, common to all biological tissues, is based on the recruitment of the embedded (load carrying) wavy collagen fibrils, which leads to the characteristic anisotropic mechanical behavior of arteries; see the classical works [49, 40]. Early works on arterial anisotropy (see, for example, [45]) considered arterial walls to be cylindrically orthotropic, which is generally accepted in the literature.

Loading beyond the (visco)elastic domain (indicated by point I in Figure 2), far outside the physiological range of deformation, often occurs during mechanical treatments such as percutaneous transluminal angioplasty. This procedure involves dilation of an artery using a balloon catheter (see [3]). In the strain range up to point II in Figure 2, the deformation process in an arterial layer is associated with inelastic effects (elastoplastic and/or damage mechanisms) leading to significant changes in the mechanical behavior (see [44, 28] and [21]). This overstretching involves dissipation, which is represented by the area between the loading and unloading curves. Hence, starting from point II, additional cyclic loading and unloading again displays stress softening, which diminishes with the number of load cycles. At point III the material exhibits a (perfectly) elastic or viscoelastic behavior. However, unloading initiated from point III returns the arterial (medial) strip to an unstressed state with non-vanishing strains remaining, these being responsible for the change of shape. With preconditioning effects neglected, the thick solid line in Figure 2 indicates the (approximate) engineering response associated with the actual physical behavior.

The model proposed in Section 5 is intended to capture only the elastic portion of the curves in Figure 2, i.e. up to point I. For the remaining portions of the curves a rate-independent elastoplastic model and the associated algorithmic formulation and finite element implementation was recently proposed in [20].

3. Continuum-Mechanical Framework

In this section we summarize the equations that provide the general continuum description of the deformation and the hyperelastic stress response of the material. As a basis for reporting the performance of different constitutive models for arteries we consider the mechanical response of a thick-walled circular cylindrical tube under various boundary loads. We specify the strain measures to be used and discuss the equilibrium equation which arises in the considered problem. We also give expressions for the torsional couple and the reduced axial force acting on the tube, these being crucial for the subsequent comparative study of constitutive models.

3.1. FINITE HYPERELASTICITY

3.1.1. *Description of the Deformation*

Let Ω_0 be a (fixed) reference configuration of the continuous body of interest (assumed to be stress-free). We use the notation $\chi : \Omega_0 \to \mathbb{R}^3$ for the deformation, which transforms a typical material point $\mathbf{X} \in \Omega_0$ to a position $\mathbf{x} = \chi(\mathbf{X}) \in \Omega$ in the deformed configuration, denoted Ω. Further, let $\mathbf{F}(\mathbf{X}) = \partial\chi(\mathbf{X})/\partial\mathbf{X}$ be the deformation gradient and $J(\mathbf{X}) = \det\mathbf{F} > 0$ the local volume ratio.

Following [14] and [41], we consider the multiplicative decomposition

$$\mathbf{F} = (J^{1/3}\mathbf{I})\overline{\mathbf{F}} \tag{1}$$

of $\mathbf{F}$ into spherical (dilatational) and unimodular (distortional) parts. We use the right and left Cauchy–Green tensors, denoted $\mathbf{C}$ and $\mathbf{b}$ respectively, and their modified counterparts, denoted $\overline{\mathbf{C}}$ and $\overline{\mathbf{b}}$ respectively, associated with $\overline{\mathbf{F}}$. From equation (1) we then have

$$\mathbf{C} = \mathbf{F}^{\mathrm{T}}\mathbf{F} = J^{2/3}\overline{\mathbf{C}}, \qquad \overline{\mathbf{C}} = \overline{\mathbf{F}}^{\mathrm{T}}\overline{\mathbf{F}}, \tag{2}$$

$$\mathbf{b} = \mathbf{F}\mathbf{F}^{\mathrm{T}} = J^{2/3}\overline{\mathbf{b}}, \qquad \overline{\mathbf{b}} = \overline{\mathbf{F}}\,\overline{\mathbf{F}}^{\mathrm{T}}. \tag{3}$$

In addition, we introduce the Green–Lagrange strain tensor $\mathbf{E}$, and, through equation (1), its associated modified strain measure $\overline{\mathbf{E}}$. Thus,

$$\mathbf{E} = \frac{1}{2}(\mathbf{C} - \mathbf{I}) = J^{2/3}\overline{\mathbf{E}} + \frac{1}{2}(J^{2/3} - 1)\mathbf{I}, \qquad \overline{\mathbf{E}} = \frac{1}{2}(\overline{\mathbf{C}} - \mathbf{I}), \tag{4}$$

where $\mathbf{I}$ denotes the second-order unit tensor.

3.1.2. *Hyperelastic Stress Response*

In order to describe the hyperelastic stress response of arterial walls, we employ a set $\{\mathbf{A}_\alpha \mid \alpha = 1, \dots, n\}$ of (second-order) tensors which characterize the wall structure, and we postulate the existence of a Helmholtz free-energy function $\Psi(\mathbf{E}, \mathbf{A}_1, \dots, \mathbf{A}_n)$. Subsequently, we assume the *decoupled* form

$$\Psi(\mathbf{E}, \mathbf{A}_1, \dots, \mathbf{A}_n) = U(J) + \overline{\Psi}(\overline{\mathbf{E}}, \mathbf{A}_1, \dots, \mathbf{A}_n), \tag{5}$$

where the function U is a purely volumetric contribution and $\overline{\Psi}$ is a purely isochoric contribution to the free energy Ψ.

From the Clausius–Planck inequality, standard arguments lead to the well-known equation $\mathbf{S} = \partial\Psi(\mathbf{E}, \mathbf{A}_1, \dots, \mathbf{A}_n)/\partial\mathbf{E}$ for the second Piola–Kirchhoff stress. Equation (5) then gives

$$\mathbf{S} = \mathbf{S}_{\text{vol}} + \overline{\mathbf{S}} \quad \text{with } \mathbf{S}_{\text{vol}} = \frac{\partial U(J)}{\partial \mathbf{E}}, \ \overline{\mathbf{S}} = \frac{\partial \overline{\Psi}(\overline{\mathbf{E}}, \mathbf{A}_1, \dots, \mathbf{A}_n)}{\partial \mathbf{E}}. \tag{6}$$

We shall also require the standard results

$$\frac{\partial J}{\partial \mathbf{E}} = J\mathbf{C}^{-1} \quad \text{and} \quad \frac{\partial \overline{\mathbf{E}}}{\partial \mathbf{E}} = J^{-2/3}\left(\mathbb{I} - \frac{1}{3}\overline{\mathbf{C}} \otimes \overline{\mathbf{C}}^{-1}\right) \tag{7}$$

from tensor analysis (see, for example, [25]), where $\mathbb{I}$ denotes the fourth-order identity tensor which, in index notation, has the form $(\mathbb{I})_{IJKL} = (\delta_{IK}\delta_{JL} + \delta_{IL}\delta_{JK})/2$, δ_{IJ} being the Kronecker delta. With these results, equations $(6)_2$ and $(6)_3$ become, after some straightforward tensor manipulations and the introduction of the hydrostatic pressure $p = \mathrm{d}U/\mathrm{d}J$ as in [25],

$$\mathbf{S}_{\text{vol}} = pJ\mathbf{C}^{-1}, \qquad \overline{\mathbf{S}} = J^{-2/3}\mathrm{Dev}\left(\frac{\partial \overline{\Psi}}{\partial \overline{\mathbf{E}}}\right). \tag{8}$$

The operator $\mathrm{Dev}(\bullet)$ in (8) is defined by

$$\mathrm{Dev}(\bullet) = (\bullet) - \frac{1}{3}[(\bullet) : \overline{\mathbf{C}}]\overline{\mathbf{C}}^{-1}, \tag{9}$$

and furnishes the physically correct deviatoric operator in the Lagrangian description, so that $\mathrm{Dev}(\bullet) : \mathbf{C} = 0$. Note that in the description of an incompressible material (which an artery is assumed to be) the hydrostatic pressure p becomes an indeterminate *Lagrange multiplier*.

A Piola transformation of equations (8) enables the Cauchy stress tensor $\boldsymbol{\sigma} = J^{-1}\mathbf{FSF}^{\mathrm{T}}$ to be put in the *decoupled* form

$$\boldsymbol{\sigma} = \boldsymbol{\sigma}_{\text{vol}} + \overline{\boldsymbol{\sigma}} \quad \text{with } \boldsymbol{\sigma}_{\text{vol}} = p\mathbf{I}, \ \overline{\boldsymbol{\sigma}} = J^{-1}\mathrm{dev}\left(\overline{\mathbf{F}}\frac{\partial \overline{\Psi}}{\partial \overline{\mathbf{E}}}\overline{\mathbf{F}}^{\mathrm{T}}\right), \tag{10}$$

analogously to equation (8), where the operator $\mathrm{dev}(\bullet)$ is defined by

$$\mathrm{dev}(\bullet) = (\bullet) - \frac{1}{3}[(\bullet) : \mathbf{I}]\mathbf{I}. \tag{11}$$

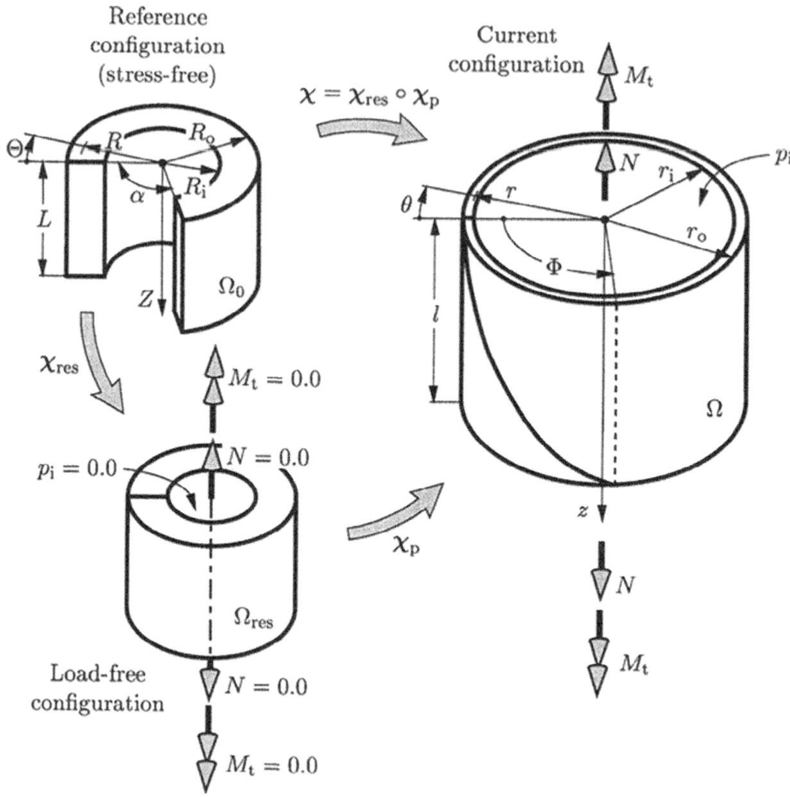

Figure 3. Arterial ring in the the (stress-free) reference configuration Ω_0, the load-free configuration Ω_{res} and the current configuration Ω.

It furnishes the physically correct deviatoric operator in the Eulerian description, so that $\text{dev}(\bullet) : \mathbf{I} = 0$.

3.2. COMBINED BENDING, INFLATION, EXTENSION AND TORSION OF A TUBE

3.2.1. *Basic Kinematics*

We consider the artery as an incompressible thick-walled cylindrical tube subjected to various loads. It is known that the load-free configuration, Ω_{res} say, in which the artery is excised from the body and *not* subjected to any loads is not a stress-free (or strain-free) reference configuration Ω_0. Thus, the arterial ring springs open when cut in a radial direction. It appears that Vaishnav and Vossoughi [61] were the first to publish this finding. Bearing in mind the statement in Section 1, we *assume* that the open sector is the undeformed (stress-free and fixed) reference configuration Ω_0, as depicted in Figure 3.

Thus, in terms of cylindrical polar coordinates (R, Θ, Z), the geometrical region Ω_0 of the tube is defined by

$$R_{\mathrm{i}} \leqslant R \leqslant R_{\mathrm{o}}, \qquad 0 \leqslant \Theta \leqslant (2\pi - \alpha), \qquad 0 \leqslant Z \leqslant L, \tag{12}$$

where R_{i}, R_{o}, α and L denote the inner and outer radii, the opening angle and length of the undeformed (split) tube, respectively. Note that the opening angle α identified in Figure 3 differs from the definition normally used (see, for example, Fung and Liu [19]).

The deformation χ takes Ω_0 into the current configuration Ω. For the considered problem $\chi = \chi_{\mathrm{p}} \circ \chi_{\mathrm{res}}$ is the composition of the deformations χ_{res} and χ_{p}, as indicated in Figure 3, where χ_{res} generates the load-free configuration Ω_{res} associated with residual stresses, while χ_{p} is associated with inflation, axial elongation and torsion of the tube, and leads to the final configuration Ω. It is important to note that the residually stressed configuration Ω_{res} of an artery arises from certain growth mechanisms of the different layers, and hence, in general, the residually stressed state is more complex than considered here. For discussion of stress-dependent growth and remodeling we refer to, for example, Rodriguez et al. [50] and Rachev [46]. It may also be noted that it has been found that residual stress accumulates due to cyclic loading in two-phase models of soft tissue *without* growth (see, for example, Huyghe et al. [34]).

In terms of cylindrical polar coordinates (r, θ, z), the geometry of the deformed configuration Ω is given by

$$r_{\mathrm{i}} \leqslant r \leqslant r_{\mathrm{o}}, \qquad 0 \leqslant \theta \leqslant 2\pi, \qquad 0 \leqslant z \leqslant l, \tag{13}$$

where r_{i}, r_{o} and l denote the inner and outer radii and the length of the deformed tube, respectively.

The deformation χ, which is taken to be isochoric, may then be written in the form

$$\chi = r\mathbf{e}_r + z\mathbf{e}_z \tag{14}$$

with reference to the (unit) basis vectors $\{\mathbf{e}_r, \mathbf{e}_\theta, \mathbf{e}_z\}$ associated with the cylindrical polar coordinates (r, θ, z), where

$$r = \sqrt{\frac{R^2 - R_{\mathrm{i}}^2}{k\lambda_z} + r_{\mathrm{i}}^2}, \qquad \theta = k\Theta + Z\frac{\Phi}{L}, \qquad z = \lambda_z Z, \tag{15}$$

λ_z is the (constant) axial stretch, the parameter k, defined by $k = 2\pi/(2\pi - \alpha)$, is a convenient measure of the tube opening angle in the unstressed configuration, r_{i} is the inner radius in the deformed configuration and Φ is the angle of twist of the tube arising from the torsion.

In addition to λ_z, it is convenient to introduce the notations defined by

$$\lambda_r(R) = \frac{\partial r}{\partial R} = \frac{R}{rk\lambda_z}, \qquad \lambda_\theta(R) = \frac{r}{R}\frac{\partial \theta}{\partial \Theta} = \frac{kr}{R},$$
$$\gamma(R) = r\frac{\partial \theta}{\partial z} = r\frac{\Phi}{l}. \tag{16}$$

Here $\lambda_r(R)$, $\lambda_\theta(R)$ and λ_z are the principal stretches of the deformation associated with the radial, circumferential and axial directions *when there is no twist*, while $\gamma(R)$, which is associated with the twist, represents locally the amount of shear in a (θ, z)-plane. Since each of these quantities depends only on the radius R, the one-dimensional character of the problem is apparent. When $\gamma \neq 0$, λ_r is the principal stretch in the radial direction but λ_θ and λ_z are not then principal stretches. The condition that the volume is preserved during the deformation is independent of γ and requires simply that

$$\lambda_r \lambda_\theta \lambda_z = 1. \tag{17}$$

Note that

$$r_{\mathrm{i}} = \lambda_{\theta\mathrm{i}} \frac{R_{\mathrm{i}}}{k}, \tag{18}$$

where $\lambda_{\theta\,\mathrm{i}}$ denotes the value of λ_θ at the inner surface of the tube.

The first term $k\Theta$ in $(15)_2$ represents the deformation from configuration Ω_0 to Ω_{res} while the second term $Z\Phi/L$ describes the influence of the torsion. In terms of the parameters k, $\lambda_{\theta\mathrm{i}}$, λ_z and Φ, equations (15), (18) define the combined bending, inflation, axial extension and torsion of a thick-walled tube.

On use of (17), the deformation gradient $\mathbf{F}$, referred to cylindrical polar coordinates, may be expressed in the form

$$\mathbf{F} = \overline{\mathbf{F}} = (\lambda_\theta \lambda_z)^{-1} \mathbf{e}_r \otimes \mathbf{E}_R + \lambda_\theta \, \mathbf{e}_\theta \otimes \mathbf{E}_\Theta + \gamma \lambda_z \, \mathbf{e}_\theta \otimes \mathbf{E}_Z + \lambda_z \, \mathbf{e}_z \otimes \mathbf{E}_Z, \tag{19}$$

where $\{\mathbf{E}_R, \mathbf{E}_\Theta, \mathbf{E}_Z\}$ is the set of unit cylindrical polar basis vectors associated with (R, Θ, Z). Note that $\mathbf{E}_Z = \mathbf{e}_z$.

Use of equations (2), (3) enables the Cauchy–Green tensors to be given in terms of cylindrical polar coordinates. Thus,

$$\begin{aligned} \mathbf{C} = \overline{\mathbf{C}} &= (\lambda_\theta \lambda_z)^{-2} \mathbf{E}_R \otimes \mathbf{E}_R + \lambda_\theta^2 \, \mathbf{E}_\Theta \otimes \mathbf{E}_\Theta + \lambda_z^2 (1 + \gamma^2) \, \mathbf{E}_Z \otimes \mathbf{E}_Z \\ &\quad + \gamma \lambda_\theta \lambda_z (\mathbf{E}_\Theta \otimes \mathbf{E}_Z + \mathbf{E}_Z \otimes \mathbf{E}_\Theta), \end{aligned} \tag{20}$$

$$\begin{aligned} \mathbf{b} = \overline{\mathbf{b}} &= (\lambda_\theta \lambda_z)^{-2} \mathbf{e}_r \otimes \mathbf{e}_r + (\lambda_\theta^2 + \gamma^2 \lambda_z^2) \mathbf{e}_\theta \otimes \mathbf{e}_\theta + \lambda_z^2 \, \mathbf{e}_z \otimes \mathbf{e}_z \\ &\quad + \gamma \lambda_z^2 (\mathbf{e}_\theta \otimes \mathbf{e}_z + \mathbf{e}_z \otimes \mathbf{e}_\theta). \end{aligned} \tag{21}$$

The deformation gradient (19) and the Cauchy–Green tensors (20) and (21) play a crucial role in the derivation of the state of stress in an arterial wall. A more general deformation including azimuthal and axial shear is discussed in the paper by Guccione et al. [22].

3.2.2. *Equilibrium Equations*

In the absence of body forces the equilibrium equations are

$$\operatorname{div} \boldsymbol{\sigma} = \mathbf{0}, \tag{22}$$

where div(•) denotes the spatial divergence of the spatial tensor field (•). Note that in cylindrical polar coordinates (r, θ, z), because of the geometrical and constitutive symmetry, the only non-trivial component of (22) is

$$\frac{\mathrm{d}\sigma_{rr}}{\mathrm{d}r} + \frac{(\sigma_{rr} - \sigma_{\theta\theta})}{r} = 0 \tag{23}$$

(see, for example, [42]). From this equation and the boundary condition $\sigma_{rr}|_{r=r_\mathrm{o}} = 0$ on the outer surface of the tube, the radial Cauchy stress σ_{rr} may be calculated as

$$\sigma_{rr}(\xi) = \int_{\xi}^{r_\mathrm{o}} (\sigma_{rr} - \sigma_{\theta\theta})\frac{\mathrm{d}r}{r}, \qquad r_\mathrm{i} \leqslant \xi \leqslant r_\mathrm{o}. \tag{24}$$

The internal pressure $p_\mathrm{i} = -\sigma_{rr}|_{r=r_\mathrm{i}}$ is then obtained in the form

$$p_\mathrm{i} = \int_{r_\mathrm{i}}^{r_\mathrm{o}} (\sigma_{\theta\theta} - \sigma_{rr})\frac{\mathrm{d}r}{r}. \tag{25}$$

This equation plays an important role in the numerical solution of the problem considered.

When the state of deformation is known, expressions for the axial force N and the torsional couple M_t can be calculated via the definitions

$$N = 2\pi \int_{r_\mathrm{i}}^{r_\mathrm{o}} \sigma_{zz}\, r \mathrm{d}r, \qquad M_\mathrm{t} = 2\pi \int_{r_\mathrm{i}}^{r_\mathrm{o}} \sigma_{\theta z} r^2\, \mathrm{d}r. \tag{26}$$

In view of the additive split of the Cauchy stress tensor $\boldsymbol{\sigma}$ into volumetric and isochoric parts introduced in $(10)_1$, we may recast equations (25) and $(26)_2$ by using the decompositions $\sigma_{\theta\theta} = p + \overline{\sigma}_{\theta\theta}$ and $\sigma_{rr} = p + \overline{\sigma}_{rr}$ to obtain

$$p_\mathrm{i} = \int_{r_\mathrm{i}}^{r_\mathrm{o}} (\overline{\sigma}_{\theta\theta} - \overline{\sigma}_{rr})\frac{\mathrm{d}r}{r}, \qquad M_\mathrm{t} = 2\pi \int_{r_\mathrm{i}}^{r_\mathrm{o}} \overline{\sigma}_{\theta z} r^2\, \mathrm{d}r, \tag{27}$$

where $\overline{\sigma}_{\theta\theta}$, $\overline{\sigma}_{rr}$ denote the isochoric parts of the normal components of (Cauchy) stress in the circumferential and radial directions, while $\overline{\sigma}_{\theta z} = \sigma_{\theta z}$ is the shear component of (Cauchy) stress acting tangentially to the cross-section of the tube.

Use of the additive split (10) and equation (24) enables the axial force N in equation $(26)_1$ to be expressed as

$$N = 2\pi \int_{r_\mathrm{i}}^{r_\mathrm{o}} \left[\int_{r_\mathrm{o}}^{\xi} (\overline{\sigma}_{\theta\theta} - \overline{\sigma}_{rr})\frac{\mathrm{d}r}{r} - \overline{\sigma}_{rr} + \overline{\sigma}_{zz} \right] \xi\, \mathrm{d}\xi. \tag{28}$$

Reversal of the order of integration in (28) and use of the expression $(27)_1$ leads to the general formula

$$F = \pi \int_{r_\mathrm{i}}^{r_\mathrm{o}} (2\overline{\sigma}_{zz} - \overline{\sigma}_{\theta\theta} - \overline{\sigma}_{rr}) r\, \mathrm{d}r \tag{29}$$

for the *reduced axial force* $F = N - r_{\mathrm{i}}^2 \pi p_{\mathrm{i}}$. This expression for F is very important since it gives precisely the force that is measured during inflation tests on arteries. A specific form of equation (29) is given in equation (15) of [5]. In subsequent sections of the present paper equations (27) and (29) will be specialized for several constitutive models that have been used in the literature to represent the mechanical response of arteries.

REMARK 3.1. For a thin-walled cylindrical tube we make the simplification $\sigma_{rr} = p + \overline{\sigma}_{rr} = 0$ for the radial stress (the membrane approximation). On use of $\sigma_{\theta\theta} = p + \overline{\sigma}_{\theta\theta}$ and $\overline{\sigma}_{\theta z} = \sigma_{\theta z}$, equations $(27)_1$, $(27)_2$ and (29) then enable the reduced equations for the internal pressure p_{i}, the torsional couple M_{t} and the reduced axial force F to be given simply as

$$p_{\mathrm{i}} = \frac{h}{r}\sigma_{\theta\theta}, \qquad M_{\mathrm{t}} = 2\pi r^2 h \sigma_{\theta z}, \qquad F = \pi r h(2\sigma_{zz} - \sigma_{\theta\theta}), \tag{30}$$

where r and h denote the radius and wall thickness of the deformed tube, respectively. It is important to note that with this membrane approximation the contribution χ_{res} to the deformation is inadmissible and residual stresses cannot be included.

REMARK 3.2. Here we describe briefly the numerical technique used in the solution of the problem of bending, axial extension, inflation and torsion of a thick-walled cylindrical tube.

By assuming a particular state of residual strain (characterized by the parameter k), the fixed axial stretch λ_z and fixed angle of twist Φ of the tube, the isochoric part of the strain (and hence the stress) can always be expressed in terms of the two variables $\lambda_{\theta\mathrm{i}}$ and r, i.e. the circumferential stretch at the inner surface of the tube and the radius, respectively. Hence, the equation of equilibrium $(27)_1$ may be written in the general form

$$p_{\mathrm{i}} = \int_{r_{\mathrm{i}}}^{r_{\mathrm{o}}} \mathcal{F}(\lambda_{\theta\mathrm{i}}, r)\frac{\mathrm{d}r}{r}, \tag{31}$$

where r_{i} is given in terms of $\lambda_{\theta\mathrm{i}}$ by (18). Since closed-form evaluation of equation (31) is only possible for very simple constitutive equations, we employ a *Gaussian integration scheme* [31], i.e.

$$p_{\mathrm{i}} \approx \sum_{j=1}^{n} \mathcal{F}(\lambda_{\theta\mathrm{i}}, r_j)\frac{w_j}{r_j}, \tag{32}$$

where w_j and r_j $(j = 1, \ldots, n)$, denote the weights and the Gaussian points, and n is the order of integration. Equation (32) is, in general, nonlinear in the single unknown $\lambda_{\theta\mathrm{i}}$, and, for given p_{i}, can be solved for $\lambda_{\theta\mathrm{i}}$ using, for example, a standard *Newton iteration* with the initial value $\lambda_{\theta\mathrm{i}} = 1.0$.

Since the deformation is now determined, the torsional couple M_t and the reduced axial force F follow directly from equations $(27)_2$ and (29), respectively. This computation is carried out by employing another *Gaussian integration.* It turns out that for the considered range of deformations a three-point integration ($n = 3$) with the accuracy of order five gives sufficiently accurate solutions.

REMARK 3.3. The theory described above is designed to capture the deformation behavior in the central part of a tube so as to exclude end effects. Therefore, axial dependence of the deformation is not considered. This reflects the typical setting used in experiments (see, for example, [18] or [53], amongst others).

4. Some Constitutive Models for Arterial Walls

The *active* mechanical behavior of arterial walls is governed mainly by the intrinsic properties of elastin and collagen fibers and by the degree of activation of smooth muscles. An adequate constitutive model for arteries which incorporates the active state (contraction of smooth muscles) was proposed recently by Rachev and Hayashi [47].

The *passive* mechanical behavior of arterial walls is quite different and is governed mainly by the elastin and collagen fibers (see, for example, [8]). The passive state of the smooth muscles may also contribute to the passive arterial behavior but the extent of this contribution is not yet known. Most constitutive models proposed for arteries are valid for the passive state of smooth muscles and are based on a phenomenological approach which describes the artery as a macroscopic system. Furthermore, most of these models were designed to capture the response near the physiological state and in this respect they have been successfully applied in fitting experimental data. The most common potentials (strain-energy functions) are of exponential type, although polynomial and logarithmic forms are also used. For a review of a number of constitutive models describing the overall passive behavior, see Humphrey [32].

Some of the constitutive models proposed use the biphasic theory to describe arterial walls as hydrated soft tissues; see, for example, the works by Simon and co-workers [56, 55]. Less frequently used are models which account for the specific morphological structure of arterial walls. One attempt to model the helically wound fibrous structure is provided by Tözeren [60], which is based on the idea that the only wall constituent is the fiber structure. However, this is a significant simplification of the histological structure.

Another structural model due to Wuyts et al. [67] assumes that the wavy collagen fibrils are embedded in concentrically arranged elastin/smooth-muscle membranes, which is in agreement with the histological situation [49]. The model in [67] assumes that the collagen fibrils have a statistically distributed initial length. Each fibril may be stretched initially with a very low force but thereafter its behavior is linearly elastic. Only the media is considered as (solid) mechanically relevant. Although the model proposed in [67] attempts to incorporate histological information, which is a very promising approach, it is only possible to represent the deformation behavior of axially-symmetric thick-walled vessels. Another drawback is the fact that the artery is considered as a tube reinforced by circularly oriented collagen and elastin fibers, which does not model the real histological situation.

Most of the constitutive models treat the arterial wall as a single layer, but a number of two-layer models have been proposed in the literature. Two-layer models which include anisotropy are those due to, for example, Von Maltzahn et al. [38], Demiray [11] and Rachev [46]. However, the emphasis of the latter paper is on stress-dependent remodeling.

In this section we evaluate and compare some prominent potentials of the exponential, polynomial and logarithmic type which are often used to characterize the overall *passive* behavior of arterial walls. The study aims to illustrate the performance of the potentials and their reliability for the prediction of the state of deformation. In particular, we use a systematic analysis to examine the inflation of a cylindrical tube at various axial stretches λ_z and to compute the evolution of the inner radius r_i with the internal pressure p_i and the reduced axial force F. In addition, at an internal pressure of $p_i = 13.33$ [kPa] (100 [mm Hg]), the approximate physiological pressure, we evaluate the effect of twist (torsion) on the tube and determine the dependence of the shear γ_i at the inner surface on the torsional couple M_t and the reduced axial force F. However, as will be shown, the mechanical behavior of an anisotropic cylindrical tube under torsion can in general only be investigated if the constitutive model is based on a fully three-dimensional formulation.

4.1. THREE-DIMENSIONAL FORMULATION

This section is concerned with three-dimensional strain-energy functions appropriate for the analysis of *thick-walled* tubes, which is a necessary point of departure for the study of the mechanical behavior of arterial walls.

4.1.1. *Strain-Energy Function Proposed by Delfino et al. [10]*

As already mentioned, the different layers of arterial walls are highly anisotropic due to the organized arrangement of the load carrying (collagen) components. However, there are many *isotropic* strain-energy functions proposed in the literature and used in practice to characterize the mechanical response of arterial walls (see, for example, the oversimplified rubber-like potential used in [30] which cannot represent the strong stiffening effect of arteries in the large strain domain).

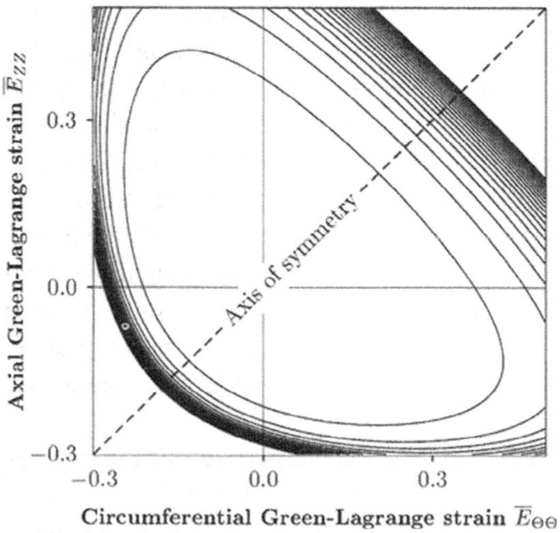

Figure 4. Contour plot of the convex potential (33) with material parameters $a = 44.2$ [kPa] and $b = 16.7$ [–] (see [10]).

Recently, Delfino et al. [10] proposed an (isotropic) rubber-like potential for carotid arteries which is able to model the typical stiffening effects in the high pressure domain. The strain-energy function $\overline{\Psi}$ has the form

$$\overline{\Psi} = \frac{a}{b}\left\{\exp\left[\frac{b}{2}(\bar{I}_1 - 3)\right] - 1\right\} \tag{33}$$

[10], where $a > 0$ is a stress-like material parameter and $b > 0$ is a non-dimensional parameter. The first invariant of the modified right Cauchy–Green tensor $\overline{\mathbf{C}}$ is defined as $\bar{I}_1 = \overline{\mathbf{C}} : \mathbf{I}$. Since the exponential function increases monotonically with $\bar{I}_1$ it is easy to show that strict local convexity of the potential (33) as a function of $\overline{\mathbf{C}}$ (or equivalently $\overline{\mathbf{E}}$) is guaranteed, bearing in mind that because of the incompressibility condition the components of $\overline{\mathbf{E}}$ are not independent.

In the present context, strict local convexity means that the second derivative of Ψ with respect to $\mathbf{E}$ is positive definite, with appropriate modifications to account for incompressibility. This fundamental physical requirement in hyperelasticity ensures that undesirable material instabilities are precluded (for a general discussion of convexity in hyperelasticity the reader is referred to, for example, [42], Section 6, and [7]). It can be shown that strict local convexity of Ψ implies that the contours of constant Ψ are convex, and, in particular, that the projections of these contours in the $(\overline{E}_{\Theta\Theta}, \overline{E}_{ZZ})$-plane are convex. On the other hand, if the contours are not convex then it can be deduced that the potential Ψ is not strictly locally convex. The consequences of this will be seen in the following sections, which show results contrasting with those in Figure 4, in which the (convex) contours are illustrated. For this figure we have used material parameters $a = 44.2$ [kPa] and $b = 16.7$ [–]

Table I. Material and geometrical data of a human carotid artery [10].

Material	Geometry	
$a = 44.2$ [kPa]	$\alpha = 0.0°$	$\alpha = 100.0°$
$b = 16.7$ [–]		
	$R_i = 3.1$ [mm]	$R_i = 4.46$ [mm]
	$R_o = 4.0$ [mm]	$R_o = 5.36$ [mm]

as proposed in [10]. Since the potential is isotropic the contour plots are symmetric in the line which bisects the axes.

For the function (33) the isochoric contribution $\overline{\boldsymbol{\sigma}}$ to the Cauchy stress tensor $\boldsymbol{\sigma}$ is obtained from $(10)_3$ as

$$\overline{\boldsymbol{\sigma}} = 2\overline{\Psi}_1 \operatorname{dev} \overline{\mathbf{b}}, \tag{34}$$

where $\overline{\Psi}_1 = \partial\overline{\Psi}/\partial\bar{I}_1$ and $\overline{\mathbf{b}}$ is the modified left Cauchy–Green tensor. Note that $J = 1$.

Hence, with the definition (3) of the left Cauchy–Green tensor, the isochoric Cauchy stress components, which are used in equations $(27)_1$, $(27)_2$ and (29), are given by (34). In order to investigate the specific arterial response we use the material constants for a human carotid artery given in [10]. The values are given in Table I. For the description of the stress-free configuration we have taken $\alpha = 100.0°$, which is based on the value in [10] adjusted for the different definitions of opening angle. For consistency we take the geometry in the unloaded configuration for the residually stressed case to be the same as that for the case without residual stress. The geometry in the unstressed configuration is obtained using the incompressibility condition, which furnishes a connection between the inner and outer radii in this configuration, together with the simplifying assumption of unchanged wall thickness. The resulting values are given in Table I. In addition, for purposes of comparison of the shear γ_i at the inner surface of the tube, the undeformed length L of the arterial tube was taken to be equal to the value of the inner radius R_i corresponding to $\alpha = 0.0°$. This was followed for all the cylinder models studied subsequently.

This basis for the computation of the radii in the stress-free configuration is also adopted in the following sections. The values of the material constants that we use for our investigations are those given in the papers in which the energy functions were introduced. We note that these values were determined under the assumption that the unloaded configuration is stress free (it is not clear if the same assumption was used in [10]).

The *in vivo* axial pre-stretch is based on *in situ* measurements prior to removal of the artery and is given as $\lambda_z = 1.1$ (see [10]). The mechanical response of the human carotid artery during inflation and torsion is shown in Figure 5. The internal

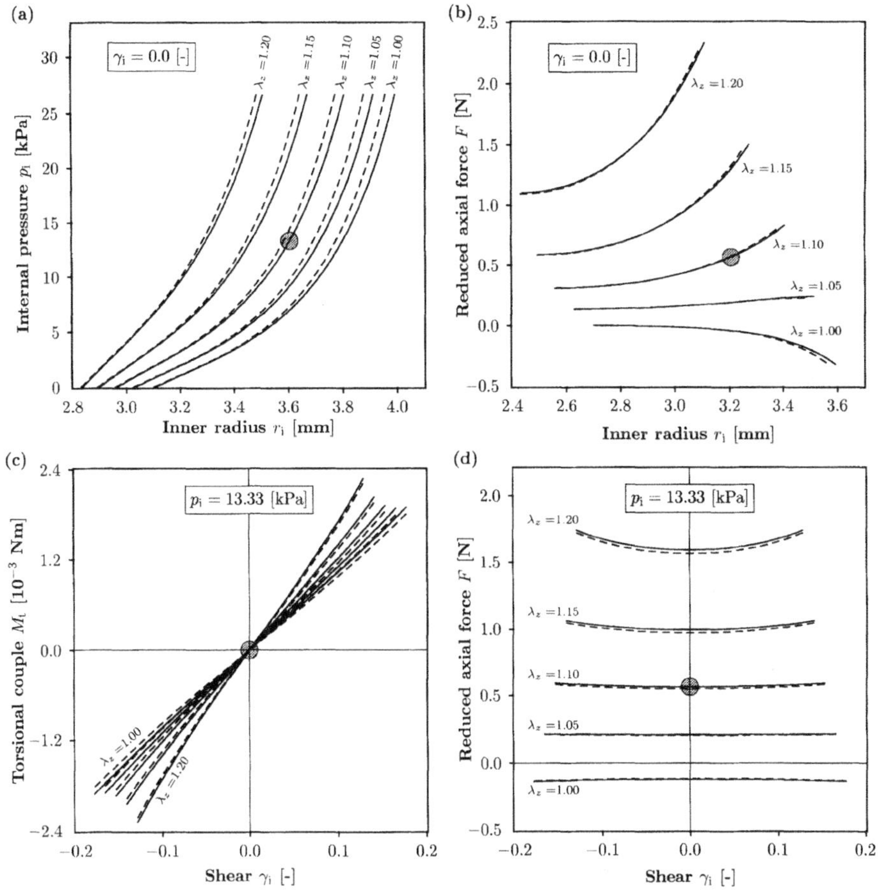

Figure 5. Deformation behavior of a human carotid artery during inflation and torsion using the constitutive model proposed in [10]. Solid lines are numerical results with residual strains included ($\alpha = 100.0°$) while the dashed lines are results without residual strains ($\alpha = 0.0°$). Dependence of (a) the internal pressure p_i and (b) the reduced axial force F on the inner radius r_i in the absence of shear deformation ($\gamma_i = 0$). Dependence of (c) the torsional couple M_t and (d) the reduced axial force F on the shear γ_i at fixed internal pressure $p_i = 13.33$ [kPa]. The shaded circles indicate the approximate central region of the physiological state.

pressure p_i and the angle of twist Φ are varied within the ranges

$$0 \leqslant p_i \leqslant 26.67 \text{ [kPa]} \quad \text{and} \quad -0.15 \leqslant \Phi \leqslant 0.15 \text{ [rad]}. \tag{35}$$

These loadings are applied at fixed axial stretches of the artery varying between $\lambda_z = 1.0$ and $\lambda_z = 1.2$.

The predicted response is in good qualitative agreement with the experimentally observed mechanical behavior of arteries; see, for example, the survey article [32] or compare with the (rare) data on shear tests of arteries provided in [12]. With the potential (33) the typical stiffening effect at high pressures can be replicated, as can be seen in Figure 5(a). Remarkably, the reduced axial force F is hardly influenced by the changes of internal pressures p_i at $\lambda_z = 1.05$, which is close to the physiological pre-stretch of $\lambda_z = 1.1$; see Figure 5(b). In Figure 5(c) and (d) respectively the torsional couple M_t and the reduced axial force F are plotted against the amount of shear at the inner surface, i.e. $\gamma_\mathrm{i} = \Phi r_\mathrm{i}/l = \Phi r_\mathrm{i}/\lambda_z L$.

The solid lines show numerical results based on a load-free, but not stress-free, configuration ($\alpha = 100.0°$), while the dashed lines are based on a load-free and stress-free configuration ($\alpha = 0.0°$). As illustrated in Figure 5(a), residual stresses influence the internal pressure/inner radius behavior moderately, while Figure 5(b)–(d) shows a very minor influence of the residual stresses on the global mechanical response of the artery. Note that incorporation of the residual stresses in the load-free configuration softens the material of the artery in the sense that a given inner radius r_i is achieved with a lower internal pressure p_i, a finding which is in agreement with the analytical studies of Humphrey [32], p. 101.

4.1.2. *Strain-Energy Function of Fung's Type*

The strain-energy function used most extensively for arteries appears to be the two-dimensional exponential form proposed by Fung et al. [18]. A generalization to the three-dimensional regime, presented by Chuong and Fung [5], assumes that the principal directions of the stress tensor coincide with the radial, circumferential and axial directions of the artery. Shear deformations due to, for example, torsion of the artery, were not considered. To incorporate shear deformations $\overline{E}_{\Theta Z}$ in the shear planes $z =$ constant, Deng et al. [12] proposed an extension of the classical two-dimensional function given in [18].

Many modifications of these strain-energy functions have been published subsequently. For example, a combined polynomial-exponential form of the strain-energy function incorporating shear deformations was given by Kas'yanov and Rachev [35]. The most general strain-energy function of *Fung's* type is formulated by Humphrey [32]. It is suitable for arbitrary (three-dimensional) states of deformations and has the form

$$\overline{\Psi} = \frac{1}{2}c[\exp(Q) - 1], \tag{36}$$

where c is a material parameter and Q is given by

$$\begin{aligned} Q = {} & b_1\overline{E}^2_{\Theta\Theta} + b_2\overline{E}^2_{ZZ} + b_3\overline{E}^2_{RR} + 2b_4\overline{E}_{\Theta\Theta}\overline{E}_{ZZ} + 2b_5\overline{E}_{ZZ}\overline{E}_{RR} \\ & + 2b_6\overline{E}_{RR}\overline{E}_{\Theta\Theta} + b_7\overline{E}^2_{\Theta Z} + b_8\overline{E}^2_{RZ} + b_9\overline{E}^2_{R\Theta}. \end{aligned} \tag{37}$$

Here b_i, $i = 1, \ldots, 9$, are non-dimensional material parameters, while $\overline{E}_{IJ}$, for $I, J = R, \Theta, Z$, are the components of the modified Green–Lagrange strain tensor referred to cylindrical polar coordinates (R, Θ, Z).

In the work of Fung et al. [18] and Chuong and Fung [5] there is no *a priori* restriction on the material parameters presented. However, it is important to note that in order for the (anisotropic) function $\overline{\Psi}$ to be convex in the sense discussed

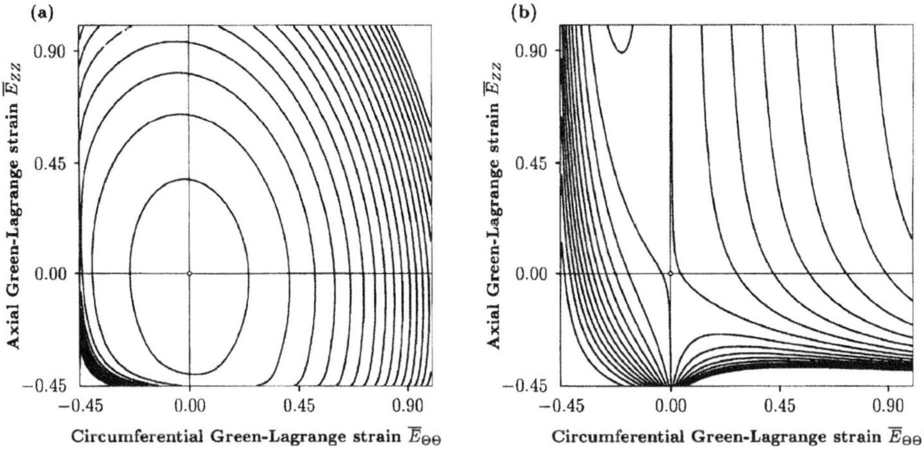

Figure 6. Contour plots of the potential (36) with (a) material parameters $c, b_1, \ldots, b_7$ in Table II, and (b) a set of parameters chosen to illustrate non-convexity.

Table II. Material and geometrical data for a rabbit carotid artery in respect of (37) (see experiment 71 in [5]; b_7 is based on the study in [12]).

Material	Geometry	
$c = 26.95$ [kPa]	$\alpha = 0.0°$	$\alpha = 160.0°$
$b_1 = 0.9925$ [–]		
$b_2 = 0.4180$ [–]	$R_i = 0.71$ [mm]	$R_i = 1.43$ [mm]
$b_3 = 0.0089$ [–]	$R_o = 1.10$ [mm]	$R_o = 1.82$ [mm]
$b_4 = 0.0749$ [–]		
$b_5 = 0.0295$ [–]		
$b_6 = 0.0193$ [–]		
$b_7 = 5.0000$ [–]		

in Section 4.1.1 the material parameters $b_1, \ldots, b_9$ must not be chosen arbitrarily. Figure 6(a) shows a contour plot of the potential (36) with the material parameters proposed in [5] (see the summary in Table II), while the contour plot in Figure 6(b) uses an alternative set of parameters chosen to illustrate non-convexity of the strain-energy function. This is one of many possible choices which lead to non-convexity. The 'physical' meanings of the individual parameters are unclear (see the discussion in the book by Fung [17], Section 8.6.2). Hence, if this strain-energy function is used care must be taken to select appropriate restrictions on the values of c and b_i, since unconstrained parameter optimization does not, in general, guarantee convexity. It is therefore important to be sure that the optimization process is performed within a range of parameters for which convexity is assured.

Furthermore, in the computational context, in order to obtain solutions of complex nonlinear (initial boundary-value) problems, incremental/iterative solution techniques of Newton's type are frequently applied to solve a sequence of linearized problems. These techniques, often employed in computational biomechanics, use predicted states of deformation which might be far from the range of deformation for which the experimental tests were conducted. This might lead to numerical problems within the solution procedure when strain-energy functions are used which are not convex *a priori*. To be specific, parameters used outside the range for which the fitting process was performed might induce a non-convex potential. A further comment on parameter identification for the strain-energy function (36) used in combination with relation (37) is appropriate. Because of the large number of material parameters b_i a least-square procedure can lead to problems of non-uniqueness associated with their sensitivity to small changes in the data, as pointed out, for example, by Fung [17], Section 8.6.1.

With equations (36) and $(10)_3$ the components of the isochoric part of the Cauchy stress tensor may be obtained in the cylindrical polar coordinate system (r, θ, z), so that it is a straightforward task to solve the equilibrium equation $(27)_1$ and to calculate the torsional couple M_t and the reduced axial force F, as given by equations $(27)_2$ and (29), respectively. The material and geometrical data used for this computation are from a carotid artery of a rabbit and are summarized in Table II. The values of the parameters c and b_i, $i = 1, \ldots, 6$, are taken from [5], while b_7 is an estimated value (measuring the resistance to distortion) based on the shear moduli of arteries as presented in [12]. Since the associated shear strains $\overline{E}_{ZR}$ and $\overline{E}_{R\Theta}$ vanish in the considered problem, the parameters b_8 and b_9 need not be considered. In order to investigate the influence of residual stresses on the global response of the artery, two different stress-free states are considered, namely $\alpha = 0.0°$ and $\alpha = 160.0°$. The (mean) value 160.0° is based on the study [23].

The *in vivo* axial pre-stretch of the artery is given as $\lambda_z = 1.695$ (calculated from the axial component of the Green–Lagrange strain given in [18], Table I, experiment 71) and the internal pressure p_i and the angle of twist Φ are varied within the ranges

$$0 \leqslant p_\mathrm{i} \leqslant 21.33\ [\mathrm{kPa}] \quad \text{and} \quad -0.10 \leqslant \Phi \leqslant 0.10\ [\mathrm{rad}] \tag{38}$$

(compare with [12, 18]). These loads are applied at fixed axial stretches of the artery varying between $\lambda_z = 1.5$ and $\lambda_z = 1.9$. Figure 7 shows the computed deformation behavior of the artery under various loading conditions. The influence of residual strain on the deformation field is much larger than was the case in the study of Section 4.1.1 (see Figure 5). This is because the opening angle α and the ratio of the wall thickness to the diameter were larger than the values used in Section 4.1.1. Note that at $\lambda_z = 1.9$ the reduced axial force F first increases with the inner radius r_i and then, at high pressures, it tends to decrease; see Figure 7(b). This characteristic is not observed experimentally.

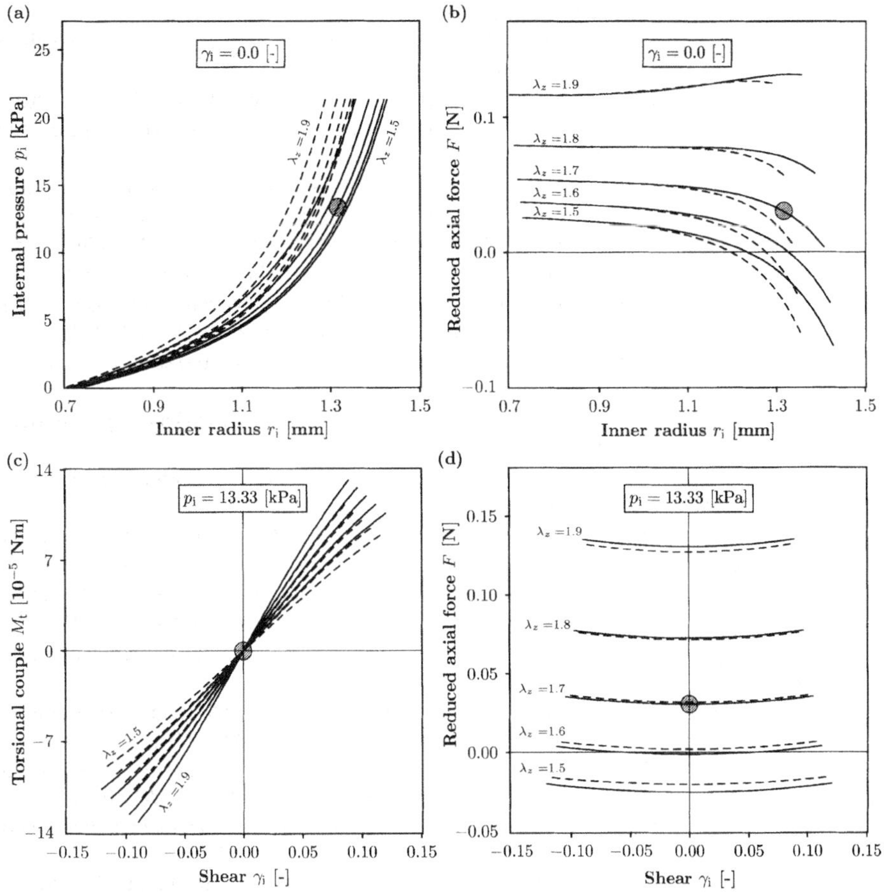

Figure 7. Deformation behavior of a rabbit carotid artery during inflation and torsion for the constitutive model (36), (37) with $b_8 = b_9 = 0$. Solid lines are numerical data with residual strains included ($\alpha = 160.0°$) while the dashed lines are results without residual strains ($\alpha = 0.0°$). Dependence of (a) the internal pressure p_i and (b) the reduced axial force F on inner radius r_i without shear deformation ($\gamma_i = 0$). Dependence of (c) the torsional couple M_t and (d) the reduced axial force F on the shear γ_i at fixed internal pressure $p_i = 13.33$ [kPa]. The shaded circles indicate the approximate central region of the physiological state.

4.2. TWO-DIMENSIONAL FORMULATION

Since it is generally accepted that it is appropriate to treat arterial walls as incompressible materials, the restriction $J = 1$ can be used to find alternative expressions for the strain-energy function $\overline{\Psi}$, which, in general, is a function of the strain components $\overline{E}_{RR}, \overline{E}_{\Theta\Theta}, \overline{E}_{ZZ}, \overline{E}_{R\Theta}, \overline{E}_{RZ}, \overline{E}_{\Theta Z}$. The alternative potential $\hat{\overline{\Psi}}(\overline{E}_{\Theta\Theta}, \overline{E}_{ZZ}, \overline{E}_{\Theta Z})$, which we refer to as the *two-dimensional* counterpart of $\overline{\Psi}$, is very popular and used frequently in the literature. Such a two-dimensional formulation is not capable of describing the three-dimensional *anisotropic* behavior of a thick-walled cylindrical tube under, for example, inflation and torsion. However, for the special case of combined bending, inflation and axial extension the potential $\hat{\overline{\Psi}}$ is suitable for predicting the three-dimensional state of stress. A comparative study of three important examples of such potentials concludes this section.

4.2.1. *The Strain-Energy Function* $\hat{\overline{\Psi}}(\overline{E}_{\Theta\Theta}, \overline{E}_{ZZ}, \overline{E}_{\Theta Z})$

We consider a thick-walled cylindrical tube of incompressible material deformed in such a way that the modified Green–Lagrange strains $\overline{E}_{R\Theta}$ and $\overline{E}_{RZ}$ are zero, as is the case for the deformation considered in Section 3.2. Using these assumptions and with elimination of $\overline{E}_{RR}$ via the incompressibility constraint an alternative two-dimensional approximation of $\overline{\Psi}$ may be given in the form

$$\overline{\Psi}(\overline{E}_{RR}, \overline{E}_{\Theta\Theta}, \overline{E}_{ZZ}, \overline{E}_{R\Theta}, \overline{E}_{RZ}, \overline{E}_{\Theta Z}) = \hat{\overline{\Psi}}(\overline{E}_{\Theta\Theta}, \overline{E}_{ZZ}, \overline{E}_{\Theta Z}), \tag{39}$$

where $\hat{\overline{\Psi}}$ is a strain energy with the three independent strain variables $\overline{E}_{\Theta\Theta}$, $\overline{E}_{ZZ}$, $\overline{E}_{\Theta Z}$. Using the chain rule, the derivatives

$$\frac{\partial \hat{\overline{\Psi}}}{\partial \overline{E}_\alpha} = \frac{\partial \overline{\Psi}}{\partial \overline{E}_\alpha} + \frac{\partial \overline{\Psi}}{\partial \overline{E}_{RR}} \frac{\partial \overline{E}_{RR}}{\partial \overline{E}_\alpha}, \qquad \alpha = \Theta\Theta, ZZ, \Theta Z, \tag{40}$$

are obtained. The constraint $\det \overline{\mathbf{C}} = \det(2\overline{\mathbf{E}} + \mathbf{I}) = 1$ and equation $(20)_2$ enable $\overline{E}_{RR}$ to be expressed in terms of the independent components $\overline{E}_{\Theta\Theta}$, $\overline{E}_{ZZ}$, $\overline{E}_{\Theta Z}$ according to

$$\overline{E}_{RR} = \frac{1}{2}\{[(2\overline{E}_{\Theta\Theta} + 1)(2\overline{E}_{ZZ} + 1) - 4\overline{E}^2_{\Theta Z}]^{-1} - 1\}. \tag{41}$$

Hence, with (41), we find from (40) that

$$\frac{\partial \overline{\Psi}}{\partial \overline{E}_{\Theta\Theta}} = \frac{\partial \hat{\overline{\Psi}}}{\partial \overline{E}_{\Theta\Theta}} + (2\overline{E}_{ZZ} + 1)(2\overline{E}_{RR} + 1)^2 \frac{\partial \overline{\Psi}}{\partial \overline{E}_{RR}}, \tag{42}$$

$$\frac{\partial \overline{\Psi}}{\partial \overline{E}_{ZZ}} = \frac{\partial \hat{\overline{\Psi}}}{\partial \overline{E}_{ZZ}} + (2\overline{E}_{\Theta\Theta} + 1)(2\overline{E}_{RR} + 1)^2 \frac{\partial \overline{\Psi}}{\partial \overline{E}_{RR}}, \tag{43}$$

$$\frac{\partial \overline{\Psi}}{\partial \overline{E}_{\Theta Z}} = \frac{\partial \hat{\overline{\Psi}}}{\partial \overline{E}_{\Theta Z}} - 4\overline{E}_{\Theta Z}(2\overline{E}_{RR} + 1)^2 \frac{\partial \overline{\Psi}}{\partial \overline{E}_{RR}}. \tag{44}$$

The aim now is to solve the equilibrium equation $(27)_1$, which requires the stress difference $\overline{\sigma}_{\theta\theta} - \overline{\sigma}_{rr}$. From the stress equation $(10)_3$ and the kinematic relation $(19)_2$ we find that $\overline{\sigma}_{\theta\theta} = \lambda_\theta^2 \partial\overline{\Psi}/\partial\overline{E}_{\Theta\Theta} + 2\gamma\lambda_z\lambda_\theta\partial\overline{\Psi}/\partial\overline{E}_{\Theta Z} + \gamma^2\lambda_z^2\partial\overline{\Psi}/\partial\overline{E}_{ZZ}$ and $\overline{\sigma}_{rr} = \lambda_r^2\partial\overline{\Psi}/\partial\overline{E}_{RR}$. Using equations (42)–(44) we find that the stress difference cannot be given in terms of the potential $\hat{\overline{\Psi}}$ alone since the expression $\partial\overline{\Psi}/\partial\overline{E}_{RR}$ cannot in general be eliminated and remains undetermined. Moreover, $\overline{\sigma}_{\theta z}$ similarly depends on $\partial\overline{\Psi}/\partial\overline{E}_{RR}$. This means, in particular, that it is not in general possible to use the alternative potential $\hat{\overline{\Psi}}$ to derive the complete state of stress in an anisotropic cylindrical tube under inflation and torsion in a three-dimensional context.

It is worth noting, however, that the particular stress combinations

$$\overline{\sigma}_{\theta\theta} - \overline{\sigma}_{rr} - 2\gamma\overline{\sigma}_{\theta z} = \lambda_\theta^2 \frac{\partial\hat{\overline{\Psi}}}{\partial \overline{E}_{\Theta\Theta}} - \gamma^2\lambda_z^2 \frac{\partial\hat{\overline{\Psi}}}{\partial \overline{E}_{ZZ}}, \tag{45}$$

$$\overline{\sigma}_{zz} - \overline{\sigma}_{rr} = \lambda_z^2 \frac{\partial\hat{\overline{\Psi}}}{\partial \overline{E}_{ZZ}} \tag{46}$$

are given in terms of $\hat{\overline{\Psi}}$, and we note that $\lambda_\theta^2 = 2\overline{E}_{\Theta\Theta} + 1$, $(1+\gamma^2)\lambda_z^2 = 2\overline{E}_{ZZ} + 1$.

Exceptionally, for an isotropic material, since the Cauchy stress tensor is coaxial with the left Cauchy–Green tensor, the universal relation

$$\gamma\lambda_z^2(\overline{\sigma}_{\theta\theta} - \overline{\sigma}_{zz}) = (\lambda_\theta^2 + \gamma^2\lambda_z^2 - \lambda_z^2)\overline{\sigma}_{\theta z} \tag{47}$$

is obtained. This can be shown as follows. Let $\cos\phi\mathbf{e}_\theta + \sin\phi\mathbf{e}_z$ denote a principal axis of the Cauchy–Green tensor $\mathbf{b}$. Then, using (21), the angle ϕ may be calculated in the form

$$\tan 2\phi = \frac{2\gamma\lambda_z^2}{\lambda_\theta^2 + \gamma^2\lambda_z^2 - \lambda_z^2}, \tag{48}$$

while, for the Cauchy stresses, we have

$$\tan 2\phi = \frac{2\overline{\sigma}_{\theta z}}{\overline{\sigma}_{\theta\theta} - \overline{\sigma}_{zz}}. \tag{49}$$

The combination of (45)–(47) enables the normal stress differences and the shear stress to be expressed in terms of $\hat{\overline{\Psi}}$. We omit the details since consideration of isotropic models is not of interest here.

REMARK 4.1. For an anisotropic material, in the special case $\gamma = 0$ (no shear deformation), equations $(27)_1$ and (28) can be solved on the basis of the two-dimensional form of strain-energy function $\hat{\overline{\Psi}}(\overline{E}_{\Theta\Theta}, \overline{E}_{ZZ}, \overline{E}_{\Theta Z})$. With $\gamma = 0$, it follows from (45) that

$$\overline{\sigma}_{\theta\theta} - \overline{\sigma}_{rr} = (1 + 2\overline{E}_{\Theta\Theta})\frac{\partial\hat{\overline{\Psi}}}{\partial \overline{E}_{\Theta\Theta}}. \tag{50}$$

Using (46) together with (50) the expression $2\overline{\sigma}_{zz} - \overline{\sigma}_{\theta\theta} - \overline{\sigma}_{rr}$ in (29) is then obtained in the form

$$2\overline{\sigma}_{zz} - \overline{\sigma}_{\theta\theta} - \overline{\sigma}_{rr} = 2(1 + 2\overline{E}_{ZZ})\frac{\partial\hat{\overline{\Psi}}}{\partial \overline{E}_{ZZ}} - (1 + 2\overline{E}_{\Theta\Theta})\frac{\partial\hat{\overline{\Psi}}}{\partial \overline{E}_{\Theta\Theta}}. \tag{51}$$

Hence, the reduced axial force F may be expressed in terms of the strain energy $\hat{\overline{\Psi}}$.

Table III. Material and geometrical data for a rabbit carotid artery based on the potential (52) (see experiment 71 in [18]).

Material	Geometry	
$c_1 = -24.385$ [kPa]	$\alpha = 0.0°$	$\alpha = 160.0°$
$c_2 = -3.589$ [kPa]		
$c_3 = -1.982$ [kPa]	$R_i = 0.71$ [mm]	$R_i = 1.43$ [mm]
$c_4 = 46.334$ [kPa]	$R_o = 1.10$ [mm]	$R_o = 1.82$ [mm]
$c_5 = 32.321$ [kPa]		
$c_6 = 3.743$ [kPa]		
$c_7 = 3.266$ [kPa]		

4.2.2. *Strain-Energy Function Proposed by Vaishnav et al. [62]*

Two-dimensional forms of strain-energy functions suitable for the description of the deformation behavior of canine thoracic aorta using polynomial expressions have been proposed by Vaishnav et al. [62]. This classic paper presents three polynomial expressions with 3, 7 or 12 material parameters. As studied in [62] the three-parameter model is too inaccurate for a serious investigation and the twelve-parameter model does not have a significant advantage over the seven-parameter model. Hence, in the present paper we focus attention on the seven-parameter model, which is written in the form

$$\begin{aligned}\hat{\overline{\Psi}} = {} & c_1\overline{E}_{\Theta\Theta}^2 + c_2\overline{E}_{\Theta\Theta}\overline{E}_{ZZ} + c_3\overline{E}_{ZZ}^2 + c_4\overline{E}_{\Theta\Theta}^3 + c_5\overline{E}_{\Theta\Theta}^2\overline{E}_{ZZ} \\ & + c_6\overline{E}_{\Theta\Theta}\overline{E}_{ZZ}^2 + c_7\overline{E}_{ZZ}^3,\end{aligned} \tag{52}$$

where $c_1, \ldots, c_7$ are stress-like material parameters and $\overline{E}_{\Theta\Theta}$ and $\overline{E}_{ZZ}$ are the components of the modified Green–Lagrange strain tensor in the circumferential and axial directions, respectively. In Fung et al. [18] the form (52) was used to fit the parameters $c_1, \ldots, c_7$ to experimental data from rabbit carotid arteries. The result is summarized in Table III. Note that with these values of the material parameters the strain-energy function (52) is not convex, as the contour plot in Figure 8 shows. In fact, because of the cubic nature of the strain-energy function (52), it is not convex for *any* set of values of the material constants.

While the performance of the constitutive law (52) is acceptable in the tensile region ($\overline{E}_{\Theta\Theta} > 0$, $\overline{E}_{ZZ} > 0$), it fails for compressive strains. Moreover, Fung [17] showed that two completely different sets of material parameters $c_1, \ldots, c_7$ are able to represent the mechanical response of the same artery quite well. This lack of uniqueness of the material parameters is problematic. Note that a polynomial expression, different from that in [62], has been proposed in [63] in order to model the *three-dimensional* anisotropic behavior of a canine carotid artery and of a rabbit aorta.

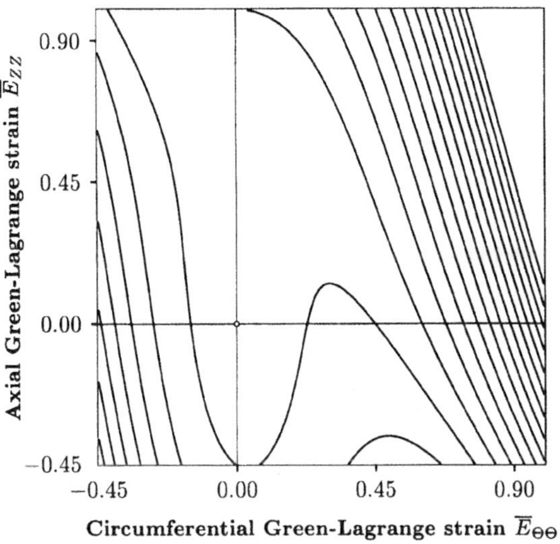

Figure 8. Contour plot of the potential (52) with material parameters given in Table III.

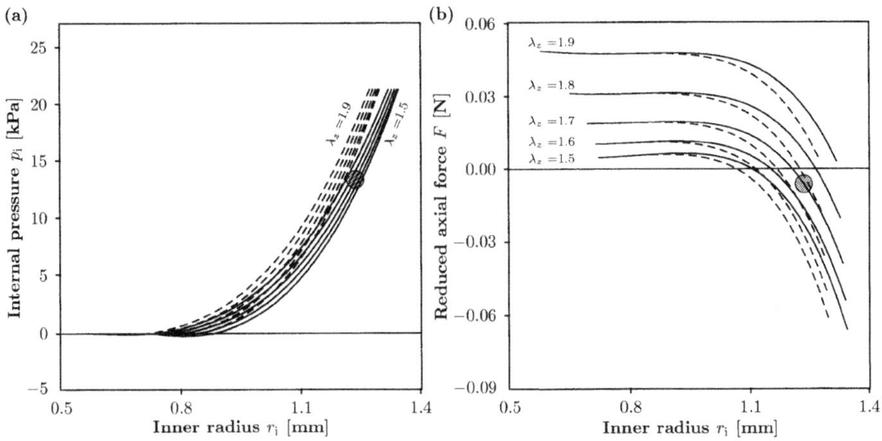

Figure 9. Deformation behavior of a rabbit carotid artery during inflation based on the constitutive model in [62]. Solid lines are numerical results with residual strains included ($\alpha = 160.0°$) while dashed lines are results without residual strains ($\alpha = 0.0°$). Dependence of (a) the internal pressure p_i, and (b) the reduced axial force F on the inner radius r_i, without shear deformation ($\gamma_i = 0$). The shaded circles indicate the approximate central region of the physiological state.

Using (52) we may solve the equilibrium equation $(27)_1$ and calculate the reduced axial force F from equation (29). For this purpose, we use the formulas (50) and (51) and apply the procedure described in Remark 3.2. The material and geometrical data for a rabbit carotid artery are as summarized in Table III and we use the same set of geometrical data and the same range of loading as in Section 4.1.2.

Figure 9 shows the predicted mechanical response of the considered artery (experiment 71 in [18]). The dependence of the internal pressure p_i on the inner radius

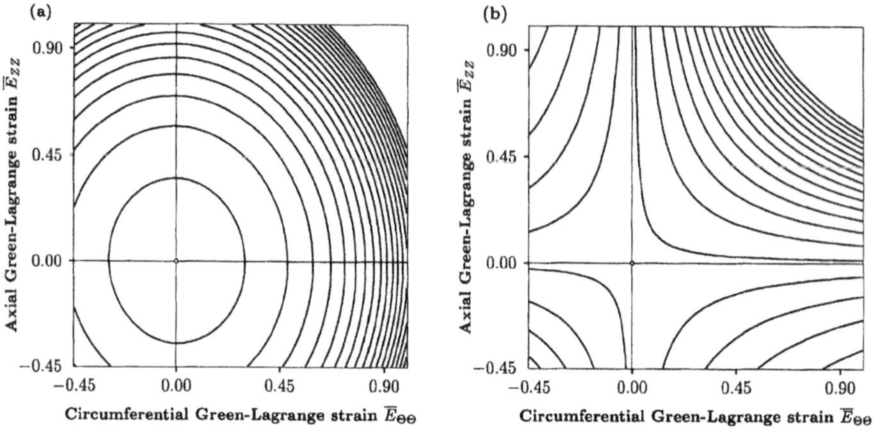

Figure 10. Contour plots of the potential (53) with (a) material parameters given in Table IV, adopted from [59], and (b) with a set of parameters chosen to illustrate non-convexity.

r_i is plotted in Figure 9(a). Note that the solution of the equilibrium equation $(27)_1$ is not unique, since, for example, $p_i = 0$ corresponds to two different values of r_i. This is a consequence of the non-convexity of the potential. The dependence of the reduced axial force F on the inner radius r_i is plotted in Figure 9(b). As can be seen, for axial stretches higher than the typical physiological stretch, the characteristic increase of F with the inner radius r_i illustrated in Figures 5(b) and 7(b) is not predicted by the potential (52) with the set of material parameters given in Table III.

The potential (52) represents the first attempt to describe the *anisotropic* mechanical response of arteries. However, as indicated above, its applicability is limited.

4.2.3. *Strain-Energy Function Proposed by Fung et al. [18]*

The well-known exponential strain-energy function due to Fung et al. [18] has been proposed in the two-dimensional form

$$\hat{\overline{\Psi}} = \frac{1}{2}c[\exp(\hat{Q}) - 1], \qquad \hat{Q} = b_1\overline{E}^2_{\Theta\Theta} + b_2\overline{E}^2_{ZZ} + 2b_4\overline{E}_{\Theta\Theta}\overline{E}_{ZZ}, \tag{53}$$

where c is a stress-like material parameter and b_1, b_2, b_4 are non-dimensional parameters. However, as discussed in Section 4.1.2, the material parameters cannot be chosen arbitrarily if convexity of the function (53) is desired. Contour plots of the potential (53) are shown in Figure 10. The material parameters proposed in [59], and given in Table IV, are used in Figure 10(a), in which case the contours are convex. On the other hand, non-convexity is illustrated in Figure 10(b) in respect of a specific choice of parameters. As in the case of Section 4.1.2, the non-convexity can be demonstrated for a wide range of parameter values. This is easy to check because of the quadratic nature of $\hat{Q}$ in equation $(53)_2$. In fact, it can be shown, for example, that if $c > 0$, then (53) is strictly locally convex *if and only if* $b_1 > 0$, $b_2 > 0$ and $b_1 b_2 > b_4^2$.

Table IV. Material and geometrical data for a dog carotid artery based on the potential (53) (see experiment D850815C in [59]).

Material	Geometry	
$c = 28.58$ [kPa]	$\alpha = 0.0°$	$\alpha = 160.0°$
$b_1 = 0.8329$ [–]		
$b_2 = 0.6004$ [–]	$R_i = 1.21$ [mm]	$R_i = 2.40$ [mm]
$b_4 = 0.0169$ [–]	$R_o = 1.77$ [mm]	$R_o = 2.96$ [mm]

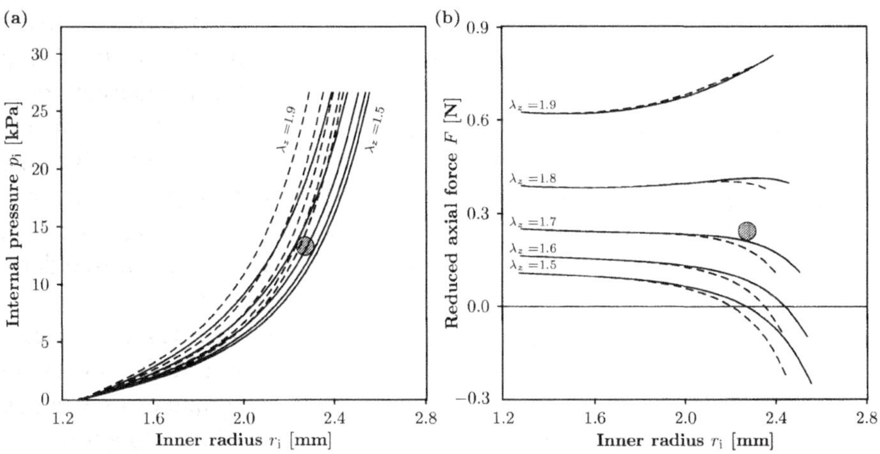

Figure 11. Deformation behavior of a dog carotid artery (see experiment D850815C in [59]) during inflation using the constitutive model in [18]. Solid lines are numerical results with residual strains included ($\alpha = 160.0°$) while the dashed lines are results without residual strains ($\alpha = 0.0°$). Dependence of (a) the internal pressure p_i and (b) the reduced axial force F on the inner radius r_i, without shear deformation ($\gamma_i = 0$). The shaded circles indicate the approximate central region of the physiological state.

By means of equations (53) and (50) we may solve the equation of equilibrium $(27)_1$ and calculate the reduced axial force F, as outlined in Remark 3.2. The resulting response is shown in Figure 11 for the material and geometrical data given in Table IV. In order to investigate the influence of residual stresses on the response of the artery, two different stress-free states are considered ($\alpha = 0.0°$ and $\alpha = 160.0°$). The *in vivo* axial pre-stretch of the artery is given as $\lambda_z = 1.72$ and the internal pressure p_i is varied within the range

$$0 \leqslant p_i \leqslant 26.67 \text{ [kPa]} \tag{54}$$

(see experiment D850815C in [59]). The inflation is considered at fixed axial stretches of the artery between $\lambda_z = 1.5$ and $\lambda_z = 1.9$.

The potential (53) is able to model the basic characteristics of the mechanical behavior of arteries except in the low pressure domain. The problem in the low

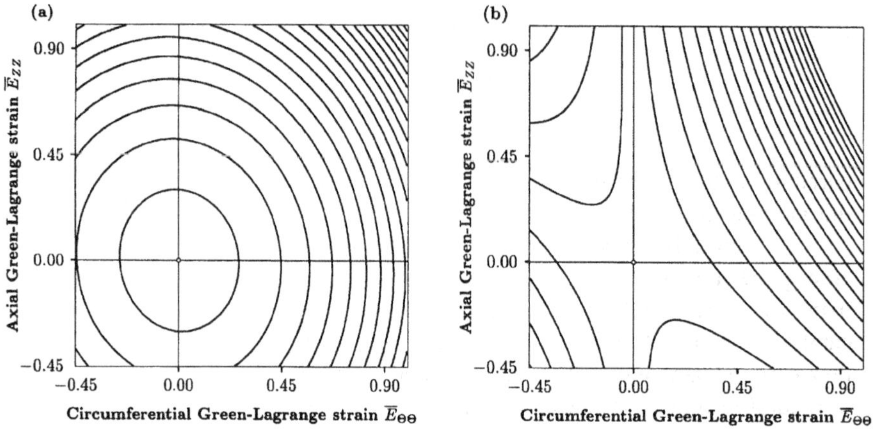

Figure 12. Contour plots of the potential (55) with (a) material parameters given in Table V, adopted from [59], and (b) a set of parameters chosen to illustrate non-convexity.

pressure domain, as can be seen in Figure 11(a), is that r_{i} is independent of the axial stretch when $p_{\mathrm{i}} = 0$. A strong influence of the residual stresses on the $p_{\mathrm{i}} - r_{\mathrm{i}}$ behavior is observed.

4.2.4. *Strain-Energy Function Proposed by Takamizawa and Hayashi [59]*

Another well-known two-dimensional form of strain-energy function for arteries was proposed by Takamizawa and Hayashi [59]. It has the logarithmic form

$$\hat{\overline{\Psi}} = -c\mathrm{Ln}(1 - \psi), \tag{55}$$

where c is a stress-like material parameter and the function ψ is given in the form

$$\psi = \frac{1}{2} b_1 \overline{E}_{\Theta\Theta}^2 + \frac{1}{2} b_2 \overline{E}_{ZZ}^2 + b_4 \overline{E}_{\Theta\Theta} \overline{E}_{ZZ}. \tag{56}$$

Here b_1, b_2, b_4 are non-dimensional material parameters and $\overline{E}_{\Theta\Theta}$, $\overline{E}_{ZZ}$ are the components of the modified Green–Lagrange strain tensor $\overline{\mathbf{E}}$ in the circumferential and axial directions, respectively.

Note that the proposed definition (56) does not, in general, preclude ψ from being 1, which leads to an infinite value of $\hat{\overline{\Psi}}$ in certain states of deformation. Furthermore, for $\psi > 1$, the argument of the logarithmic function (55) is negative and the function is not then defined. This type of strain-energy function is therefore only applicable for a limited range of states of deformation. Moreover, it is convex under the same conditions as discussed in respect to (53). The material parameters proposed in [59], and given in Table V, are used to produce the convex contours in Figure 12(a), while an alternative set of parameters is used in Figure 12(b) to illustrate non-convexity of the strain-energy function.

Using equations (55), (56) and (50) we may solve equation $(27)_1$ and calculate the reduced axial force F, as outlined in Remark 3.2. The material data for

Table V. Material and geometrical data for a dog carotid artery based on (55) (see experiment D850815C in [59]).

Material	Geometry	
$c = 57.94$ [kPa]	$\alpha = 0.0°$	$\alpha = 160.0°$
$b_1 = 0.6311$ [–]		
$b_2 = 0.4728$ [–]	$R_i = 1.21$ [mm]	$R_i = 2.40$ [mm]
$b_4 = 0.0301$ [–]	$R_o = 1.77$ [mm]	$R_o = 2.96$ [mm]

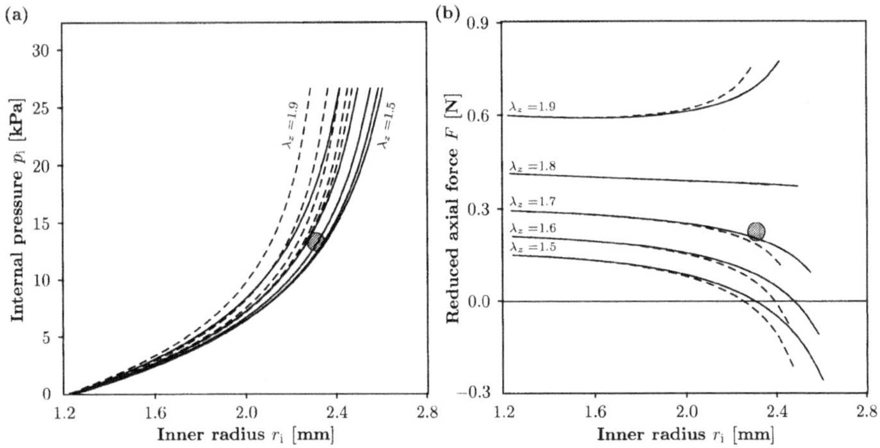

Figure 13. Deformation behavior of a dog carotid artery during inflation using the constitutive model in [59] (experiment D850815C). Solid lines are numerical results with residual strains included ($\alpha = 160.0°$) while the dashed lines are results without residual strains ($\alpha = 0.0°$). Dependence of (a) the internal pressure p_i and (b) the reduced axial force F on the inner radius r_i, without shear deformation ($\gamma_i = 0$). The shaded circles indicate the approximate central region of the physiological state.

a dog carotid artery (see [59]) and the geometrical data used are summarized in Table V. We investigated the same range of loading as described in Section 4.2.3. The resulting arterial response is shown in Figure 13.

The potential (55) is able to represent the typical response of arteries quite well [59] except in the low pressure region, as also observed in Section 4.2.3 in respect of the potential (53). The residual stresses have a strong influence on the $p_i - r_i$ behavior. Finally, we remark that the potential (55), if used within a (displacement-driven) finite element formulation, may, because of the problems mentioned above, lead to numerical difficulties.

5. A Multi-Layer Model for Arterial Walls

In this section we propose a potential that models each layer of the artery as a fiber-reinforced composite. The basic idea is to formulate a constitutive model which incorporates some histological information. Hence, the material parameters involved may be associated with the histological structure of arterial walls (i.e. fiber directions), a feature which is not possible with the phenomenological models described so far in the paper. The underlying physical background of the proposed constitutive model leads to a formulation that avoids the problems encountered with some phenomenological models. The model is based on the theory of the mechanics of fiber-reinforced composites [57] and embodies the symmetries of a cylindrically orthotropic material.

5.1. CONSTITUTIVE MODEL FOR THE ARTERY LAYERS

Since arteries are composed of (thick-walled) layers we model each of these layers with a separate strain-energy function. From the engineering point of view each layer may be considered as a composite reinforced by two families of (collagen) fibers which are arranged in symmetrical spirals.

We assume that each layer responds with similar mechanical characteristics and we therefore use the same form of strain-energy function (but a different set of material parameters) for each layer. We suggest an additive split of the isochoric strain-energy function $\overline{\Psi}$ into a part $\overline{\Psi}_{\text{iso}}$ associated with *iso*tropic deformations and a part $\overline{\Psi}_{\text{aniso}}$ associated with *aniso*tropic deformations [29]. Since the (wavy) collagen fibers of arterial walls are not active at low pressures (they do not store strain energy) we associate $\overline{\Psi}_{\text{iso}}$ with the mechanical response of the non-collagenous matrix material, which we assume to be isotropic. The resistance to stretch at high pressures is almost entirely due to collagenous fibers [49] and this mechanical response is therefore taken to be governed by the anisotropic function $\overline{\Psi}_{\text{aniso}}$. Hence, the (two-term) potential is written as

$$\overline{\Psi}(\overline{\mathbf{C}}, \mathbf{a}_{0\,1}, \mathbf{a}_{0\,2}) = \overline{\Psi}_{\text{iso}}(\overline{\mathbf{C}}) + \overline{\Psi}_{\text{aniso}}(\overline{\mathbf{C}}, \mathbf{a}_{0\,1}, \mathbf{a}_{0\,2}), \tag{57}$$

where the families of collagenous fibers are characterized by the two (reference) direction vectors $\mathbf{a}_{0\,i}$, $i = 1, 2$, with $|\mathbf{a}_{0\,i}| = 1$. Note that in (57) we use $\overline{\mathbf{C}}$ rather than $\overline{\mathbf{E}}$ as the deformation measure.

We include structure tensors in accordance with the formulation in Section 3.1.2. Specifically, we incorporate two such tensors, $\mathbf{A}_i$, $i = 1, 2$, defined as the tensor products $\mathbf{a}_{0\,i} \otimes \mathbf{a}_{0\,i}$. The integrity basis for the three symmetric second-order tensors $\overline{\mathbf{C}}, \mathbf{A}_1, \mathbf{A}_2$ then consists of the invariants

$$\bar{I}_1(\overline{\mathbf{C}}) = \operatorname{tr}\overline{\mathbf{C}}, \qquad \bar{I}_2(\overline{\mathbf{C}}) = \frac{1}{2}[(\operatorname{tr}\overline{\mathbf{C}})^2 - \operatorname{tr}\overline{\mathbf{C}}^2], \qquad \bar{I}_3(\overline{\mathbf{C}}) = \det\overline{\mathbf{C}} = 1, \tag{58}$$

$$\bar{I}_4(\overline{\mathbf{C}}, \mathbf{a}_{0\,1}) = \overline{\mathbf{C}} : \mathbf{A}_1, \qquad \bar{I}_5(\overline{\mathbf{C}}, \mathbf{a}_{0\,1}) = \overline{\mathbf{C}}^2 : \mathbf{A}_1, \tag{59}$$

$$\bar{I}_6(\overline{\mathbf{C}}, \mathbf{a}_{0\,2}) = \overline{\mathbf{C}} : \mathbf{A}_2, \qquad \bar{I}_7(\overline{\mathbf{C}}, \mathbf{a}_{0\,2}) = \overline{\mathbf{C}}^2 : \mathbf{A}_2, \tag{60}$$

$$\bar{I}_8(\overline{\mathbf{C}}, \mathbf{a}_{0\,1}, \mathbf{a}_{0\,2}) = (\mathbf{a}_{0\,1} \cdot \mathbf{a}_{0\,2})\, \mathbf{a}_{0\,1} \cdot \overline{\mathbf{C}}\mathbf{a}_{0\,2}, \qquad \bar{I}_9(\mathbf{a}_{0\,1}, \mathbf{a}_{0\,2}) = (\mathbf{a}_{0\,1} \cdot \mathbf{a}_{0\,2})^2; \tag{61}$$

see [57, 25]. Since the invariants $\bar{I}_3$, $\bar{I}_9$ are constants we may express equation (57) in the reduced form

$$\overline{\Psi}(\overline{\mathbf{C}}, \mathbf{A}_1, \mathbf{A}_2) = \overline{\Psi}_{\mathrm{iso}}(\bar{I}_1, \bar{I}_2) + \overline{\Psi}_{\mathrm{aniso}}(\bar{I}_1, \bar{I}_2, \bar{I}_4, \ldots, \bar{I}_8). \tag{62}$$

Note that the invariants $\bar{I}_4$ and $\bar{I}_6$ are the squares of the stretches in the directions of $\mathbf{a}_{0\,1}$ and $\mathbf{a}_{0\,2}$, respectively, so that they are stretch measures for the two families of (collagen) fibers and therefore have a clear physical interpretation. For simplicity, in order to minimize the number of material parameters, we consider the reduced form of (62) given by

$$\overline{\Psi}(\overline{\mathbf{C}}, \mathbf{A}_1, \mathbf{A}_2) = \overline{\Psi}_{\mathrm{iso}}(\bar{I}_1) + \overline{\Psi}_{\mathrm{aniso}}(\bar{I}_4, \bar{I}_6). \tag{63}$$

The anisotropy then arises only through the invariants $\bar{I}_4$ and $\bar{I}_6$, but this is sufficiently general to capture the typical features of arterial response.

Finally, the two contributions $\overline{\Psi}_{\mathrm{iso}}$ and $\overline{\Psi}_{\mathrm{aniso}}$ to the function $\overline{\Psi}$ must be particularized so as to fit the material parameters to the experimentally observed response of the arterial layers. We use the (classical) neo-Hookean model to determine the isotropic response in each layer, and we write

$$\overline{\Psi}_{\mathrm{iso}}(\bar{I}_1) = \frac{c}{2}(\bar{I}_1 - 3), \tag{64}$$

where $c > 0$ is a stress-like material parameter. The strong stiffening effect of each layer observed at high pressures motivates the use of an exponential function for the description of the strain energy stored in the collagen fibers, and for this we propose

$$\overline{\Psi}_{\mathrm{aniso}}(\bar{I}_4, \bar{I}_6) = \frac{k_1}{2k_2} \sum_{i=4,6} \left\{\exp[k_2(\bar{I}_i - 1)^2] - 1\right\}, \tag{65}$$

where $k_1 > 0$ is a stress-like material parameter and $k_2 > 0$ is a dimensionless parameter. An appropriate choice of k_1 and k_2 enables the histologically-based assumption that the collagen fibers do not influence the mechanical response of the artery in the low pressure domain [49] to be modeled.

All that remains is to determine an expression for the stress, which we provide here in the Eulerian description. Using $(10)_3$ and the proposed particularizations (64) and (65), we obtain, after some straightforward manipulations, the explicit isochoric contribution $\bar{\boldsymbol{\sigma}}$ to the Cauchy stress tensor, namely

$$\bar{\boldsymbol{\sigma}} = c\,\mathrm{dev}\,\bar{\mathbf{b}} + \sum_{i=4,6} 2\overline{\Psi}_i \mathrm{dev}(\mathbf{a}_i \otimes \mathbf{a}_i), \tag{66}$$

where $\overline{\Psi}_4 = \partial\overline{\Psi}_{\mathrm{aniso}}/\partial\bar{I}_4$, $\overline{\Psi}_6 = \partial\overline{\Psi}_{\mathrm{aniso}}/\partial\bar{I}_6$ denote (scalar) response functions and $\mathbf{a}_i = \overline{\mathbf{F}}\mathbf{a}_{0\,i}$, $i = 1, 2$, the Eulerian counterparts of $\mathbf{a}_{0\,i}$. For a detailed derivation

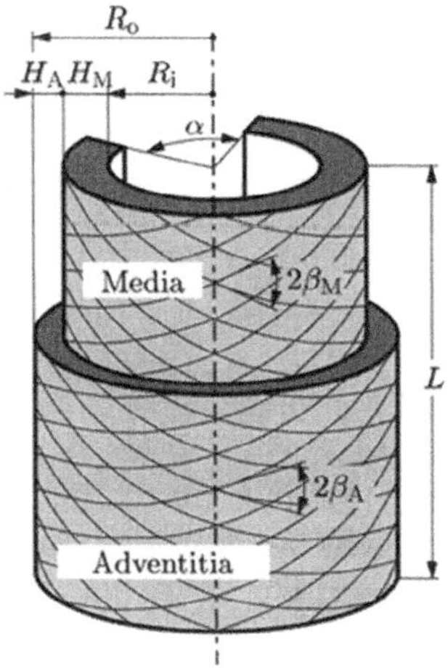

	Material	Geometry
Media	$c_M = 3.0000$ [kPa] $k_{1M} = 2.3632$ [kPa] $k_{2M} = 0.8393$ [–]	$H_M = 0.26$ [mm] $\beta_M = 29.0°$
Adventitia	$c_A = 0.3000$ [kPa] $k_{1A} = 0.5620$ [kPa] $k_{2A} = 0.7112$ [–]	$H_A = 0.13$ [mm] $\beta_A = 62.0°$

$R_i = 0.71$ [mm]	for	$\alpha = 0.0°$
$R_i = 1.43$ [mm]	for	$\alpha = 160.0°$

Figure 14. Material and geometrical data for a carotid artery from a rabbit in respect of (67) and (68) (see experiment 71 in [5]).

of equation (66) the reader is referred to the more general constitutive framework described in [20].

5.2. ARTERY MODELED AS A TWO-LAYER THICK-WALLED TUBE WITH RESIDUAL STRAINS

In order to report the performance of the proposed constitutive model we study the mechanical response of a *healthy young* arterial segment (with no pathological intimal changes). For this case the innermost layer of the artery is not of (solid) mechanical interest, and we therefore focus attention on modeling the two remaining layers, i.e. the media and the adventitia. It is then appropriate to model the artery as a two-layer thick-walled tube (with residual strains), as illustrated in Figure 14.

This model uses 6 material parameters, i.e. c_M, $k_{1\,M}$, $k_{2\,M}$ for the media and c_A, $k_{1\,A}$, $k_{2\,A}$ for the adventitia. In respect of equations (63)–(65) the free-energy functions for the considered two-layer problem may be written as

$$\overline{\Psi}_M = \frac{c_M}{2}(\bar{I}_1 - 3) + \frac{k_{1\,M}}{2k_{2\,M}} \sum_{i=4,6} \left\{\exp[k_{2\,M}(\bar{I}_{i\,M} - 1)^2] - 1\right\},$$
$$R_i \leqslant R \leqslant R_i + H_M, \tag{67}$$

$$\overline{\Psi}_A = \frac{c_A}{2}(\bar{I}_1 - 3) + \frac{k_{1\,A}}{2k_{2\,A}} \sum_{i=4,6} \left\{\exp[k_{2\,A}(\bar{I}_{i\,A} - 1)^2] - 1\right\},$$
$$R_i + H_M \leqslant R \leqslant R_o \tag{68}$$

for the media and adventitia, respectively. The constants c_M and c_A are associated with the non-collagenous matrix of the material, which describes the *isotropic* part of the overall response of the tissue. Note, however, that the matrix material is significantly less stiff than its elastin fiber constituent. The constants $k_{1\,\mathrm{M}}$, $k_{2\,\mathrm{M}}$ and $k_{1\,\mathrm{A}}$, $k_{2\,\mathrm{A}}$ are associated with the *anisotropic* contribution of collagen to the overall response. The material parameters are constants and do not depend on the geometry, opening angle or fiber angle. The internal pressure/radius response, of course, does depend on geometry, opening angle and fiber angle, but we have not included here an analysis of the effect of changes in these quantities. However, our studies have found that in the high-pressure regime the stress–strain response depends significantly on the fiber angles (as should be expected). The fiber angles are associated with the stress-free configuration, as indicated in Figure 14, and we have assumed that they are the same in the load-free configuration. The difference in angle between the unstressed and unloaded configurations for the case we considered goes from (approximately) $-3.0°$ on the inner boundary to $+2.7°$ on the outer boundary (mean value $0.2°$). This approximation has a negligible influence on the subsequent analysis.

The invariants, associated with the media M and the adventitia A, are defined by $\bar{I}_{4\,j} = \mathbf{A}_{1\,j} : \overline{\mathbf{C}}$ and $\bar{I}_{6\,j} = \mathbf{A}_{2\,j} : \overline{\mathbf{C}}$, $j = \mathrm{M, A}$, and H_{M} is the reference thickness of the media, as illustrated in Figure 14. The tensors $\mathbf{A}_{1\,j}$, $\mathbf{A}_{2\,j}$, characterizing the structure of the media and adventitia, are given by

$$\mathbf{A}_{1\,j} = \mathbf{a}_{0\,1\,j} \otimes \mathbf{a}_{0\,1\,j}, \qquad \mathbf{A}_{2\,j} = \mathbf{a}_{0\,2\,j} \otimes \mathbf{a}_{0\,2\,j}, \qquad j = \mathrm{M, A}, \tag{69}$$

where, in a cylindrical polar coordinate system, the components of the direction vectors $\mathbf{a}_{0\,1\,j}$ and $\mathbf{a}_{0\,2\,j}$ have, in matrix notation, the forms

$$[\mathbf{a}_{0\,1\,j}] = \begin{bmatrix} 0 \\ \cos\beta_j \\ \sin\beta_j \end{bmatrix}, \qquad [\mathbf{a}_{0\,2\,j}] = \begin{bmatrix} 0 \\ \cos\beta_j \\ -\sin\beta_j \end{bmatrix}, \qquad j = \mathrm{M, A}, \tag{70}$$

and β_j, $j = \mathrm{M}, A$, are the angles between the collagen fibers (arranged in symmetrical spirals) and the circumferential direction in the media and adventitia, as indicated in Figure 14. Note that Finlay et al. [13] reported that in, for example, human brain arteries the (collagenous) fiber orientations also have small components in the radial direction. However, we neglect this feature in the present work.

Because of the wavy structure of collagen it is regarded as not being able to support compressive stresses. We therefore assume that the fibers are active in extension and inactive in compression. Hence, in the proposed model the anisotropic terms in the free-energy functions (67) and (68) should only contribute when the fibers are extended, that is when $\bar{I}_{4\,j} > 1$ or $\bar{I}_{6\,j} > 1$, $j = \mathrm{M, A}$. If one or more of these conditions is not satisfied then the relevant part of the anisotropic function is omitted from the expressions (67) and (68). If, for example, $\bar{I}_{4\,\mathrm{A}}$ and $\bar{I}_{6\,\mathrm{A}}$ are less than or equal to 1, then the response of the adventitia is purely isotropic. When these conditions are taken into account, convexity is guaranteed by the form of the free-energy functions (67) and (68).

Contour plots for the two arterial layers (media and adventitia) based on the material parameters given in Figure 14 are depicted in Figure 15. As can be seen,

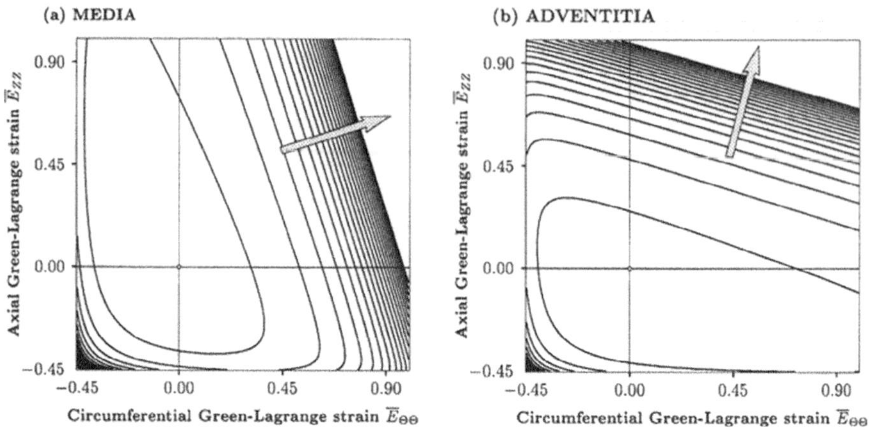

Figure 15. Contour plots of the potentials (67) for the media (a) and (68) for the adventitia (b) using material parameters given in Figure 14. The arrows show the directions of greatest ascent in the regions $\overline{E}_{\Theta\Theta} \geqslant 0, \overline{E}_{ZZ} \geqslant 0$.

the two potentials are convex and anisotropic. In addition, the directions of greatest ascent (illustrated by arrows in Figure 15) in the region $\overline{E}_{\Theta\Theta} \geqslant 0, \overline{E}_{ZZ} \geqslant 0$ are clearly different for the two layers. These directions are determined mainly by the orientations of the collagen fibers, which tend to be nearly circumferential in the media and nearly axial in the adventitia. Note that where $\overline{E}_{\Theta\Theta} < 0$ and $\overline{E}_{ZZ} < 0$ the (symmetric) contours reflect the isotropy in this region.

Experimental tests performed by Von Maltzahn et al. [39], Yu et al. [69] and Xie et al. [68] indicate that the elastic properties of the media and adventitia are different. Their results show that the media is much stiffer than the adventitia. In particular, it was found that in the neighborhood of the reference configuration the mean value of Young's modulus for the media, for several pig thoracic aortas, is about an *order of magnitude higher* than that of the adventitia [69]. For our proposed constitutive model this observation implies that for these materials the neo-Hookean parameters are such that the ratio $c_{\mathrm{M}}/c_{\mathrm{A}}$ is typically in the range of 6 to 14. This effectively reduces the number of material parameters, and for definiteness we therefore set $c_{\mathrm{M}} = 10c_{\mathrm{A}}$ for purposes of numerical calculation. In general, however, this ratio depends on the topographical site.

We use geometrical data from [5] for a carotid artery from a rabbit (experiment 71 in [18]) and make the assumptions that the media occupies $2/3$ of the arterial wall thickness and that the wall thickness of each layer in the unloaded configuration ($\alpha = 0.0°$) is the same as for the case without residual stress ($\alpha = 160.0°$). In order to identify the material parameters of the two-layer model for healthy arterial walls, we fitted the parameters to the experimental data from experiment 71 in [18] and used the standard nonlinear *Levenberg–Marquardt* algorithm. The material parameters obtained are summarized in Figure 14. For more explanation of the underlying fitting process the reader is referred to [21]. For purposes of

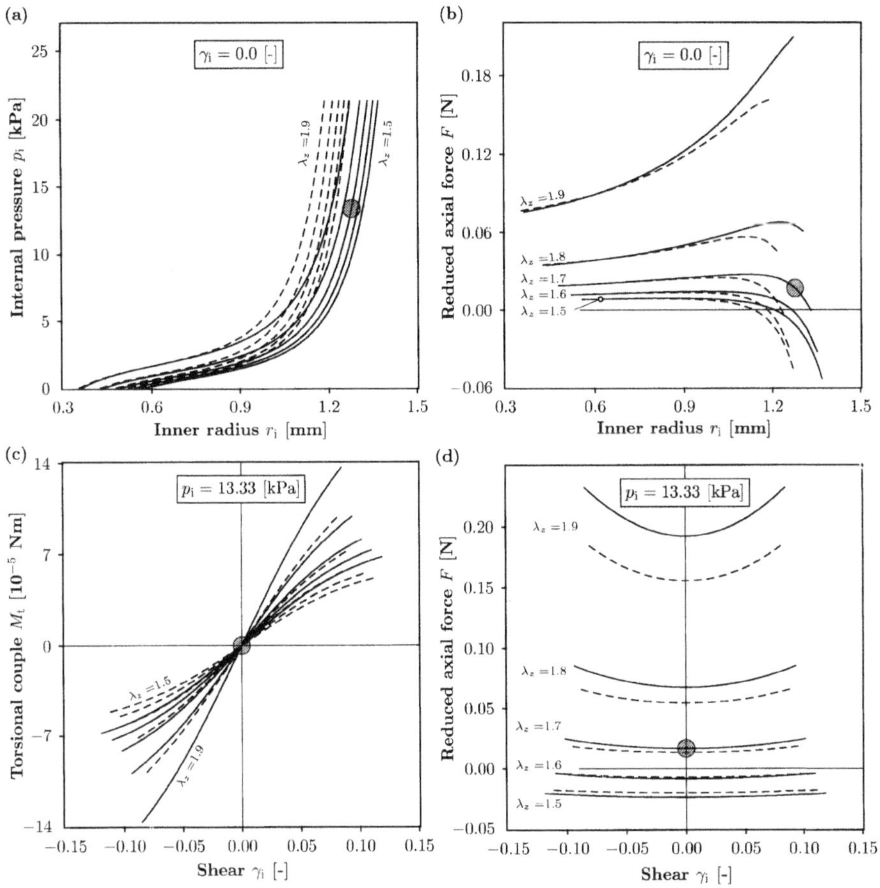

Figure 16. Deformation behavior of a carotid artery during inflation and torsion using the constitutive model (67)–(68). Solid lines are numerical results with residual strains included ($\alpha = 160.0°$) and the dashed lines are results without residual strains ($\alpha = 0.0°$). Dependence of (a) the internal pressure p_i and (b) the reduced axial force F on inner radius r_i, without shear deformation ($\gamma_i = 0$). Dependence of (c) the torsional couple M_t and (d) the reduced axial force F on the shear γ_i at fixed internal pressure $p_i = 13.33$ [kPa]. The shaded circles indicate the approximate central region of the physiological state.

comparison, we make the assumption that the configuration shown in Figure 14 is stress-free. But, bearing in mind the discussion in Section 3.2.1 concerning a single layer, this simplifying assumption must be regarded as an approximation. In practice, the opening angles and the stress-free configurations for the separate layers would be different.

The mechanical response of the carotid artery during bending, inflation, axial extension and torsion is shown in Figure 16. The internal pressure p_i and the angle of twist Φ are varied within the ranges

$$0 \leqslant p_i \leqslant 21.33 \text{ [kPa]} \quad \text{and} \quad -0.10 \leqslant \Phi \leqslant 0.10 \text{ [rad]}. \tag{71}$$

The internal pressure versus radius behavior in the low pressure domain shown in Figure 16(a) clearly differs from the pressure/radius curves discussed in previous sections, as comparison with Figures 5(a), 7(a), 9(a), 11(a) and 13(a) shows. The proposed model (67)–(68) is able to describe the salient features of arterial elasticity, such as the experimentally observed 'sigma-shaped' form of the pressure/radius relationship; see Figure 2(a) in [66]. Note that residual strains have a strong influence on the global pressure/radius response of the artery, which is similar to what was observed for the other potentials treated in this paper except for that in [10], as discussed in Section 4.1.1, for which $\alpha = 100.0°$ and a smaller value of the wall thickness to diameter ratio was used.

Figure 16(b) shows that the proposed potential (67)–(68) is also able to model the typical evolution of the reduced axial force F with inflation (increase of the inner radius) of the artery; see Figure 2(b) in [66]. This means that F is a decreasing function of r_i at axial stretches λ_z less than some value above the physiological stretch and an increasing function for λ_z greater than this value (this effect is also evident in Figure 7(b)). This characteristic behavior can also be replicated with the other potentials described in Section 4 except for the constitutive model in [62]; see Figure 9(b).

The response of the artery during torsion at the internal (physiological) pressure $p_\mathrm{i} = 13.33$ [kPa] is plotted in Figure 16(c), (d). As can be seen from Figure 16(c), the torsional couple M_t increases more slowly than the shear $\gamma_\mathrm{i} = \Phi r_\mathrm{i}/l = \Phi r_\mathrm{i}/\lambda_z L$ on the inner boundary increases (i.e. the slope of the curve decreases). One possible explanation of this interesting phenomenon is as follows: since the artery is inflated with the internal physiological pressure, the (collagen) fiber reinforcement is activated and the fibers are much stiffer than the matrix material. During torsion from this state of deformation the nearly inextensible fibers cause the arterial diameter to decrease, which leads to a reduction in the torsional couple M_t given by equation $(26)_2$. This realistic diameter-shrinking behavior of the artery during torsion seems to be a consequence of the considered fiber reinforcement (orthotropy). However, this effect may also be predicted by a non-convex isotropic strain-energy function. Of the potentials discussed in Section 4 only that reviewed in Section 4.1.2 can predict this phenomenon (see Figure 7(c)).

In Figure 16(d) the reduced axial force F during torsion is plotted against the shear γ_i. For an axial pre-stretch $\lambda_z = 1.5$ the reduction in the inner radius r_i due to torsion is about 5.8% ($\gamma_\mathrm{i} = 0.119$) and 7.8% for $\lambda_z = 1.9$ ($\gamma_\mathrm{i} = 0.085$). This behavior is in qualitative agreement with experimental observations presented in [12] and may also be reproduced with the potentials described in Sections 4.1.1 and 4.1.2; see Figures 5(d), 7(d). Note the relatively strong influence of residual strains at high axial stretches compared with that shown in Figure 7(d).

REMARK 5.1. The fully three-dimensional formulation of the convex potential (67)–(68) allows the characteristic anisotropic behavior of healthy arteries under combined bending, inflation, axial extension and torsion to be predicted. It is not, however, restricted to a particular geometry such as axisymmetry, and is

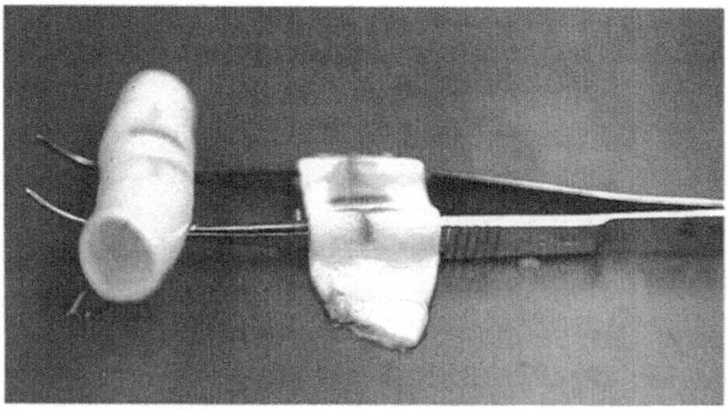

Figure 17. Mechanical separation of the layers of a human external iliac artery into a stiff media-intima tube (on the left-hand side) and a limp adventitia (from *C. Schulze-Bauer, MD*, Computational Biomechanics, Graz University of Technology, Austria, with permission).

accessible to approximation techniques such as the finite element method. For an extension of the anisotropic model to the finite viscoelastic domain see [26] and for an extension of the constitutive framework to the elastoplastic domain see [20] and [21]. All these recent works focus on implementation of the models in a finite element program.

The proposed constitutive model has the merit that it is based partly on histological information. It therefore allows the material parameters to be associated with the constituents (matrix and collagen) of each solid mechanically-relevant layer. Since the media and adventitia have different physiological functions, a two-layer model using different strain-energy functions for the media and adventitia seems to be essential. This idea goes back to von Maltzahn et al. [38], who proposed a cylindrical two-layer model using an isotropic (polynomial) function (with one coefficient) for the media and an anisotropic strain-energy function for the adventitia (with three coefficients). This approach is indispensable for the study of stress distributions across the arterial wall and allows a histomechanical investigation of the arterial layers and their underlying physiological functions. Extension to a three-layer constitutive model incorporating pathological intimal changes is a straightforward task.

Figure 17, in which mechanically separated media-intima and adventitia layers are shown, provides a graphic illustration of the effect of the different properties and (possibly) different residual stresses in the two tubes. In the unloaded configuration the relevant geometrical quantities are – media-intima: inner radius, 4.07 [mm]; thickness 0.98 [mm]; adventitia: thickness 0.4 [mm]. The given geometry should be taken into account in interpreting the figure.

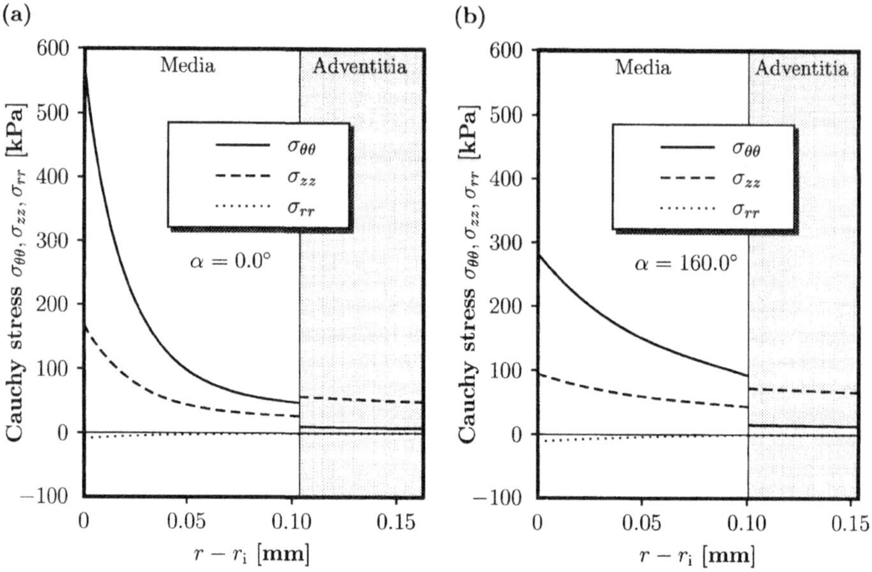

Figure 18. Plots of the principal Cauchy stresses $\sigma_{\theta\theta}$, σ_{zz}, σ_{rr} in the circumferential, axial and radial directions through the deformed media and adventitia layers in the physiological state with $p_{\mathrm{i}} = 13.33$ [kPa], $\lambda_z = 1.7$, $\gamma_{\mathrm{i}} = 0$: (a) without residual stress ($\alpha = 0.0°$); (b) with residual stress ($\alpha = 160.0°$). The abscissa is $r - r_{\mathrm{i}}$. The numerical results are obtained for the constitutive models (67) and (68) with geometrical data and material constants as in Figure 14.

5.3. STRESS DISTRIBUTION THROUGH THE DEFORMED ARTERIAL WALL

One important aspect of the influence of residual stress is the effect that it has on the stress distribution through the arterial wall in the physiological state. Whilst we have seen from, for example, Figure 16 its effect on the overall pressure/radius response its effect on the stress distribution through the arterial wall is more pronounced. This is illustrated in Figure 18, in which the distributions of the principal Cauchy stress components $\sigma_{\theta\theta}$, σ_{zz} and σ_{rr} through the deformed wall thickness (media and adventitia layers) are plotted against $r - r_{\mathrm{i}}$, where r is the deformed radial coordinate and r_{i} the deformed inner radius. The geometrical data and material constants shown in Figure 14 are again used in conjunction with the material models (67) and (68). The physiological state is taken to correspond to $p_{\mathrm{i}} = 13.33$ [kPa] and $\lambda_z = 1.7$, with no torsion ($\gamma_{\mathrm{i}} = 0$). The calculation can be carried out by using any numerical tool. However, in order to solve the three-dimensional boundary-value problem for the stress components $\sigma_{\theta\theta}$, σ_{zz} and σ_{rr} (rather than for $\overline{\sigma}_{\theta\theta}$, $\overline{\sigma}_{zz}$ and $\overline{\sigma}_{rr}$, as used throughout the text) it seems to be convenient to employ the (mixed) finite element method. Details of the computational aspects are described in Holzapfel and Gasser [26].

Figure 18(a) shows the Cauchy stress distributions for the case in which there are no residual stresses ($\alpha = 0.0°$), while Figure 18(b) shows the corresponding plot with residual stresses included ($\alpha = 160.0°$). The tangential stresses $\sigma_{\theta\theta}$ and

σ_{zz} are discontinous at the media/adventitia interface, while the radial stress σ_{rr} is continuous. Note that the magnitude of σ_{rr} is much smaller than that of the tangential stresses. The behavior shown in Figure 18(a) is similar to that found by Von Maltzahn and Warriyar [39] for bovine carotid arteries. Figure 18 demonstrates the relatively high values of the circumferential stress in the media compared with that in the adventitia, which was also found, for example, in [39]. Within the media there is a significant difference between the distributions of $\sigma_{\theta\theta}$ and σ_{zz} in the two plots. Interestingly, although the maximum circumferential stress $\sigma_{\theta\theta}$ (which occurs at the inner wall) is reduced significantly, the mean circumferential stress through the wall is increased by the presence of residual stress. Note that the residual stress also influences the deformed wall thickness and the strains in the physiological state. In particular, at the inner wall, for example, the circumferential strain (measured relative to the stress-free configuration) is reduced by the residual stress.

The most important influence of the residual stress is the reduction in the maximum stress values $\sigma_{\theta\theta}$ and σ_{zz} (which occur at the inner side of the media) and the gradients of $\sigma_{\theta\theta}$ and σ_{zz} in the media, an effect which has been reported previously for a single layer (see, for example, Fung [16], Section 11.3). Stress gradients would be reduced further by larger values of α, as has also been described in [16]. Indeed, it is often assumed that the arterial wall adapts itself so that the circumferential stresses are uniform within each layer. Moreover, the assumption of uniform strain is sometimes adopted (see, for example, Takamizawa and Hayashi [59]). Some consequences of these assumptions have been discussed in a recent paper by Ogden and Schulze-Bauer [43].

6. Summary and Concluding Remarks

For a deeper understanding of the highly nonlinear deformation mechanisms and stress distributions in arteries under different loading conditions and the improvement of diagnostics and therapeutical procedures that are based on mechanical treatments, a reliable constitutive model of arteries is an essential prerequisite.

For the description of the nonlinear elastic behavior of arterial walls, there are essentially polynomial, exponential and logarithmic forms of strain-energy function available in the literature. A representative selection of models in common use has been investigated in this paper and evaluated in detail. This comparative study was conducted in respect of the mechanical response of a thick-walled tube under combined bending, inflation, axial extension and torsion and with reference to fundamental continuum mechanical principles. It is hoped that this simple study will offer some guidance for the evaluation of alternative forms of strain-energy function for arteries.

The constitutive model of Delfino et al. [10], which is based on an isotropic description, is not able to reproduce the pronounced anisotropic mechanical behavior of arteries observed in several experimental investigations. Nevertheless, it is worth noting that its predictions for the restricted kinematics and loading conditions considered here are in some respects qualitatively similar to those for the anisotropic energy functions discussed in Section 4.

The two-dimensional formulation discussed in Section 4.2.1 does not, in general, permit the stress response under certain combined loadings (such as inflation and torsion) to be modeled. Exceptionally, if the material is isotropic or if it describes a membrane model, such combined loadings can be analyzed, as shown in Section 4.2.1. However, such a formulation, because it omits $\overline{E}_{R\Theta}$, $\overline{E}_{RZ}$ and $\overline{E}_{RR}$, is inherently limited to specific kinematics or to a membrane description. For example, if $\overline{E}_{RR}$ is omitted then the inflation/torsion problem cannot be solved for a thick-walled tube. The three well-established anisotropic models developed by Vaishnav et al. [62], Fung et al. [18] and Takamizawa and Hayashi [59], which we have discussed in Section 4.2, are special cases of this two-dimensional formulation. *A fortiori*, their applicability is limited, but, as we have seen, they do predict qualitatively reasonable response for restricted geometry and loadings. Moreover, they have contributed to our current level of understanding of arterial wall mechanics. Thus, isotropic or two-dimensional anisotropic energy functions may be valuable under some conditions, but, bearing in mind the limitations discussed above, they should be used with caution.

In the theory of elasticity the notion of convexity of the strain-energy function (which is dependent on the choice of deformation measure used) has an important role in ensuring physically meaningful and unambiguous mechanical behavior. It also induces desirable mathematical features in the governing equations, which are important from the point of view of numerical computations. A problem detected in the potentials considered in this comparative study is the general lack of convexity. For example, the anisotropic potential of Vaishnav et al. [62] is not convex for *any* set of material parameters. The strain-energy functions of Fung et al. [18], Chuong and Fung [5] and Takamizawa and Hayashi [59] are not convex for all possible sets of material parameters, and restrictions on these parameters are therefore needed to ensure convexity and, therefore, to avoid material instabilities.

The three-dimensional anisotropic mechanical response of arteries points to the need for three-dimensional constitutive models, and suitable generalizations of the above-mentioned models must therefore be employed. All the models discussed above are based on a phenomenological approach in which the *macroscopic* nature of the biological material is modeled. This approach, which is concerned mainly with fitting the constitutive equations to experimental data, is not capable of relating the deformation mechanism to the known architectural structure of the arterial wall. The material parameters have no direct physical meaning and are therefore treated as numbers without clear physical interpretation.

From this comparative study and the experience gained, it may be concluded that there is a need for an alternative form of constitutive model which avoids the limitations discussed. It is for this reason that we have proposed an approach in which arterial walls are approximated as two-layer thick-walled tubes, with each layer modeled as a highly deformable fiber-reinforced composite. This leads to a fully three-dimensional anisotropic material description of the artery incorporating histological information. The proposed two-layer model uses a set of *six* material parameters whose interpretations can be partly based on the underlying histological structure.

The new model discussed in this paper is consistent with both mechanical and mathematical requirements and is suitable for use within the context of finite element methods (see, for example, [26, 20] and [21]). It is also applicable for arbitrary geometries so that more complex boundary-value problems can be solved. As described in Section 5.3, this approach enables insight into the nature of the stress distribution across the arterial wall to be gained, and therefore offers the potential for a detailed study of the mechanical functionality of arteries.

The importance of including residual strains (and stresses), which was shown previously by scientists such as Chuong and Fung [6], has been emphasized. As we have seen, incorporation of residual strains in the load-free configuration changes not only the overall pressure/radius response of the artery but also the stress distribution through the deformed arterial wall (see also [5]). Thus, in order to predict reliable stress distributions, the parameter identification process must incorporate residual strains in the load-free configuration.

Acknowledgements

Financial support for this research was provided by the *Austrian Science Foundation* under *START-Award Y74-TEC* and *FWF-Project No. P11899-TEC*. This support is gratefully acknowledged. Ray Ogden is grateful to Graz University of Technology for support during several visits and to the British Council in Vienna for financial support for this research. The authors are indebted to *C.A.J. Schulze-Bauer, MD* for his contributions to this work and for valuable discussions.

References

1. H. Abè, K. Hayashi and M. Sato (eds), *Data Book on Mechanical Properties of Living Cells, Tissues, and Organs*, Springer-Verlag, New York (1996).
2. H. Bader, Dependence of wall stress in the human thoracic aorta on age and pressure. *Circ. Res.* **20** (1967) 354–361.
3. P.C. Block, Mechanism of transluminal angioplasty. *Am. J. Cardiology* **53** (1984) 69C–71C.
4. T.E. Carew, R.N. Vaishnav and D.J. Patel, Compressibility of the arterial wall. *Circ. Res.* **23** (1968) 61–68.
5. C.J. Chuong and Y.C. Fung, Three-dimensional stress distribution in arteries. *J. Biomech. Engr.* **105** (1983) 268–274.
6. C.J. Chuong and Y.C. Fung, Residual stress in arteries. In: G.W. Schmid-Schönbein, S.L-Y. Woo and B.W. Zweifach (eds), *Frontiers in Biomechanics*, Springer-Verlag, New York (1986), pp. 117–129.
7. P.G. Ciarlet, *Mathematical Elasticity. Volume I: Three-Dimensional Elasticity*, North-Holland, Amsterdam (1988).

8. R.H. Cox, Regional variation of series elasticity in canine arterial smooth muscles. *Am. J. Physiol.* **234** (1978) H542–H551.
9. R.H. Cox, Comparison of arterial wall mechanics using ring and cylindrical segments. *Am. J. Phys.* **244** (1983) H298–H303.
10. A. Delfino, N. Stergiopulos, J.E. Moore and J.-J. Meister, Residual strain effects on the stress field in a thick wall finite element model of the human carotid bifurcation. *J. Biomech.* **30** (1997) 777–786.
11. H. Demiray, A layered cylindrical shell model for an aorta. *Int. J. Engr. Sci.* **29** (1991) 47–54.
12. S.X. Deng, J. Tomioka, J.C. Debes and Y.C. Fung, New experiments on shear modulus of elasticity of arteries. *Am. J. Physiol.* **266** (1994) H1–H10.
13. H.M. Finlay, L. McCullough and P.B. Canham, Three-dimensional collagen organization of human brain arteries at different transmural pressures. *J. Vasc. Res.* **32** (1995) 301–312.
14. P. Flory, Thermodynamic relations for high elastic materials. *Trans. Faraday Soc.* **57** (1961) 829–838.
15. R.F. Fuchs, Zur Physiologie und Wachstumsmechanik des Blutgefäßsystems. *Archiv für die gesamte Physiologie* **28** (1900).
16. Y.C. Fung, *Biomechanics: Motion, Flow, Stress, and Growth*, Springer-Verlag, New York (1990).
17. Y.C. Fung, *Biomechanics: Mechanical Properties of Living Tissue*, 2nd edn, Springer-Verlag, New York (1993).
18. Y.C. Fung, K. Fronek and P. Patitucci, Pseudoelasticity of arteries and the choice of its mathematical expression. *Am. J. Physiol.* **237** (1979) H620–H631.
19. Y.C. Fung and S.Q. Liu, Change of residual strains in arteries due to hypertrophy caused by aortic constriction. *Circ. Res.* **65** (1989) 1340–1349.
20. T.C. Gasser and G.A. Holzapfel, Rate-independent elastoplastic constitutive modeling of biological soft tissues: Part I. Continuum basis, algorithmic formulation and finite element implementation. Submitted (2000).
21. T.C. Gasser, C.A.J. Schulze-Bauer, E. Pernkopf, M. Stadler and G.A. Holzapfel, Rate-independent elastoplastic constitutive modeling of biological soft tissues: Part II. Percutaneous transluminal angioplasty. Submitted (2000).
22. J.M. Guccione, A.D. McCulloch and L.K. Waldman, Passive material properties of intact ventricular myocardium determined from a cylindrical model. *ASME J. Biomech. Engr.* **113** (1991) 42–55.
23. H.C. Han and Y.C. Fung, Species dependence of the zero-stress state of aorta: Pig versus rat. *J. Biomech. Engr.* **113** (1991) 446–451.
24. K. Hayashi, Experimental approaches on measuring the mechanical properties and constitutive laws of arterial walls. *J. Biomech. Engr.* **115** (1993) 481–488.
25. G.A. Holzapfel, *Nonlinear Solid Mechanics. A Continuum Approach for Engineering*, Wiley, Chichester (2000).
26. G.A. Holzapfel and T.C. Gasser, A viscoelastic model for fiber-reinforced composites at finite strains: Continuum basis, computational aspects and applications. *Comput. Methods Appl. Mech. Engr.* In press (2000).
27. G.A. Holzapfel, T.C. Gasser, M. Stadler and C.A.J. Schulze-Bauer, A multi-layer structural model for the elastic and viscoelastic behavior of arterial walls: Continuum formulation and finite element analysis. Submitted (2000).
28. G.A. Holzapfel, C.A.J. Schulze-Bauer and M. Stadler, Mechanics of angioplasty: Wall, balloon and stent. In: J. Casey and G. Bao (eds), *Mechanics in Biology*, AMD-Vol. 242/BED-Vol. 46, The American Society of Mechanical Engineers, New York (2000), pp. 141–156.
29. G.A. Holzapfel and H.W. Weizsäcker, Biomechanical behavior of the arterial wall and its numerical characterization. *Comp. Biol. Med.* **28** (1998) 377–392.

30. W.H. Hoppmann and L. Wan, Large deformation of elastic tubes. *J. Biomech.* **3** (1970) 593–600.
31. T.J.R. Hughes, *The Finite Element Method: Linear Static and Dynamic Finite Element Analysis*, Prentice-Hall, Englewood Cliffs, NJ (1987).
32. J.D. Humphrey, Mechanics of arterial wall: Review and directions. *Critical Reviews in Biomed. Engr.* **23** (1995) 1–162.
33. J.D. Humphrey, An evaluation of pseudoelastic descriptors used in arterial mechanics. *J. Biomech. Engr.* **121** (1999) 259–262.
34. J.M. Huyghe, D.H. van Campen, T. Arts and R.M. Heethaar, A two-phase finite element model of the diastolic left ventricle. *J. Biomech.* **24** (1991) 527–538.
35. V.A. Kas'yanov and A.I. Rachev, Deformation of blood vessels upon stretching, internal pressure, and torsion. *Mech. Comp. Mat.* **16** (1980) 76–80.
36. Y. Lanir and Y.C. Fung, Two-dimensional mechanical properties of rabbit skin-I. Experimental system. *J. Biomech.* **7** (1974) 29–34.
37. B.M. Learoyd and M.G. Taylor, Alterations with age in the viscoelastic properties of human arterial walls. *Circ. Res.* **18** (1966) 278–292.
38. W.-W. Von Maltzahn, D. Besdo and W. Wiemer, Elastic properties of arteries: A nonlinear two-layer cylindrical model. *J. Biomech.* **14** (1981) 389–397.
39. W.-W. Von Maltzahn and R.G. Warriyar, Experimental measurments of elastic properties of media and adventitia of bovine carotid arteries. *J. Biomech.* **17** (1984) 839–847.
40. W.W. Nichols and M.F. O'Rourke, *McDonald's Blood Flow in Arteries*, 4th edn, Arnold, London (1998), chapter 4, pp. 73–97.
41. R.W. Ogden, Nearly isochoric elastic deformations: Application to rubberlike solids. *J. Mech. Phys. Solids* **26** (1978) 37–57.
42. R.W. Ogden, *Non-linear Elastic Deformations*, Dover Publication, New York (1997).
43. R.W. Ogden and C.A.J. Schulze-Bauer, Phenomenological and structural aspects of the mechanical response of arteries. In: J. Casey and G. Bao (eds), *Mechanics in Biology*, AMD-Vol. 242/BED-Vol. 46, The American Society of Mechanical Engineers, New York (2000), pp. 125–140.
44. H.S. Oktay, T. Kang, J. D. Humphrey and G.G. Bishop, Changes in the mechanical behavior of arteries following balloon angioplasty. In: *1991 ASME Advances in Bioengineering*, New York (1991).
45. D.J. Patel and D.L. Fry, The elastic symmetry of arterial segments in dogs. *Circ. Res.* **24** (1969) 1–8.
46. A. Rachev, Theoretical study of the effect of stress-dependent remodeling on arterial geometry under hypertensive conditions. *J. Biomech.* **30** (1997) 819–827.
47. A. Rachev and K. Hayashi, Theoretical study of the effects of vascular smooth muscle contraction on strain and stress distributions in arteries. *Ann. Biomed. Engr.* **27** (1999) 459–468.
48. J.A.G. Rhodin, Architecture of the vessel wall. In: H.V. Sparks Jr., D.F. Bohr, A.D. Somlyo and S.R. Geiger (eds), *Handbook of Physiology, The Cardiovascular System*, Vol. 2, American Physiologial Society, Bethesda, Maryland (1980), pp. 1–31.
49. M.R. Roach and A.C. Burton, The reason for the shape of the distensibility curve of arteries. *Canad. J. Biochem. Physiol.* **35** (1957) 681–690.
50. E.K. Rodriguez, A. Hoger and A.D. McCulloch, Stress-dependent finite growth in soft elastic tissues. *J. Biomech.* **27** (1994) 455–467.
51. C.S. Roy, The elastic properties of the arterial wall. *J. Physiol.* **3** (1880–1882) 125–159.
52. B.S. Schultze-Jena, Über die schraubenförmige Struktur der Arterienwand. *Gegenbauers Morphol. Jahrbuch* **83** (1939) 230–246.
53. C.A.J. Schulze-Bauer, C. Mörth and G.A. Holzapfel, Passive biaxial mechanical response of aged human iliac arteries. Submitted (2000).

54. F.H. Silver, D.L. Christiansen and C.M. Buntin, Mechanical properties of the aorta: A review. *Critical Reviews in Biomed. Engr.* **17** (1989) 323–358.
55. B.R. Simon, M.V. Kaufmann, M.A. McAfee and A.L. Baldwin, Porohyperelastic finite element analysis of large arteries using ABAQUS. *J. Biomech. Engr.* **120** (1998) 296–298.
56. B.R. Simon, M.V. Kaufmann, M.A. McAfee, A.L. Baldwin and L.M. Wilson, Identification and determination of material properties for porohyperelastic analysis of large arteries. *J. Biomech. Engr.* **120** (1998) 188–194.
57. A.J.M. Spencer, Constitutive theory for strongly anisotropic solids, In: A.J.M. Spencer (ed.), *Continuum Theory of the Mechanics of Fibre-Reinforced Composites*, CISM Courses and Lectures No. 282, International Centre for Mechanical Sciences, Springer-Verlag, Wien (1984), pp. 1–32.
58. J. Staubesand, Anatomie der Blutgefäße. I. Funktionelle Morphologie der Arterien, Venen und arterio-venösen Anastomosen. In: M. Ratschow (ed.), *Angiology*, Thieme, Stuttgart (1959), Chapter 2, pp. 23–82.
59. K. Takamizawa and K. Hayashi, Strain energy density function and uniform strain hypothesis for arterial mechanics. *J. Biomech.* **20** (1987) 7–17.
60. A. Tözeren, Elastic properties of arteries and their influence on the cardiovascular system. *J. Biomech. Engr.* **106** (1984) 182–185.
61. R.N. Vaishnav and J. Vossoughi, Estimation of residual strains in aortic segments. In: C.W. Hall (ed.), *Biomedical Engineering II: Recent Developments*, Pergamon Press, New York (1983), pp. 330–333.
62. R.N. Vaishnav, J.T. Young and D.J. Patel, Distribution of stresses and of strain-energy density through the wall thickness in a canine aortic segment. *Circ. Res.* **32** (1973) 577–583.
63. D.A. Vorp, K.R. Rajagopal, P.J. Smolinsky and H.S. Borovetz, Identification of elastic properties of homogeneous orthotropic vascular segments in distension. *J. Biomech.* **28** (1995) 501–512.
64. J. Vossoughi, Z. Hedjazi and F.S.I. Boriss, Intimal residual stress and strain in large arteries. In: *1993 ASME Advances in Bioengineering*, New York (1993), pp. 434–437.
65. J. Vossoughi and A. Tözeren, Determination of an effective shear modulus of aorta. *Russian J. Biomech.* **1–2** (1998) 20–35.
66. H.W. Weizsäcker and J.G. Pinto, Isotropy and anisotropy of the arterial wall. *J. Biomech.* **21** (1988) 477–487.
67. F.L. Wuyts, V.J. Vanhuyse, G.J. Langewouters, W.F. Decraemer, E.R. Raman and S. Buyle, Elastic properties of human aortas in relation to age and atherosclerosis: A structural model. *Phys. Med. Biol.* **40** (1995) 1577–1597.
68. J. Xie, J. Zhou and Y.C. Fung, Bending of blood vessel wall: Stress–strain laws of the intima-media and adventitia layers. *J. Biomech. Engr.* **117** (1995) 136–145.
69. Q. Yu, J. Zhou and Y.C. Fung, Neutral axis location in bending and Young's modulus of different layers of arterial wall. *Am. J. Physiol.* **265** (1993) H52–H60.

Journal of Elasticity **61:** 49–81, 2000.

Structure, Mechanical Properties, and Mechanics of Intracranial Saccular Aneurysms

J.D. HUMPHREY[1] and P.B. CANHAM[2]
[1]*Biomedical Engineering, Texas A&M University, College Station, TX, U.S.A.*
E-mail: jdh@acs.tamu.edu
[2]*Medical Biophysics, University of Western Ontario, London, Ontario, Canada*

Received 20 April 2000; in revised form 13 February 2001

Abstract. Intracranial saccular aneurysms remain an enigma; it is not known why they form, why they enlarge, or why only some of them rupture. Nonetheless, there is general agreement that mechanics plays an essential role in each aspect of the natural history of these potentially deadly lesions. In this paper, we review recent findings that discount limit point instabilities under quasi-static increases in pressure and resonance under dynamic loading as possible mechanisms of enlargement of saccular aneurysms. Indeed, recent histopathological data suggest that aneurysms enlarge due to a stress-mediated process of growth and remodeling of collagen, the primary load-bearing constituent within the wall. We submit that advanced theoretical, experimental, and numerical studies of this process are essential to further progress in treating this class of pathologies. The purpose of this review is to provide background and direction that encourages elasticians to contribute to this important area of research.

Key words: collagen structure, stability, rupture criterion, growth mechanics.

1. Introduction

Intracranial aneurysms are focal dilatations of the arterial wall that usually occur in or near the circle of Willis, the primary network of vessels that supplies blood to the brain. In general, these aneurysms occur in one of two forms: *fusiform lesions*, which are elongated dilatations of an artery, and *saccular lesions*, which are local sac-like out-pouchings. This paper focuses on the more common saccular form, which usually develops at the apex of a bifurcation (Figure 1). Rupture of saccular aneurysms is the leading cause of spontaneous subarachnoid hemorrhage (*SAH*), which despite advances in neurosurgery and neuroradiology continues to result in a high mortality rate (35–50%) and severe morbidity among the survivors [20, 26, 102]. Fortunately, with advances in medical imaging, greater numbers of unruptured aneurysms are being detected. There are two primary methods of treating these lesions: intracranial surgery, wherein the lesion is isolated from the blood flow by placing a small metal clip at its neck, and catheter-based interventions, which include the deployment of metallic coils that promote the formation of clots within the lesion that again isolate it from the blood flow [10, 71]. Conservative

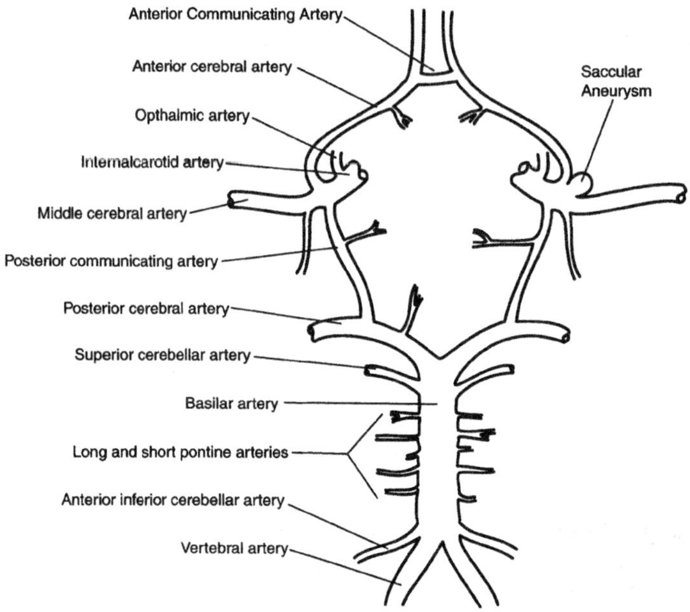

Figure 1. Schema of the cerebral vasculature illustrating the circle of Willis and surrounding arteries; shown, too, is a typical aneurysm at a bifurcation (where the fundus = pole).

management is also a clinical option, however, for it is thought that many saccular aneurysms will not rupture [10, 26, 80]. For example, a recent international trial [103] reported a small risk of rupture ($\sim$0.1% per year) for aneurysms less than 10 mm in maximum dimension; these results have generated considerable controversy, however (e.g., [4]). The primary clinical dilemma, therefore, is whether a patient should be subjected to a prophylactic procedure that has associated risks given that it is unlikely that the aneurysm will rupture, or if it is better to monitor periodically the patient for changes in the lesion while accepting the devastating consequences associated with *SAH* should a rupture occur.

The goals of this paper are threefold: to review our current knowledge of the biology and structure of saccular aneurysms, to assess recent developments that address the associated mechanics, and to identify questions about aneurysmal development, enlargement, and rupture that require further biomechanical study. One goal of histo-mechanical analysis, for example, is to predict better the likelihood of enlargement of a given lesion and its rupture-potential, the former of which may occur over periods from weeks to decades. It is hoped, therefore, that this paper stimulates experimental, theoretical, and computational research that will complement that in the basic and clinical sciences and thereby contribute to improved treatment strategies.

2. Natural History and Significance

Two-to-five percent of the general population in the Western World likely harbors a saccular aneurysm, ruptured or unruptured. Histopathological and clinical studies reveal further that these lesions are more prevalent in women (55–65%), and that they occur predominantly in the anterior and middle portions of the cerebral vasculature (cf. Figure 1). For example, a review by Ferguson [26] reports distributions of 37% in the internal carotid artery, 31% in the anterior cerebral and anterior communicating arteries, 13% in the middle cerebral artery, 9% in the basilar artery, 5% in the vertebral artery, and 5% other; different reports indicate a slightly higher percentage in the middle cerebral artery, but are otherwise similar [92, 98]. Multiple lesions occur in ~15–30% of aneurysm patients. Although saccular aneurysms may remain dormant for years to decades, the small percentage that rupture tend to do so during the 5th–7th decades of life (mean age ~52 years old). For more detail, see Sekhar and Heros [80], Hashimoto and Handa [38], Kassel and Torner [52], and Wiebers et al. [102].

The natural history of saccular aneurysms consists of at least three phases: pathogenesis, enlargement, and rupture. The initiation of saccular aneurysms is the subject of considerable debate, but it is generally accepted that unique structural features of the cerebral vasculature contribute to the pathogenesis (e.g., [7, 80]). Cerebral arteries do not have an external elastic lamina, they have sparse medial elastin, they lack supporting perivascular tissue, and they have structural irregularities at the apex of their bifurcations [27, 40, 91]. It is thought that these factors may render the cerebral artery susceptible to a local weakening of the wall under the persistent action of hemodynamic loads, particularly in hypertension [92]. One theme, in particular, is that the internal elastic lamina and muscular media must become markedly fragmented or degraded in order for a saccular aneurysm to form [11, 24, 80]. Other risk factors may include heavy alcohol consumption, cigarette smoking, and the long term use of analgesics or oral contraceptives, although these are thought to play a lesser role (see [20, 73, 103]). Increased familial incidence in some populations suggests that genetics is important. It has been hypothesized, for example, that a genetic defect may disrupt the normal synthesis of certain types of collagen (e.g., types III and V) within the cerebral vasculature, which in turn may weaken the arterial wall [41, 74]. Similarly, it has been hypothesized that an asymmetrically formed circle of Willis may be of genetic origin, and may increase the hemodynamic load on portions of the vasculature [20, 92]. There is a pressing need for much more research on the roles of genetics, risk factors, cellular responses to mechanical stresses, and hemodynamics in the pathogenesis.

Aneurysms typically enlarge from the initially small out-pouching or dilatation of the arterial wall, which can result in lesions having diameters up to 30 mm as well as complex shapes and composition. Unfortunately, little is known about the mechanisms by which this enlargement occurs, or its time-course. Some recent studies suggest that slower rates of enlargement are associated with a lower risk of rupture [51, 66]. Regardless, among other hypotheses, it has been suggested that lesions may enlarge rapidly due to structural instabilities, that is via either a limit point instability or resonance. These two hypotheses are discussed below based on more recent nonlinear analyses and shown unlikely, at least for particular classes of lesions. As it will be seen, there is a pressing need for an increased understanding of this critical phase of the natural history.

Rupture of an aneurysm implies one of two outcomes: a catastrophic tearing of a portion of the lesion, with significant bleeding that is often fatal, or a small "leak", with minimal bleeding but clinical symptoms. Small leaks may be sealed by a fibrin patch and followed by the formation of an intraluminal or intramural thrombus; such repair may render the lesion more susceptible to subsequent enlargement or catastrophic rupture. Although histomechanical failure mechanisms are unknown, rupture usually occurs at the fundus (Figure 1) despite the neck often being thinner [18, 80]. Moreover, in the case of coexisting aneurysms, the larger one usually ruptures first, or if of nearly the same size, the proximal one will usually rupture first [18, 50].

Several studies have associated various physical factors with rupture-potential. Asari and Ohmoto [3] suggested that it is the combination of lesion location (e.g., middle cerebral artery), shape (i.e., multilobular or not), and the presence of hypertension that best indicates a high risk of rupture. Hademenos et al. [35] similarly reported that multilobular lesions are more prone to rupture, but they suggested further that the less prevalent posterior lesions have a higher chance to rupture. The vast majority of other studies draw conclusions based primarily on the size of the lesion, however, with estimates of the critical maximum dimension ranging from 3 to 10 mm [18, 52, 98, 102, 103]. From a mechanical perspective, of course, shape and wall thickness are more important contributors to rupture-potential than overall size [84]. Ujiie et al. [99] reported that 59% of lesions are round, 24% oval, and 22% barlike; Parlea et al. [75] reported that aneurysms of the anterior communicating artery (recall Figure 1) tended to be pear-shaped, and, although a clear pattern could not be established, most aneurysms tend toward a spherical shape. That shape has not been considered more is particularly surprising since Crompton [17] showed long ago that lesions in women tend to have a greater neck : height ratio and they are more likely to rupture. Lesion thickness can range from 30 to 500 μm in the unloaded configuration [96]. It is believed that increasing thickness corresponds primarily, but not exclusively, with continued enlargement [93]. Based on a study of 23 unruptured lesions, Suzuki and Ohara [96] suggested further that saccular aneurysms fall into one of four categories: uniformly thin (22%), thick at the fundus but thin at the neck (17%), thin at the neck but variable elsewhere (43%), or thick at the neck but variable elsewhere (18%). Asari and Ohmoto [3] report similar findings, including that the uniformly thin lesions tended to be the smallest (less than 4 mm in diameter). Whereas most previous mathematical models have assumed idealized shapes and uniform wall thickness (in the undeformed configuration), there is a clear need for more complete data upon which more realistic models can be based.

3. Histopathology

Early studies using light microscopy showed relatively little structural organization within saccular aneurysms as compared to the nearby parent arteries [28, 39, 79, 89]. Like other muscular arteries, cerebral arteries consist of three main layers: the tunica adventitia (outer layer) consists primarily of collagen, the tunica media consists primarily of smooth muscle with interspersed collagen and some elastin, and the tunica intima (inner layer) consists primarily of a basement membrane bordered on the luminal side by a single layer of endothelial cells. The transition from parent vessel to aneurysm is characterized by a sharp break in the media [7, 39, 80], thus it is generally thought that the aneurysm stems from the adventitia and intima [69, 79]. Stehbens [89] suggested, however, that the appearance of the aneurysmal wall does not indicate that part of the arterial wall from which the sac came for many of its features seem to be acquired during the later stages of enlargement. Regardless, a general finding is that the aneurysmal wall consists primarily of collagen, with small patches of smooth muscle of the stellate form, as is usually found in the intima of vessels. The elastic lamina also tends to split into several laminae at the neck of the aneurysm and to be fragmented or absent in the aneurysm wall [39, 89]. Stehbens [90] reported that in those portions of an aneurysm that resemble an intimal proliferation, the collagen fibers are sparse, of variable length and caliber, and arranged haphazardly, whereas when the wall is fibrotic, the collagen is arranged in distinct laminae. That the fundus is generally thicker than the neck [80, 89], and yet the site at which rupture is most likely to occur [18, 79, 92], is one of the unresolved paradoxes of aneurysm structure.

Electron microscopy has confirmed this general histology, with additional detail regarding the presence of monocytes, fibroblasts, macrophages, and cellular debris [60, 72, 92]. Spatial variations in these cell types provide possible clues with regard to lesion heterogeneity, including local weakening of the wall, and dynamic changes therein [53, 54]. Recent immunocytochemical and immunofluorescence studies reveal further detail on aneurysm composition. Austin et al. [7] found type I collagen and fibronectin to be distributed uniformly throughout the aneurysmal wall, and Mimata et al. [69] identified the fibrillar types I and III collagens, which are mainly responsible for the tensile strength, throughout the wall. The microfibrillar type VI collagen is also distributed throughout the wall, albeit mostly in the outer region, and the basement membrane type IV collagen is localized around the sparsely distributed smooth muscle cells. Kosierkiewicz et al. [55] examined lesions with atherosclerotic involvement. They identified an intimal type thickening even in some small aneurysms and advanced plaques with smooth muscle cells and lipid-laden macrophages in many large aneurysms. They also reported that it was often difficult to separate the atherosclerotic region from the rest of the wall.

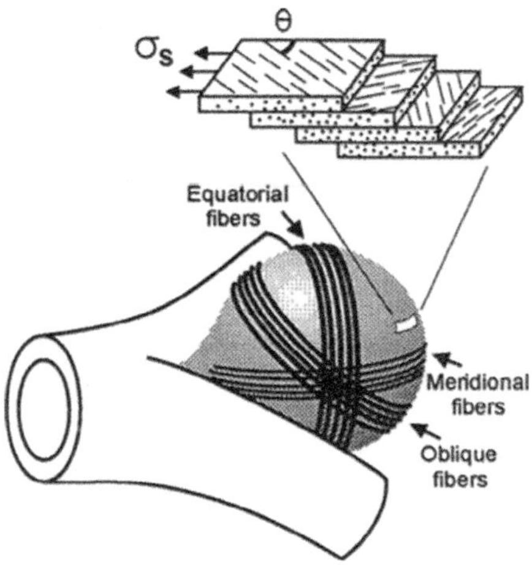

Figure 2. Schema of a saccular lesion showing overlapping layers of collagen fibers of different strengths. The inset illustrates how the alignments in adjacent layers combine to give tensile strength σ_S (modified from [66]).

The main characteristic of the aneurysm wall is its multidirectional collagen fibers – at physiological pressures they become straight and thereby govern the overall stiffness of the lesion (Figure 2). As the aneurysm enlarges, collagen is repeatedly synthesized and degraded; that is, the architecture evolves "continuously". Three important parameters are changes in the orientation, cross-linking, and volume fraction of the various types of collagen. For example, type I collagen is substantially stiffer than type III, and the alignment of the collagen fibers is fundamental to the strength of the tissue that must bear biaxial loading. Thus, there is a need for combined histo-mechanical analyses.

Most general microscopic studies of the aneurysmal wall have been on lesions that were not fixed at arterial pressure. In the unloaded state, the collagen fibers are wavy, and their preferred orientations are not discerned easily when viewed using normal stains for light microscopy. Since collagen is birefringent, however, polarized light can be used to identify the collagen and to assess its orientation [104, 106] Results by our group that exploit these observations are presented below in the sub-section on collagen architecture.

4. Assessments of Mechanical Behavior

Scott et al. [83] performed *in vitro* pressure-volume tests on seven human saccular aneurysms obtained at autopsy. Data were reduced assuming that the lesions were perfect spheres, having deformed volumes of $4\pi a^3/3$; this allowed them to estimate the deformed radii a and wall tension $T = Pa/2$ (using Laplace's relation for a thin-walled pressure vessel). Scott et al. reported that aneurysms exhibit a nonlinear behavior over finite strains, and suggested that they are stiffer than normal vessels. They also reported that two aneurysms exhibited a critical breaking (Cauchy) stress σ_c on the order of 2 to 3 MPa. Because they measured global volumes, not local strains, the results are both averaged and effectively one-dimensional and consequently not sufficient for quantifying the requisite multiaxial constitutive relations.

Steiger et al. [93] reported results from uniaxial extension tests on thin strips of tissue excised from six human saccular lesions. Important findings are that lesion behavior differs at the fundus and neck: tearing occurred at a stretch $\lambda = 1.37$ and a stress $\sigma_c = 0.5$ MPa in the fundus and $\lambda = 1.57$ and $\sigma_c = 1.2$ MPa in the neck. That strain, not stress, was a more consistent metric of failure is consistent with results on the failure of arteries (see [45]). Nevertheless, Steiger's study is limited because the data were reduced using the linearized measure of strain, and based on the overall length of the specimen rather than a central gage length. Moreover, the 1-D data did not discriminate between the meridional and circumferential behaviors. Toth et al. [97] recently reported similar uniaxial data from 22 human aneurysms, including 17 harvested at surgery. The characteristics of the latter group includes a mean diameter of 11.6 mm (ranging from 5 to 23 mm) and a mean patient age of 47 years old (range from 32 to 63). Twelve of the 17 lesions were from females, and 11 of the 17 patients had a history of *SAH*. Similar to Steiger et al., it was found that (circumferentially oriented) specimens from the fundus tore at lower stretches (i.e., $\lambda = 1.23$) than those from the neck ($\lambda = 1.55$). Moreover, the strength near the fundus was greater in the meridional than the circumferential direction. Additional results were presented in terms of moduli for a Kelvin–Voigt linear viscoelastic model; use of a linearized measure of strain is clearly inappropriate given the reported stretches up to 55%. Remarkably, these three studies represent the entirety of the data up to 1999 on the mechanical behavior of human saccular aneurysms. The need for experimental data and the associated constitutive formulations is clear therefore.

In summary, histopathology and mechanical tests reveal the following general characteristics of "non-complicated" saccular aneurysms: they are thin-walled shell-like structures that consist primarily of a 2-D plexus of collagen, they appear to have negligible bending stiffness, they exhibit nonlinear anisotropic pseudoelastic responses over finite strain, and their properties vary regionally. Clearly, therefore, a nonlinear membrane theory is a reasonable starting point for analysis. Although Scott's data are not sufficient for detailed quantification of multiaxial behavior, including anisotropy and heterogeneity, Kyriacou and Humphrey [57] showed that they are well described by a Fung-type pseudostrain-energy function w, which is defined per undeformed surface area consistent with the direct membrane approach [47]. That is, consider a w of the form

$$w = c[\mathrm{e}^{Q} - 1], \qquad Q = c_1 E_{11}^2 + c_2 E_{22}^2 + 2c_3 E_{11} E_{22}, \tag{1}$$

where E_{AB} are the principal (in-plane) Green strains and c and c_i are material parameters. For the quasi-static inflation of a perfectly spherical membrane (as

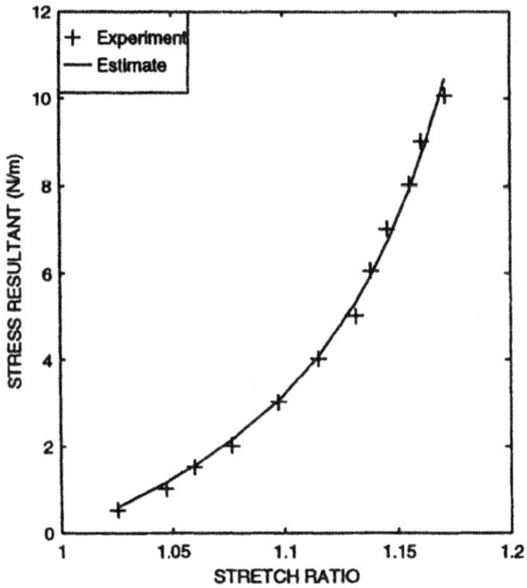

Figure 3. Best-fit of equation (1) to the tension-stretch data of Scott et al. (from Kyriacou and Humphrey, with permission).

assumed by Scott et al.), the 2-D deformation gradient is $\mathbf{F} = \text{diag}[\lambda, \lambda]$ where $\lambda = a/A$ (with A the undeformed radius) and the principal Green strains are $E_{11} = E_{22} = \frac{1}{2}(\lambda^2 - 1)$. Recalling the general constitutive relation for a membrane,

$$T_{\alpha\beta} = \frac{1}{\det \mathbf{F}} F_{\alpha\Delta} F_{\beta\Xi} \frac{\partial w}{\partial E_{\Delta\Xi}}, \quad \alpha, \beta, \Delta, \Xi = 1, 2, \tag{2}$$

with $c_2 \equiv c_1$ due to the implicit assumption of in-plane isotropy in Scott's analysis, we see that the uniform tension $T (= T_1 \equiv T_2)$ is

$$T = c\Gamma(\lambda^2 - 1)\exp[0.5\Gamma(\lambda^2 - 1)^2], \tag{3}$$

where $\Gamma \equiv (c_1 + c_3)$. Kyriacou and Humphrey determined the best-fit values of the two independent material parameters c and Γ via a Marquardt–Levenberg regression of the data presented by Scott et al., that is by minimizing the sum-of-the-squares of the error between the calculated and measured tensions (i.e., stress resultants). The best-fit values were $c = 0.88$ N/m and $\Gamma = 12.99$, which yielded the fit to data shown in Figure 3. Despite this good fit, the inadequacy of the data is evident: they do not allow separate determination of c_1 and c_2, which embody the material symmetry, they do not separate the contributions due to c_1 and c_3, and they do not provide information on possible heterogeneities. Based on results on arteries reported by Fung and colleagues, Kyriacou and Humphrey assumed that $c_3 \sim c_1/10$, and hence $c = 0.88$ N/m, $c_1 \equiv c_2 = 11.82$ and $c_3 = 1.18$. Likewise, various anisotropies can be explored by varying the values of c_1 and c_2 (see Section 7).

5. Hypothesized Mechanisms of Enlargement and Rupture

5.1. LIMIT POINT INSTABILITIES

A longstanding question with regard to saccular aneurysms has been, how can a structure that consists of collagen, which exhibits high stiffness and low extensibility, continue to enlarge and eventually rupture? In an effort to address this question, Austin et al. [6] and Akkas [1] suggested that saccular aneurysms suffer limit point instabilities, that is (mathematical) bifurcations in their quasi-static response to increases in distension pressure. Note, however, that Austin et al. based their conclusions on *in vitro* experiments on a "model lesion" that they constructed by gluing a 0.8 mm thick collagen patch onto the center of a 0.175 mm thick elastomeric membrane that was fixed around its periphery and inflated from underneath. Because of the use of the elastomeric membrane, it is to be expected that this model exhibited a limit point instability [8]. Akkas, on the other hand, reported computational results for the inflation of a neo-Hookean model (i.e., $W = c(\mathrm{tr}\,\mathbf{C} - 3)$, where $\mathbf{C} = \mathbf{F}^{\mathrm{T}}\mathbf{F}$) of a saccular aneurysm, which also exhibited a limit point as expected. Because aneurysms, like most collagenous soft tissues, tend to exhibit an exponential rather than rubber-like behavior, it is clear that these studies needed to be revisited.

Consider an idealized spherical lesion* having an undeformed radius A, a uniform initial thickness H, and subjected to a uniform distension pressure P. With $\mathbf{F} = \mathrm{diag}[\lambda, \lambda]$, is it easy to show that the pressure-stretch relation for the Fung-type form of w (equation (1)) is

$$P(\lambda) = \frac{2c\Gamma}{A}(\lambda - 1/\lambda)\exp[0.5\Gamma(\lambda^2 - 1)^2], \tag{4}$$

which is easily non-dimensionalized by multiplying $P(\lambda)$ by A/c. Regardless, a limit point exists if $\mathrm{d}P/\mathrm{d}\lambda = 0$ for any $\lambda > 1$ (note: a membrane cannot support compression, thus the restriction on λ). It can be shown numerically that the Fung material does not admit a limit point, with $\Gamma = 12.99$ from Scott's data. Inasmuch as Kyriacou and Humphrey [57] found a similar result for a more general case of an axisymmetric lesion (using finite elements), it appears that certain sub-classes of saccular aneurysms probably do not enlarge or rupture via a limit point instability. This finding re-emphasizes the importance of basing one's analysis on appropriate constitutive relations.

5.2. DYNAMIC INSTABILITIES

Richardson and Kofman [77] reported bruits in cerebral aneurysms – that is, audible tones at frequencies ~400 Hz. Ferguson [23] suggested that these bruits

* Shah et al. [84] showed that the spherical assumption is reasonable for a small sub-class of aneurysms.

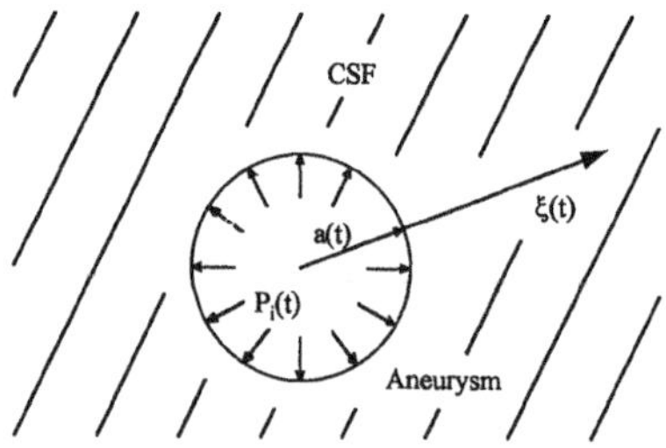

Figure 4. Schema of a spherical aneurysm surrounded by cerebral spinal fluid. The radial coordinate defining the fluid domain is $\xi \in [a(t), \infty)$ (from Humphrey, with permission).

resulted from turbulence within the lesion while others suggested that they indicated that aneurysms are excited at their natural frequency [50, 81, 86]. Resonance implies large wall motions, indeed violent vibrations, and thus was hypothesized by some as a potential mechanism of enlargement or rupture. There have been but a few analyses of the associated elastodynamics, however, most of which are based on classical shell theory and thus linearized strains and material behavior (e.g., [49, 86]). Moreover, none of these studies account for the observation that many saccular aneurysms are surrounded by cerebral spinal fluid (*CSF*). Below, we summarize a recent study by Shah and Humphrey [85] that is based on finite elasticity and accounts for the *CSF*. First, however, it is important to note the following findings from hemodynamic studies. Experimental and computational results both reveal that flow-induced wall shear stresses τ_w are small in all classes of saccular aneurysms studied to date: maximum values are ~5 to 13 Pa [65, 94], which are less than the 40 Pa needed to induce endothelial cell damage [30] and orders of magnitude less than the pressure-induced in-plane wall stresses which can be 1 to 10 MPa [12, 57]. These findings, coupled with observations that the maximum wall shear stress typically occurs at the neck, not the fundus where rupture tends to occur, suggest that intra-aneurysmal pressures are the dominant hemodynamic loads governing stress-induced rupture [29, 80, 94]. This is not to say that wall shear stresses are not important; they likely signal the endothelium to express various molecules, including growth factors that may regulate the intramural collagen. This latter role has not been explored in detail, though it ought to be. It also appears that intra-aneurysmal pressures are similar in magnitude to those in the parent vessel [16, 25, 67, 81] and that they vary little with position within the lesion [33, 37]. Consequently, it appears to be reasonable to assume (to first order) that saccular aneurysms are loaded primarily by a uniform, time-varying distension pressure.

For purposes of examining the elastodynamics, consider a thin-walled, spherical aneurysm of initial radius A and wall thickness H, but now let it be subjected to a time-varying distension pressure $P_i(t)$ and surrounded by *CSF* (Figure 4). The 2-D deformation gradient tensor $\mathbf{F} = \text{diag}[\lambda(t), \lambda(t)]$, where $\lambda(t) = a(t)/A$ and $a(t)$ are the deformed radii. In addition, let the lesion exhibit an isotropic, Fung-type behavior (equation (1)). From Kraus [56], it can be shown that the three equations of motion for a membrane reduce to a single ordinary differential equation for a pulsating sphere:

$$\rho h \frac{\mathrm{d}^2 u_r}{\mathrm{d}t^2} = -2T\kappa - t_{rr}(r_i) + t_{rr}(r_o), \tag{5}$$

where ρ is the mass density of the aneurysm, $h(t)$ is the deformed thickness (which equals $H/\lambda(t)^2$ if incompressible), $u_r(t) = a(t) - A$ is the radial displacement, $T = T(\lambda)$ is the constitutively determined wall tension, $\kappa(t) = 1/a(t)$ is the curvature in the deformed configuration, and t_{rr} are radial stresses on the inner and outer surfaces of the lesion. Assuming a prescribed time-varying uniform luminal pressure, $t_{rr}(r_i) = -P_i(t)$, Milnor [68] shows that arterial pressures are well described by a Fourier series representation of the form,

$$P_i(t) = P_m + \sum_{n=1}^{N} \big(A_n \cos(n\omega t) + B_n \sin(n\omega t)\big), \tag{6}$$

where P_m is the mean blood pressure, A_n and B_n are Fourier coefficients for N harmonics, and ω is the circular frequency. Ferguson [25] reported micro-catheter measured intra-aneurysmal blood pressures in humans. From these data, it can be shown that specific values of A_n and B_n, for the first 5 harmonics, are $A_1 = -7.13$, $B_1 = 4.64$, $A_2 = -3.08$, $B_2 = -1.18$, $A_3 = -0.130$, $B_3 = -0.564$, $A_4 = -0.205$, $B_4 = -0.346$, $A_5 = -0.0662$, and $B_5 = -0.120$, all in mmHg, with $P_m = 65.7$ mmHg. Note that these lower pressures were recorded in supine, anesthetized patients.

The cerebral spinal fluid (*CSF*) could similarly be assumed to exert a uniform time-varying pressure $P_o(t)$ on the outer surface of the membrane that is a reaction to the pressure-induced distension of the lesion. This is tantamount to treating the *CSF* as an ideal fluid (i.e., inviscid and incompressible). In this case, $t_{rr}(r_o) = -P_o(t)$, where P_o can be determined by solving the pressure field in the fluid domain $\xi \in [a, \infty)$. For an ideal fluid, the governing differential equations are the balance of mass and linear momentum (i.e., Euler) equations. In the absence of body forces, they can be written as $\nabla \cdot \mathbf{v} = 0$ and $-\nabla P = \rho_f \mathbf{a}$, respectively, where $\mathbf{v}$ and $\mathbf{a}$ are the fluid velocity and acceleration and ρ_f is the mass density of the *CSF*. It is probably better to assume that the *CSF* is viscous (e.g., Newtonian), however. In this case, the governing differential equations are the balance of mass and the incompressible Navier–Stokes form of the linear momentum equations ($-\nabla P + \mu \nabla^2 \mathbf{v} = \rho_f \mathbf{a}$), and the outer stress boundary condition is $t_{rr}(r_o) = -P_o(t) + 2\mu D_{\xi\xi}(\xi_o)$ where $\mathbf{D}$ is the stretching tensor ($\mathbf{D} = \frac{1}{2}(\mathbf{L} + \mathbf{L}^{\mathrm{T}})$, where $\mathbf{L}$ is the velocity gradient tensor). Because the solution for the ideal fluid can be recovered from that for the Newtonian fluid, we consider the latter here.

Assuming a flow in the radial direction ξ in a spherical domain (Figure 4), mass balance requires that

$$\frac{1}{\xi^2}\frac{\partial}{\partial \xi}(\xi^2 v_\xi) = 0 \rightarrow v_\xi(\xi, t) = \frac{g(t)}{\xi^2}, \tag{7}$$

where the function $g(t)$ is determined by requiring a material particle on the membrane to have the same velocity as the adjacent fluid particle v_ξ. Hence, at $\xi = a$,

$$\frac{\mathrm{d}}{\mathrm{d}t}(u_r) = \frac{\mathrm{d}\lambda}{\mathrm{d}t}A = \frac{g(t)}{a^2} \rightarrow g(t) = A^3\lambda^2\frac{\mathrm{d}\lambda}{\mathrm{d}t}. \tag{8}$$

The requisite component of **D** is thus computed easily.

For this radial flow, the meridional and circumferential Navier–Stokes equations require that the fluid pressure $P = P(\xi, t)$ alone. Hence, the only "non-trivial" equation of motion is the radial one, which can be integrated over $\xi \in [a, \infty)$ to yield the pressure $P_o(t)$ exerted on the outer surface of the membrane by the surrounding CSF:

$$P_o(t) = P_\infty(t) + \rho_f A^2\left(\lambda\frac{\mathrm{d}^2\lambda}{\mathrm{d}t^2} + \frac{3}{2}\left(\frac{\mathrm{d}\lambda}{\mathrm{d}t}\right)^2\right), \tag{9}$$

wherein we have used v_ξ from mass balance and the matching condition at the solid-fluid interface. It is interesting to note that this is the same pressure field as that for an ideal fluid (i.e., as that obtained by integrating the unsteady Bernoulli equation along a radial streamline).

Taken together, these equations yield the final governing differential equation [85]:

$$\left(\frac{\rho HA}{\lambda^2} + \rho_f A^2\lambda\right)\frac{\mathrm{d}^2\lambda}{\mathrm{d}t^2} + \frac{3}{2}\rho_f A^2\left(\frac{\mathrm{d}\lambda}{\mathrm{d}t}\right)^2 + \frac{4\mu}{\lambda}\frac{\mathrm{d}\lambda}{\mathrm{d}t} + \frac{2T(\lambda)}{A\lambda} = P_i(t) - P_\infty(t) \tag{10}$$

with $T(\lambda)$ given by equation (3) and $P_i(t)$ by equation (6). This nonlinear second-order ordinary differential equation can be solved using numerical techniques such as Runge–Kutta, which is facilitated by transforming it into a system of two first-order equations (this is simplified by first non-dimensionalizing the equation). Shah and Humphrey [85] solved this system of equations for the following values of parameters, which they suggested define a representative lesion: $\rho = 1050$ kg/m^3, $A = 3 \times 10^{-3}$ m, $H = 27.8 \times 10^{-6}$ m, $\rho_f = 1000$ kg/m^3, $\mu = 1.26 \times 10^{-3}$ Ns/m^2, $P_\infty = 3$ mmHg, and $c = 0.88$ N/m, $c_1 = c_2 = 11.82$, and $c_3 = 1.18$. See the original paper for complete results. Figure 5, panel B shows that equilibrium initial conditions yield a periodic solution as expected (i.e., a closed path in the phase-plane); panel C reveals further that, for the case of perturbed initial conditions, this periodic solution serves as an attractor (i.e., the oscillations tend to dissipate

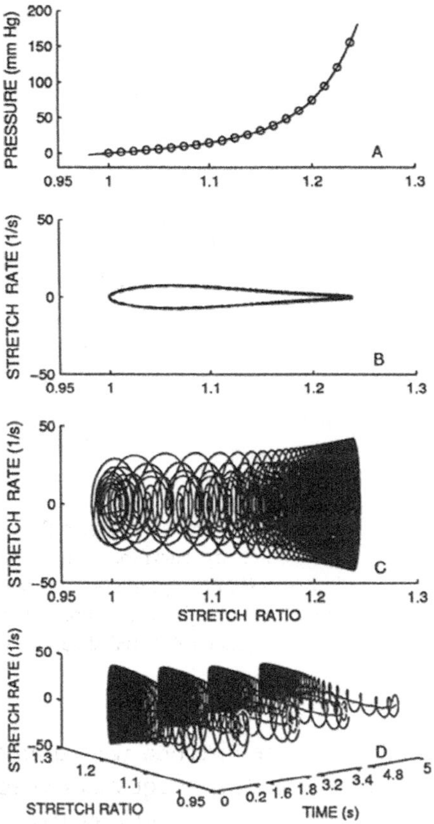

Figure 5. Results from Shah and Humphrey on the elastodynamics of a saccular aneurysm subjected to a sinusoidal forcing (pressure) function. Panel B shows the periodic solution in the unperturbed case, and panels C and D show that this solution is an attractor in the case of perturbed initial conditions, thus suggesting dynamic stability (from Shah and Humphrey, with permission).

and the perturbed solution returns to the periodic solution) and hence the solution is dynamically stable. Additional results suggest that the time-dependent solution can be treated quasi-statically as a series of equilibria. Whether this observation holds for other situations (e.g., different material parameters, different geometries, different forcing functions, etc.) must be examined individually, and remains an open problem. Based on this simple analysis, however, it appears that at least one sub-class of (nearly) spherical saccular aneurysms is dynamically stable both when $P_i(t)$ is a periodic function having a fundamental frequency less than 5–10 Hz (the non-autonomous system) and when it is a constant (the autonomous system; not shown). It appears reasonable, therefore, to emphasize quasi-static stress analyses for insight into the mechanics, a conclusion supported by Steiger [95]. Indeed, because saccular aneurysms are known to be thin, membranous tissues subject to low frequency pulsatile pressures, intuition suggests that the inertial effects would be small.

5.3. STRESS ANALYSES

There have been few rigorous studies of the quasi-static response of the aneurysmal wall to applied loads. Overly simplified analyses have been based on electrical analog models of latex balloons [5, 19] or constitutive relations that describe the behavior of rubber-like [1] or linear materials [34, 86] – (note: [34] is based on an incorrect equilibrium solution using a so-called modified Laplace's equation).

Laplace's equation $T = Pa/2$ is a universal solution for a thin-walled sphere and thus is applicable to saccular aneurysms (as used in the above limit point analysis). Canham and Ferguson [12] used Laplace's equation to estimate a critical diameter d_c at which a lesion may rupture. They assumed that the aneurysmal tissue volume v_T $(= 4\pi a^2 h)$ remains constant at all transmural pressures P (i.e., that these lesions suffer isochoric motions in a given state of enlargement), and showed that

$$d_c = \left(\frac{4\sigma_c v_T}{\pi P}\right)^{1/3}, \tag{11}$$

where σ_c is a critical wall strength. Rough estimates of $v_T = 1$ mm^3, $\sigma_c = 10$ MPa (recall that Scott et al. reported a $\sigma_c = 1$–2 MPa and Steiger et al. reported a $\sigma_c = 0.5$–1.2 MPa, both from uniaxial studies), and $P = 150$ mmHg suggested a $d_c = 8.6$ mm, a reasonable value. Limitations of this approach are the same as those in the work of Humphrey and Kyriacou [46] and Shah and Humphrey [85] – assumption of homogeneous and in-plane isotropic tissue behavior as well as homogeneity of the calculated stress and strain fields. The latter suggests that each material point is equally likely to fail, which does not account for the propensity of rupture at the fundus [18, 80, 92].

Despite the usefulness of simple spherical models, more realistic analyses are needed to account for the complex geometry, material properties, and applied loads that characterize the mechanics of an intracranial saccular aneurysm. Towards this end, Kyriacou and Humphrey [57] and Shah et al. [84] solved the equilibrium problem in weak form using the finite element method. For example, one can solve nonlinear axisymmetric and nonaxisymmetric inflation problems using the principle of virtual work,

$$\int_{\Omega_o} \delta w \, \mathrm{d}A - \int_{\Omega} P\mathbf{n} \cdot \delta\mathbf{x} \, \mathrm{d}a = 0, \tag{12}$$

where w is the 2-D strain-energy function, P the distension pressure, $\mathbf{n}$ an outward unit normal to the membrane in the current configuration Ω, $\delta\mathbf{x}$ the virtual changes in position, and Ω_o the original domain. After introduction of suitable interpolation functions and numerical integration via appropriate quadrature rules, equation (12) reduces to a system of nonlinear algebraic equations of the form $\mathbf{g}(\mathbf{q}) = \mathbf{0}$, where $\mathbf{q}$ represents the vector of (unknown) nodal positions. This equation admits an iterative Newton–Raphson solution, viz.

$$\mathbf{K}(\mathbf{q}^i)[\mathbf{q}^{i+1} - \mathbf{q}^i] = -\mathbf{g}(\mathbf{q}^i), \tag{13}$$

where $\mathbf{K} = \partial\mathbf{g}/\partial\mathbf{q}$ is the tangent matrix and i an iteration counter. This finite element solution thereby yields the current position of each node, from which one can compute strains and then stresses; this allows one to quantify the heterogeneous and anisotropic response of aneurysms to distention pressures. Kyriacou and Humphrey [57] and Shah et al. [84] considered a class of idealized axisymmetric saccular aneurysms having an initially uniform wall thickness H, a truncated spherical or elliptical geometry, a Fung-type constitutive behavior, and a clamped boundary condition at the neck. Not having sufficient data to quantify possible regional variations in material behavior, they considered a range of stress–strain behaviors from isotropic and homogeneous to anisotropic and heterogeneous. Using the aforementioned results for equation (3) and Scott's data, they defined isotropic behavior by $c = 0.8769$ N/m, $c_1 = c_2 = 11.82$, and $c_3 = 1.18$. For anisotropic behavior, the values of c and c_3 were the same, but values of c_1 and c_2 were modified to allow the ratio c_1/c_2 to vary linearly with the undeformed arc length $S \in [0, L]$ from $c_1/c_2 = 1$ at the fundus ($S = 0$) to either 3 or 1/3 at the neck ($S = L$). Kyriacou and Humphrey prescribed the variation in c_1/c_2 such that the value of $w\ (\lambda_1 = 1.18, \lambda_2 = 1.18)$ was the same at the neck as it was in the isotropic case; Shah et al. ensured that $w\ (\lambda_1 = 1.18, \lambda_2 = 1.18)$ remained the same at each point. Whereas the former allows regional variations in material heterogeneity (as suggested by the data of Steiger et al. [93]) and material symmetry, the latter maintains a type of homogeneity and thereby isolates effects of regional variations in symmetry. Of course, c_1/c_2 must equal 1 at the fundus due to axisymmetry, which requires $T_1 = T_2$ and $\lambda_1 = \lambda_2$ at that location. Finally, the prescribed boundary conditions were zero displacement at the neck (i.e., $u_r = 0$ and $u_z = 0$ at $z = 0$, which enforces $\lambda_2 \equiv 1$ at $z = 0$, where $\mathbf{u}$ is the displacement) and zero radial displacement at $r = 0$, the symmetry axis.

Perhaps the first question that one should address with the finite element method is the applicability of the Laplace equation $T_\alpha \equiv T = Pa/2$. Shah et al. [84] attempted this by first finding the best-fit sphere for the deformed configuration of model lesions as calculated by finite elements. They fit the deformed generator curve via $(r^j)^2 + (z^j - o)^2 = a^2$, where $j = 1, \ldots, n$ is the number of nodes used in the simulation, and o and a define the center and radius of the best-fit sphere. Next, they calculated the principal uniform Cauchy stress t, which equals T/h or $T\lambda^2/H$ where $h = H/\lambda^2$ is the deformed thickness, $\lambda = a/A$ the uniform stretch ratio, and A the best-fit undeformed radius. This requires a value for the uniform λ associated with T, which was obtained by inverting the constitutive relation (cf. equation (3)). Finite element and Laplace results were then compared as a function of undeformed arc length S.

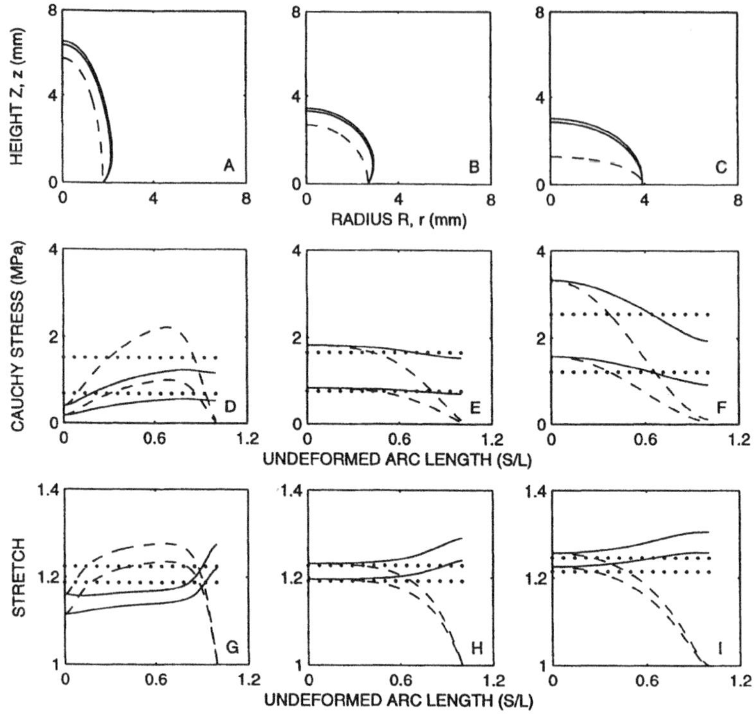

Figure 6. Finite element results for a model isotropic and axisymmetric aneurysm – $c = 0.88$ N/m, $c_1 = c_2 = 11.82$, $c_3 = 1.18$, $H = 27.8$ μm, initial volume = 0.0398 ml, and $A : B = 0.32$, 1.0, and 3.12. Panels A–C show the undeformed (dashed) and deformed (solid lines) configurations at 80 and 160 mmHg pressure. Panels D–F show the associated stresses in the meridional and circumferential directions, with the dotted lines showing the Laplace approximation. Panels G–I show the associated stretches (from Shah et al., with permission).

Figure 6 shows results for lesions having three different initial geometries (i.e., values of A/B, the ratio of the initial R and Z major axes) but otherwise the same initial lesion volume, thickness, isotropic material behavior, quasi-static distension pressure, and boundary conditions; specific values are in the figure legend. The undeformed generator curves (dashed lines in panels A to C) reveal that the prescribed geometry was one half of a complete ellipse or sphere. Panels A to C show how the initially elliptical or spherical geometry was distorted upon loading (see solid lines) due, in part, to the fixed boundary condition at the neck. Despite equal increments in pressure from 0 to 80 and then 80 to 160 mmHg, most deformation occurred at lower pressures as expected of a material that exhibits an exponential stress–strain behavior. Panels D to F reveal a number of important observations with regard to the distributions of the principal Cauchy stresses t_α: the meridional stress (solid curve) was higher than the circumferential stress (dashed curve) in lesions when $A/B \geqslant 1$, but the converse was true when $A/B < 1$; the highest multiaxial stresses occurred at the fundus in the lesions with the highest ratio of A/B; the highest multiaxial stresses occurred near $S/L \sim 0.7$ (with $S/L = 0$

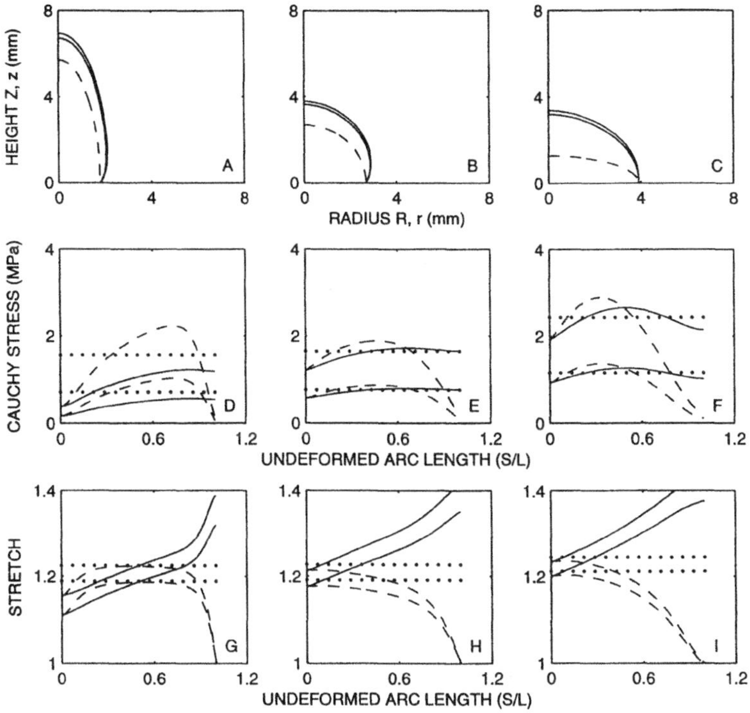

Figure 7. Similar to Figure 6 except for lesions having an increasingly greater stiffness in the circumferential direction as you move from the fundus to the neck.

at the fundus and 1 at the neck) for lesions with $A/B < 1$; and, as expected, stresses were uniform over the largest domain in lesions having $A/B = 1$ (i.e., an initially spherical geometry). Panels G to I show the associated distributions of the principal stretch ratios λ_α. Note that $\lambda_2 = 1$ at $S/L = 1$ as required by boundary conditions. The horizontal dotted lines in panels D to I show the uniform stress and stretch values predicted by the Laplace solution; recall that these were calculated based on the best-fit sphere for the deformed configuration. As expected, the Laplace approximation was best for the initially spherical geometry although one may argue that a reasonable mean value for $T_\alpha(S)$ was obtained in each case. Details on the stress field provide much greater information, however.

Additional results were reported for the same three lesions and loading conditions with the exception that the material properties varied linearly in S from isotropic at the pole (i.e., $c_1 = c_2 = 11.8$ at $S/L = 0$) to meridionally stiffer at the neck (i.e., $c_1 = 17.79$ and $c_2 = 5.93$ at $S/L = 1$). The deformed configurations were similar to those in Figure 6, and so too for the stress and stretch fields with two exceptions: the maximum stresses increased at the fundus and the meridional stretch decreased at the neck when $A/B \geqslant 1$, and the maximum values of circumferential stress and stretch (i.e., near $S/L = 0.7$) increased slightly with respect to those in Figure 6 when $A/B < 1$. Figure 7 shows results for the same three

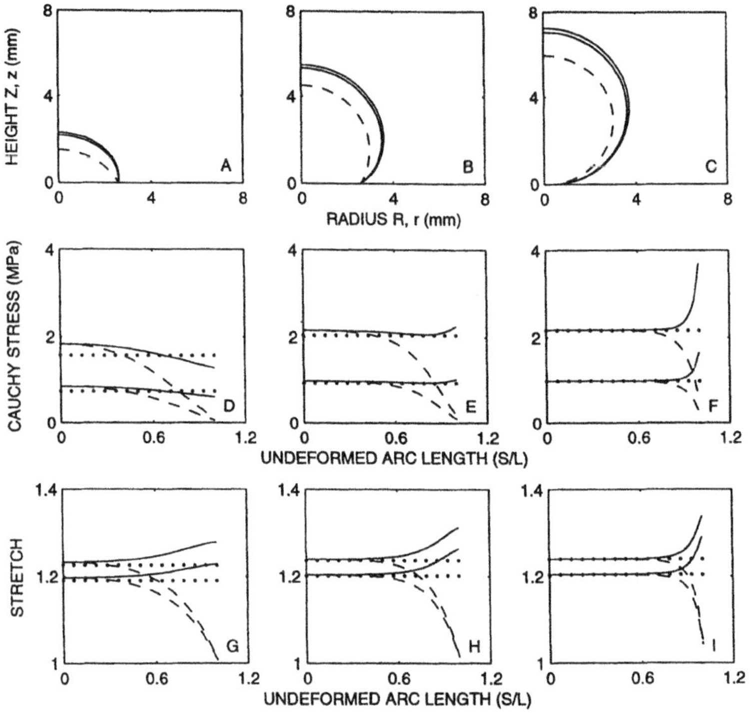

Figure 8. Similar to Figure 6 except for lesions of three different initial sizes (from Shah et al., with permission).

lesions with the exception that the material properties varied linearly in S from isotropic at the fundus to circumferentially stiffer at the neck (i.e., $c_1 = 5.93$ and $c_2 = 17.79$ at $S/L = 1$). In contrast to changes associated with the meridionally stiffer lesion, this circumferentially stiffer behavior resulted in marked differences in the stress and stretch fields, particularly when $A/B \geqslant 1$. For example, panels E and F in Figure 7 reveal that an increased circumferential stiffness resulted in a decreased equibiaxial stress at the fundus, a maximum multiaxial stress away from the fundus, and an increased meridional stretch near the neck. Increasing the circumferential stiffness when $A/B > 1$ thus tended to homogenize the stress field.

Figure 8 shows results for lesions having three different initial sizes (i.e., luminal volume) but otherwise the same initial spherical shape (i.e., the same undeformed radius, but truncated at different locations), thickness, isotropic material behavior, quasi-static distension pressures, and boundary conditions. As expected, panels A to C show that the more completely spherical geometry (panel C) yielded the most sphere-like behavior; that is, the stress and stretch fields are uniform and equibiaxial over a large portion of the lesion, the only variations being due to the boundary condition at the fixed neck (a boundary layer effect). Despite marked differences in size (undeformed and deformed), the magnitude of the stresses and stretches were nearly the same at the fundus and similar over the entire domain.

This was anticipated for the general solution for the inflation of an axisymmetric membrane is [47],

$$T_1 = \frac{P}{2\kappa_2}, \qquad T_2 = \frac{P}{\kappa_2}\left(1 - \frac{\kappa_1}{2\kappa_2}\right), \tag{14}$$

which reveals that it is the principal curvatures (κ_α), not the size per se, that controls the stress resultants – this has not yet been appreciated clinically as noted earlier with regard to the ongoing search for the "critical size". Finally, the dotted lines reveal, as expected, that a Laplace approximation is reasonable for nearly complete sphere-like lesions (panels B and C) but less good for cap-like lesions.

Despite longstanding reliance on the maximum dimension as a predictor of rupture-potential, this metric has failed to answer the most important clinical questions: Why do some lesions expand whereas others remain dormant for long periods? Why do some lesions rupture whereas most do not? Why does rupture tend to occur at the fundus even when the neck is thinner? Although based on incomplete data, the biomechanical analyses presented here reveal important insights that address these questions in part. It appears that lesions do not enlarge because of material or dynamic instabilities; it appears that the local curvature and anisotropic material properties, not lesion size, govern the distribution of intramural stress; and it appears that the stresses are greatest at the fundus if the material behavior is either isotropic or meridionally stiffer (recall that Toth et al. [97] found the latter experimentally).

6. Need for a Structurally-Based Constitutive Relation

The above results demonstrate that biomechanics can and must play a role in understanding better the natural history of saccular aneurysms and their treatment. Yet, analyses are only as good as the data upon which they are founded. A pressing need in aneurysm research is the identification of an improved stress–strain relation for the tissue. It is axiomatic that material behavior results from the internal composition of the material, hence an appropriate starting point is quantitative histology.

6.1. COLLAGEN ARCHITECTURE

Collagen is a highly structured cross-linked biopolymer, able to withstand high tensile loads. Tendon, having type I fibers, is perhaps the simplest and most thoroughly studied collagenous tissue; it has a breaking strength of 60–100 MPa and a stiffness of 1.0–2.5 GPa at its maximum extension [22, 101]. A collagenous framework provides strength and stiffness to blood vessels as well. Many arteries are able to withstand more than 10× the normal blood pressure. Even veins, which may be tested to pressures approaching 600 mmHg prior to use as bypass vessels for the heart, are exceptionally strong [2]. That intracranial saccular aneurysms rupture suggests that they are, in contrast, much less able to withstand the persistent loads due to normal physiological or pathological blood pressures (e.g., up to ~200 mmHg). It appears that two aspects of collagen structure may be key to the reduced strength of saccular aneurysms – the layered organization of the fibers within the wall and the existence of a fiber complement that is substantially weaker than normal type I collagen [7, 14, 15, 31, 69].

The general spherical shape of saccular aneurysms provides a basis for understanding the load carrying capability of the wall. Arterial blood pressure stresses the wall of a spherical lesion equally in all tangential directions, with "modest" differences in the stresses for elongated or flattened lesions [57]. Of course, more varied forms of saccular aneurysms, with secondary lobes on the primary lesion or biloculations, will have a correspondingly varied distribution of wall stresses. There is a need to correlate structure and mechanics in saccular aneurysms as attempted in other collagenous tissues (e.g., see [22, 63, 64]). Moreover, new results are needed to understand better how the aneurysmal wall behaves in the short term as a stiff elastic collagenous fabric, whereas over the longer term, it behaves *in vivo* as a remodeling, irreversibly enlarging structure. Progress has been made by applying multi-dimensional polarized light (MDPL) microscopy, which has the potential to complement the tissue mechanics. The key to assessing directional tissue properties is an evaluation of the directional distribution of the constituent fibers, layer by layer, and quantification of the proportion of fibers in each direction as well as their uniaxial strength within each layer.

Given that the aneurysmal wall consists primarily of collagen, it is useful to exploit the birefringent optical properties of individual collagen fibers – wherein the morphological, mechanical, and optical axes coincide – to study both orientation and distribution with the polarizing microscope [9]. The histological processing for such studies is standard (i.e., fixation, dehydration, paraffin infiltration and embedding, and subsequent sectioning via a microtome), except for the staining of tissue after sectioning. Birefringent enhancement stains make the measurement of the alignment of individual fibers more precise [87] and they make possible a measure of the strength of birefringence from the same local region. Thus, there are two polarized light techniques of importance in studying aneurysmal collagen: the Universal Stage attachment enables measurement of orientation in three dimensions and the Senarmont compensator, which is a 45 degree aligned quarter-wave filter, allows measurement of the birefringence of individual fibers [9].

The Universal Stage is an effective instrument for measuring two fiber angles: the azimuth, in the plane of the tissue section, and the elevation, measured out of the plane of the section. Thus, the tissue section is viewed optically as a thick transparent section and the birefringent collagen fibers within that section reveal their 3-D alignment directly on the polarizing microscope. A key point is that the inner stage of the instrument containing the tissue slide is mounted so that it can be rotated freely in three dimensions up to a tilt angle of 50°. Measurements of alignment are made at extinction, which has a precision of approximately ±1°

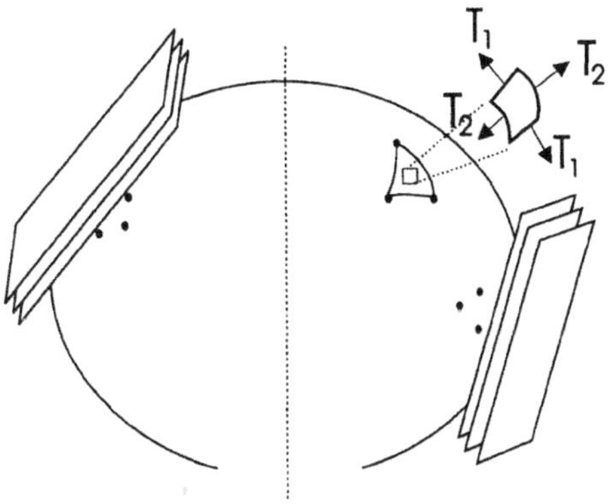

Figure 9. Schema to illustrate several tangential sectioning planes on the surface of an aneurysm. The local stretches and stress resultants can be assessed separately in each region prior to perfusion fixation.

[87]. Graphical presentation and analysis of data have been accomplished using the Lambert equal area projection.

Plane polarized light traversing a birefringent material is resolved into two rays, the *ordinary ray* and the *extraordinary ray*, which vibrate at right angles to each other and travel at different velocities. The velocity difference introduces a phase difference between the two rays that depends directly upon the strength of birefringence and the thickness of birefringent fabric. By means of the Senarmont compensator, one is able to measure, fiber by fiber, the phase difference (or phase retardation) directly on the microscope. The rationale behind the phase retardation method is that both the mechanical strength of the tissue and its birefringence depend on the cross-linked molecular structure and size of the fibers [21, 100].

An important requirement for combining both phase retardation and 3-D orientation is that the elevation angles of the measured fibers be relatively low, less than 15°, which ensures that the phase retardation is not biased toward lower values. This requirement has been met in studies on saccular aneurysms by confining analysis to tissue sections cut tangentially or near tangentially. By the method of repeated embedding, after each short series of tangential sectioning from the surface, it has been possible to retrieve collagen-related microscopic data from several regions of interest on the aneurysm surface [15]. This approach has set the stage for matching, in the future, wall structure and material behavior, region by region on individual lesions (Figure 9).

Figure 10 shows a tangentially cut section close to the luminal edge of a 2.6 mm diameter human aneurysm. Superimposed on the section are four radially aligned corridors that provide a gray-scale comparison of the strength (i.e., level) of birefringence as a function of position across the wall. The layering of the wall is evident by the concentric rings, with a parallel alignment of fibers within individual layers. The azimuthal direction of the fibers is generally coherent within layers with the mean direction varying widely from layer to layer (cf. Figure 2).

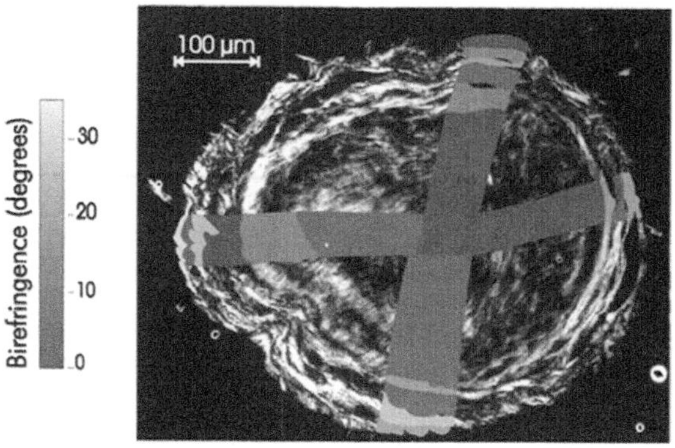

Figure 10. Polarized light micrograph of a tangential section from a 2.6 mm diameter aneurysm revealing the layered structure of the aneurysm wall (wall thickness 70 μm). Four measurement corridors are shown, with the birefringence levels of collagen within each of the layers being gray-scaled (from MacDonald et al. [66], with permission).

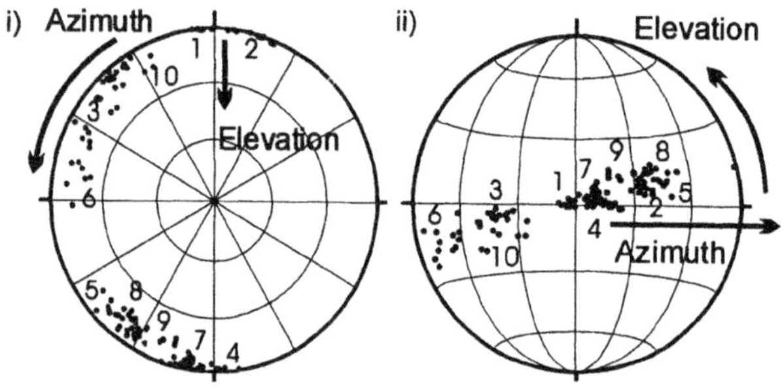

Figure 11. Lambert projections to show the orientation of collagen fibers for one of the measurement corridors from the aneurysm shown in Figure 10: (i) primary data with each layer identified by number, and (ii) data rotated to show the great circle distribution, indicating that there is a full range of directional alignments of the collagen fibers at that region on the aneurysm wall (with permission).

This distinctly layered organization has been characteristic of each of the several saccular aneurysms studied, regardless of size, provided the wall is relatively free of atherosclerosis [13, 14]. The defining feature of each layer has been primarily the mean fiber orientation, although fiber size and birefringence also contribute to the distinctiveness among layers.

Graphical presentation of directional data on Lambert projections provides a quantitative and comprehensive overview of the directional organization of the fibers in a tissue section. It is the preferred graphical method for three-dimensional data because it is an equal area projection, preserving the total area of the data regardless of its mean orientation and thus its projection position on the graph [76, 88]. Figure 11(i) is an example of a Lambert projection showing a single corridor of measurements with 10 layers across the wall of the aneurysm from Figure 10. This projection shows the clustering of the primary data around the

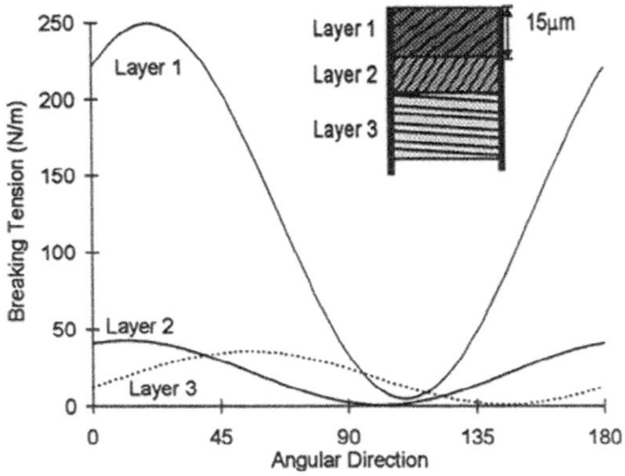

Figure 12. Plot of the calculated breaking tension in the three outer layers of an aneurysm against the angular direction of the wall, where 0° on the abscissa is the meridional direction (neck to fundus) and 90° is the equatorial direction. The inset is a schema of the three outer layers with the mean collagen orientation shown, and the gray scale indicating the strength of birefringence (from MacDonald et al., with permission).

perimeter (azimuthal directions ranging from 0 to 360°) with the elevation angle varying nonlinearly toward the center of the plot. The layers are numbered from the lumen, as layer 1, to the outer edge of the aneurysm. The computer rotated projection (Figure 11(ii)) shows the distribution of the same data relative to a great circle girdle distribution, and reveals how well these data span all orientations relative to the aneurysm surface.

Together, fiber direction and distribution provide a basis for estimating tissue strength (which also depends on cross-linking, etc.), and they can be used to develop microstructural methods such as those of Lanir [62]. The observed variation of birefringence across the aneurysm wall has suggested that it is reasonable to combine retardation and orientation data to assess directional strength of the tissue [15]. Early findings suggest that enlargement of an aneurysm requires a reorganization of the higher strength outer fibers while new collagen is added to the inner layers. Note, therefore, that several studies have linked collagen birefringence, mechanics, and the healing process – for example, in skin wound healing and maturation of a scar post myocardial infarction [21, 105, 106]. It was the study of Doillon, however, that provided the first quantitative measure of tissue strength directly from birefringence. Recently, MacDonald et al. [66] reanalyzed the data of Doillon by plotting tensile strength σ_s (kPa) versus fiber birefringence B (phase retardation in units of nm), which revealed a nonlinear relation $\sigma_s = 0.304B^{2.33}$, for which the correlation coefficient $r = 0.99$. This strength was defined as the tensile stress at which the dermal scar began to fail due to damage and/or micro-tearing. Figure 12 shows the contribution to directional strength of three outer layers of a 9 mm diameter aneurysm, and the marked differences in strength contribution because of layer thickness and fiber birefringence; included in the calculations were aneurysm radius, layer thickness, and number of layers, and the variables of fiber alignment, birefringence, and estimated local wall thickness [66]. The data revealed a tissue anisotropy of approximately $2\times$ for the direction of least to greatest strength, with a tensile strength (in the weakest direction) ranging from 0.73 to 1.8 MPa over 4 aneurysms. Recall that Steiger et al. and Toth et al. reported direct measurements of tensile strength from aneurysmal tissue strips 0.5 MPa ± 0.26 (SE) and 0.5 to 1.45 MPa, respectively – this shows a good correspondence with the few aneurysms that have been studied microscopically to date.

6.2. THEORETICALLY-MOTIVATED EXPERIMENTS

There are five general steps in the formulation of a constitutive relation (*DEICE*): *D*elineating general characteristics, *E*stablishing a theoretical framework, *I*dentifying a specific form of the relation, *C*alculating best-fit values of the material parameters, and *E*valuating the predictive capability of the final relation. Hsu et al. [42, 43] presented both a new theoretical framework and a multiaxial experimental system for accomplishing steps 2–5 for thin, axisymmetric, non-complicated (i.e., no atherosclerosis and no prior bleeds or repairs) saccular aneurysms. Briefly, the framework exploits two results noted above: equations (14), which show that the in-plane stress-resultants can be determined directly from experimental data, and equation (2), which is a general membrane constitutive relation. Taken together, it is easy to see that, in principle, one can glean information about the response functions $\partial w/\partial E_{\alpha\beta}$ for the material via,

$$\frac{\partial w}{\partial E_{11}} = \frac{\lambda_2}{\lambda_1}\left(\frac{P}{2\kappa_2}\right), \qquad \frac{\partial w}{\partial E_{22}} = \frac{\lambda_1}{\lambda_2}\left(\frac{P}{\kappa_2}\right)\left(1 - \frac{\kappa_1}{2\kappa_2}\right), \tag{15}$$

where $\mathbf{F} = \mathrm{diag}[\lambda_1, \lambda_2]$ and λ_α (stretch ratios), κ_α (principal curvatures), and P (distension pressure) are all experimentally measurable. Although one would prefer to examine these response functions by maintaining one of the principal Green strains constant while the other varies, and vice versa, this is not possible in the axisymmetric inflation problem. Hsu et al. [42] showed via numerical simulations, however, that this approach can provide information on the functional form of the strain-energy function. Unfortunately, this has yet to be accomplished in large part due to the scarcity of unruptured human lesions at autopsy and in particular those having an axisymmetric shape.

6.3. INVERSE FINITE ELEMENT PARAMETER ESTIMATION

Whether the lesion is axisymmetric or non-axisymmetric, or whether the functional form of the strain-energy is identified directly from data or simply postulated, one must calculate best-fit values of the material parameters from data. Kyriacou et al. [59] suggested that the inverse finite element method would be useful in this regard. Briefly, they evaluated this approach by comparing nodal displacements that were calculated via a (forward) finite element solution (e.g., equations (12)–(13) with an assumed set of material parameters) with those that were found experimentally for an inflated rubber membrane; estimates of the best-fit values of the material parameters were determined by minimizing the difference between computed and measured displacements in a nonlinear least squares sense. Results showed that this approach is indeed feasible, at least for simple material descriptors. Nonetheless, solving the nonlinear finite element equations (iteratively using a Newton–Raphson method) within an iterative nonlinear regression algorithm (Marquardt–Levenberg) can be computationally expensive. Seshaiyer et al. [82] suggested, therefore, that one simply perform the parameter estimation over a sub-domain $\Omega_s \subset \Omega_o$, rather than over the whole domain Ω_o. Advantages are two-fold: one avoids the necessity of knowing all of the boundary conditions, which can be challenging even in a laboratory setting, and one need not solve a large number of simultaneous finite element equations.

In particular, Seshaiyer et al. reported best-fit values of the material parameters (constitutive relation similar to equation (1)) based on pressure-strain data from multiple regions from 2 non-axisymmetric human aneurysms. The estimation was based on four linear triangular elements that defined the sub-domain, with the 1 central node "free" and the 4 outer nodes prescribed as displacement boundary conditions. As expected, the results suggested anisotropy and regional differences, the most marked of which was for a lesion that was primarily collagenous but had a region that was visibly atherosclerotic (and thus stiffer). In comparison to the values of the parameters obtained from the data of Scott et al., the more recent findings suggested an overall stiffer behavior. For example, for one (representative) region, $c = 10.18$ N/m, $c_2 = 20.03$, $c_1 = 8.71$, and $c_3 = 8.81$. This was consistent with the much less extensible behavior seen experimentally in these lesions – maximum principal stretches were on the order of 8% rather than the 18% reported by Scott et al. (who assumed that the lesions were perfect spheres). This difference may well have been due to a difference in defining the stress-free states, the more recent data likely being more reliable. Figure 13 shows an illustrative stress–stretch response to equibiaxial stretches of 10% based on the new best-fit values. Note the anisotropy, albeit not marked.

7. Growth and Remodeling

7.1. MOTIVATION

Diverse research over the last 25 years has revealed the ubiquitous role of growth and remodeling within the vasculature, one that is essential to normal tissue maintenance, the process of healing, adaptation to altered conditions, and even the progression or regression of disease. Examples include arterial adaptations to hypertension, sustained alterations in flow, and balloon angioplasty to name but a few [32, 61]. Based on the recent data by Canham and colleagues [66] as well as work on protease activity in aneurysms [31, 69] it appears that stress-mediated

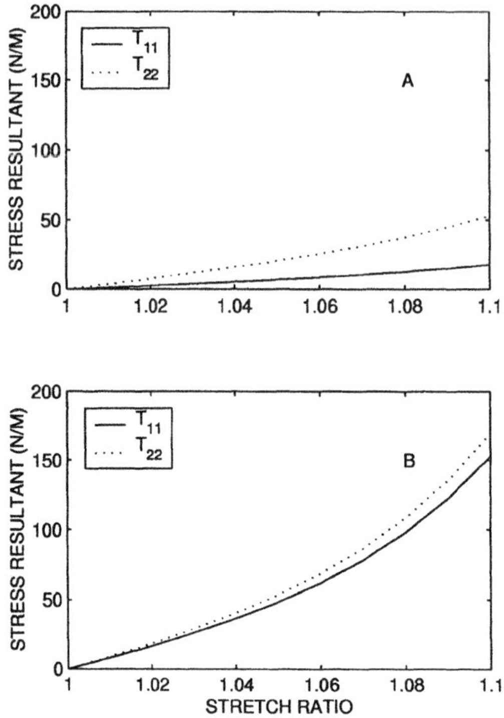

Figure 13. Calculated equibiaxial stress-stretch responses based on the best-fit material parameters determined using the sub-domain inverse finite element method for parameter estimation – results from one region on one human lesion (from Seshaiyer et al., with permission).

regulation of aneurysmal collagen may play a key role in the natural history of these lesions as well. This is largely an open problem from the perspective of mechanics, but let us briefly review recent work that illustrates its potential importance.

7.2. TOWARDS A GROWTH MODEL

There has been only one prior report of a computational model for studying the growth of saccular aneurysms, and unfortunately it is not described in detail. Steiger [95] considered a class of axisymmetric lesions (equations (14)) and stated that "tissue growth rate was set proportional to wall stress". Although there is no discussion of the constitutive or evolution equations, he reports that "sausage-shaped and disc-shaped" lesions tended to develop toward a spherical shape whereas multi-lobed lesions tended to remain complex. He suggested that localized blebs may be an attempt to stabilize a localized weakness in the wall.

Let us consider the following questions: Can fibroblasts in an enlarging aneurysm synthesize and organize collagen such that the resulting intramural stresses mimic the values experienced in the normal parent vessel? Or, does a particular distribution of material properties exist that would tend to minimize and homogenize the stress distribution within an aneurysm? In an effort to begin to examine the latter question, Ryan and Humphrey [78] studied 12 sub-classes of non-complicated lesions (maximum dimensions $\leqslant 2$ mm) defined by the triplets (A, B, Z_r) where A and B are the major/minor radii and Z_r is a "truncation level" that yields a model having a neck (cf. Figure 6). Using finite element simulations, preferred material properties were sought in terms of two parameters, $(c_2/c_1)_{\max} \in [1/11, 11]$ and $p \in [1, 6]$, where

$$\frac{c_2}{c_1} = 1 + \left[\left(\frac{c_2}{c_1}\right)_{\max} - 1\right]\left[\frac{i-1}{N-1}\right]^p \tag{16}$$

and c_2 and c_1 are the material parameters in the Fung pseudostrain-energy function (equation (1)), $(c_2/c_1)_{\max}$ is the ratio of these parameters at the neck of the lesion (arc length $S = L$), $i \in [1, N]$ is the finite element number, and p is a descriptor of how (e.g., linearly or nonlinearly) the material symmetry varies from the fundus to the neck. For example, $(c_2/c_1)_{\max} = 1$ implies isotropy at all S, $(c_2/c_1)_{\max} > 1$ yields a progressively increased circumferential stiffness, and $(c_2/c_1)_{\max} < 1$ yields a progressively increased meridional stiffness. Likewise, $p = 1$ requires the symmetry to vary linearly from the fundus to the base (as in Kyriacou and Humphrey [57], Shah et al. [84]), whereas $p > 1$ allows nonlinear variations. Preferred properties thus indicate that particular combination of $(c_2/c_1)_{\max}$ and p that minimizes and homogenizes the stress field.

Based on literally thousands of simulations (though this would be better accomplished simply as an optimization problem), it was found that the multiaxial stresses in lesions having an initially large neck : height ratio tend to be lower and nearly homogeneous if $p > 1$ and $(c_2/c_1)_{\max} > 9$. Figure 14 compares, for example, the different stress distributions for isotropic (i.e., $(c_2/c_1)_{\max} = 1$ and $p = 1$) and the preferred properties for one lesion. With the exception of the boundary layer effect (due to the imposed zero displacement boundary condition at the neck), the stresses are nearly homogeneous. Although results were different for the different geometries (see [78]), the general finding was consistent with that of Steiger [95]): small non-complicated lesions (i.e., thin and collagenous, free of atherosclerosis, fibrin patches, etc.) tended to "prefer" material properties that allowed them to become more spherical. For large neck : height ratios this requires that the lesion expand more in the z direction, which requires less stiffness in the meridional direction; for small neck : height ratios, this requires that the lesion expand more in the r direction, which requires less stiffness in the circumferential direction. These findings are teleologically reasonable as the sphere is the optimal geometry to resist a distension pressure. An unexpected finding, however, was how the lesions preferred to achieve this. For example, lesions with large neck : height ratios tended to concentrate the anisotropy near the neck (i.e., larger values of p).

Although based on idealized models, the finding that intramural stresses in aneurysms can be homogenized simply via a preferential deposition of collagen

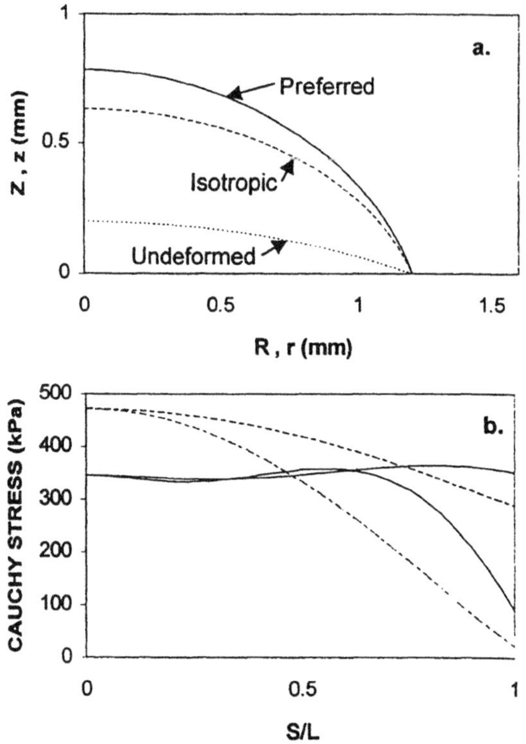

Figure 14. Stress distributions for the same model lesion except that the material behavior is isotropic in one case and "preferred" in the other. Note the tendency towards homogenization and minimization of the stresses in the latter case (from Ryan and Humphrey, with permission).

(and selective removal of old collagen) is provocative. Indeed, given recent reports that apoptosis (programmed cell death) and matrix metalloproteinase activity (matrix degradation) are both increased in saccular aneurysms, and so too for the transcription of type III collagen, it is reasonable to expect significant stress-mediated remodeling in aneurysms [15, 36, 69]. It is tempting to hypothesize, for example, that stable lesions are those which have remodeled in such a way that the stresses experienced by the fibroblasts are restored close to those in the normal parent vessel. There may be cases, of course, wherein the stresses may exceed wall strength prior to the normalization of stress due to remodeling; likewise, the insidious effects of atherosclerosis, the activation of platelets, etc. may also hinder or prevent the remodeling process and thereby lead to rupture. In such cases, the lesion could rupture prior to stabilization. There is clearly a need to explore remodeling theories (e.g., see [48]), which of course must be based on good estimates of the wall shear stress and pressure fields (and thus solid-fluid interactions), which likely serve as signals to the endothelial cells and fibroblasts to control matrix turnover, and better data on the time-course of changes in lesion geometry and microstructure. Only in this way will one be able to formulate and test various growth and remodeling theories. Much remains to be done, and finite elasticity has a clear role to play.

Acknowledgements

This work was supported in part by grants from the NIH (HL-54957) and the Medical Research Council of Canada. We also wish to acknowledge the expert technical contributions by many individuals in our labs, particularly Helen Finlay, Frank Hsu, Stelios Kyriacou, Donia McDonald, Padhu Seshaiyer, and Amit Shah, as well as our long-standing collaborations with neurosurgery colleagues, Gary Ferguson at the University of Western Ontario and Daniele Rigamonti at the Johns Hopkins Medical Institutions.

References

1. N. Akkas, Aneurysms as a biomechanical instability problem. In: F. Mosora (ed.), *Biomechanical Transport Processes*, Plenum Press (1990) pp. 303–311.
2. G.D. Angelini, S.L. Passani, I.M. Breckenridge and A.C. Newby, Nature and pressure dependence of damage induced by distension of human saphenous vein coronary artery bypass grafts. *Cardiovas. Res.* **21** (1987) 902–907.
3. S. Asari and T. Ohmoto, Growth and rupture of unruptured cerebral aneurysms based on the intraoperative appearance. *Acta Med. Okayama* **48** (1994) 257–262.
4. J.I. Ausman, The New England Journal of Medicine report on unruptured intracranial aneurysms: A critique. *Surg. Neurol.* **51** (1999) 227–229.
5. G.M. Austin, Biomathematical model of aneurysm of the Circle of Willis: The Duffing equation and some approximate solutions. *Math. Biosci.* **11** (1971) 163–172.
6. G.M. Austin, W. Schievink and R. Williams, Controlled pressure-volume factors in the enlargement of intracranial saccular aneurysms. *Neurosurg.* **24** (1989) 722–730.
7. G.M. Austin, S. Fisher, D. Dickson, D. Anderson and S. Richardson, The significance of the extracellular matrix in intracranial aneurysms. *Ann. Clin. Lab. Sci.* **23** (1993) 97–105.
8. M.F. Beatty, Topics in finite elasticity: Hyperelasticity of rubber, elastomers, and biological tissues – with examples. *Appl. Mech. Rev.* **40** (1987) 1699–1734.
9. H.S. Bennett, The microscopical investigation of biological materials with polarized light. In: *McClung's Handbook of Microscopical Technique*, 3rd edn, Hafner Publishing Co., New York (1950).
10. P.J. Camarata, R.E. Latchaw, D.A. Rufenacht and R.C. Heros, State of the art in medicine: Intracranial aneurysms. *Invest. Radiol.* **28** (1993) 373–382.
11. G.J. Campbell and M.R. Roach, Fenestrations in the internal elastic lamina at bifurcations of the human cerebral arteries. *Stroke* **12** (1981) 489–495.
12. P.B. Canham and G.G. Ferguson, A mathematical model for the mechanics of saccular aneurysms *Neurosurg.* **17** (1985) 291–295.
13. P.B. Canham, H.M. Finlay, J.G. Dixon and S.Y. Tong, Aneurysmal fabric modelled by layered great circle trajectories of collagen. *Acta Sterol.* **11** (1992) 703–711.
14. P.B. Canham, H.M. Finlay and S.Y. Tong, Stereological analysis of the layered structure of human intracranial aneurysms. *J. Microsc.* **183** (1996) 170–180.
15. P.B. Canham, H.M. Finlay, J.A. Kiernan and G.G. Ferguson, Layered structure of saccular aneurysms assessed by collagen birefringence. *Neurol. Res.* **21** (1999) 618–626.

16. A.M. Coll, J.F.D. Corral, S. Yazawa and M. Falcon, Intra-aneurysmal pressure differences in human saccular aneurysms. *Neurol.* **6** (1976) 93–96.
17. M.R. Crompton, The pathology of ruptured middle-cerebral aneurysms with special reference to differences between the sexes. *Lancet* **2** (1962) 421–425.
18. M.R. Crompton, Mechanism of growth and rupture in cerebral berry aneurysms. *Br. Med. J.* **1** (1966) 1138–1142.
19. J. Cronin, Biomathematical model of aneurysm of the Circle of Willis: A qualitative analysis of the differential equation of Austin. *Math. Biosci.* **16** (1973) 209–225.
20. S.M. de la Monte, G.W. Moore, M.A. Monk and G.M. Hutchins, Risk factors for development and rupture of intracranial berry aneurysms. *Am. J. Med.* **78** (1985) 957–964.
21. C.J. Doillon, M.G. Dunn, E. Bender and F.H. Silver, Collagen fiber formation in repair tissue: Development of strength and toughness. *Collagen Rel. Res.* **5** (1985) 481–492.
22. J.H. Evans and J.C. Barbenel, Structure and mechanical properties of tendon related to function. *Equine Vet. J.* **7** (1974) 1–8.
23. G.G. Ferguson, Turbulence in human intracranial saccular aneurysms. *J. Neurosurg.* **33** (1970) 485–497.
24. G.G Ferguson, Physical factors in the initiation, growth, and rupture of human intracranial aneurysms. *J. Neurosurg.* **37** (1972) 666–677.
25. G.G. Ferguson, Direct measurement of mean and pulsatile blood pressure at operation in human intracranial saccular aneurysms. *J. Neurosurg.* **36** (1972) 560–563.
26. G.G. Ferguson, Intracranial arterial aneurysms – a surgical perspective. In: *Handbook of Clinical Neurology* 11(55) (1989), pp. 41–87.
27. H.M. Finlay, P. Whittaker and P.B. Canham, Collagen organization in the branching region of human brain arteries. *Stroke* **29** (1998) 1595–1601.
28. W.D. Forbus, On the origin of miliary aneurysms of the superficial cerebral arteries. *Bull. Johns Hopkins Hosp.* **47** (1930) 239–284.
29. G.N. Foutrakis, H. Yonas and R.J. Sclabassi, Finite element methods in the simulation and analysis of intracranial blood flow: Saccular aneurysm formation in curved and bifurcating arteries. Tech Rep 6, Univ of Pitt Comp Neurosci (1994).
30. D.L. Fry, Acute vascular endothelial changes associated with increased blood velocity gradients. *Circ. Res.* **22** (1968) 165–167.
31. P. Gaetani, F. Tartara, F. Tancioni, R. Rodriguez y Baena, E. Casari, M. Alfano and V. Grazioli, Deficiency of total collagen content and of deoxypyridinoline in intracranial aneurysm walls. *FEBS Letters* **404** (1997) 303–306.
32. G.H. Gibbons and V.J. Dzau, The emerging concept of vascular remodeling. *Mech. of Disease* **330**(20) (1994) 1431–1438.
33. Y.P. Gobin, J.L. Counord, P. Flaud and J. Duffaux, In vitro study of hemodynamics in a giant saccular aneurysm model: Influence of flow dynamics in the parent vessel and effects of coil embolization. *Interven. NeuroRadiol.* **36** (1994) 530–536.
34. G.J. Hademenos, T. Massoud, D.J. Valentino, G. Duckwiler and F. Vinuela, A nonlinear mathematical model for the development and rupture of intracranial saccular aneurysms. *Neurol. Res.* **16** (1994) 376–384.
35. G.J. Hademenos, T.F. Massoud, F. Turjman and J.W. Sayre, Anatomical and morphological factors correlating with rupture of intracranial aneurysms in patients referred for endovascular treatment. *Neuroradiol.* **40** (1998) 755–760.
36. A. Hara, N. Yoshimi and H. Mori, Evidence for apoptosis in human intracranial aneurysms. *Neurol. Res.* **20** (1998) 127–130.
37. T. Hasimoto, Dynamic measurement of pressure and flow velocities in glass and silastic model berry aneurysms. *Neurol. Res.* **6** (1984) 22–28.
38. N. Hashimoto and H. Handa, The size of cerebral aneurysms in relation to repeated rupture. *Surg. Neurol.* **19** (1983) 107–111.

39. O. Hassler, Morphological studies on the large cerebral arteries, with reference to the aetiology of subarachnoid haemorrhage. *Acta Psychiatr. Neurol. Scand. (suppl)* **154** (1961) 1–145.
40. O. Hassler, The windows of the internal elastic lamella of the cerebral arteries. *Virchows Arch. Path. Anat.* **335** (1962) 127–132.
41. K. Hegedus, Some observations on reticular fibers in the media of the major cerebral arteries. *Surg. Neurol.* **22** (1984) 301–307.
42. F.P.K. Hsu, C. Schwab, D. Rigamonti and J.D. Humphrey, Identification of response functions for nonlinear membranes via axisymmetric inflation tests: Implications for biomechanics. *Int. J. Sol. Struct.* **31** (1994) 3375–3386.
43. F.P.K. Hsu, A.M.C. Liu, J. Downs, D. Rigamonti and J.D. Humphrey, A triplane video-based experimental system for studying axisymmetrically inflated biomembranes, *IEEE Trans. Biomed. Engr.* **42** (1995) 442–450.
44. J.D. Humphrey, R.K. Strumpf and F.C.P. Yin, A constitutive theory for biomembranes: Application to epicardium. *ASME J. Biomech. Engr.* **114** (1992) 461–466.
45. J.D. Humphrey, Arterial wall mechanics: Review and directions. *Crit. Rev. Biomed. Engr.* **23**(1/2) (1995) 1–162.
46. J.D. Humphrey and S.K. Kyriacou, The use of Laplace's equation in aneurysm mechanics. *Neurol. Res.* **18** (1996) 204–208.
47. J.D. Humphrey, Computer methods in membrane biomechanics. *Comp. Meth. Biomech. Biomed. Engr.* **1** (1998) 171–210.
48. J.D. Humphrey, Remodeling of a collagenous tissue at fixed lengths. *ASME J. Biomech. Engr.* **121** (1999) 591–597.
49. E.J.N. Hung and M.R. Botwin, Mechanics of rupture of cerebral saccular aneurysms. *J. Biomech.* **8** (1975) 385–392.
50. K.K. Jain, Mechanism of rupture of intracranial saccular aneurysms. *Surg.* (1963) 347–350.
51. H. Kamitani, H. Masuzawa, I. Kanazawa and T. Kubo, Bleeding risk in unruptured and residual cerebral aneurysms – angiographic annual growth rate in nineteen patients. *Acta Neurochir. (Wein)* **141** (1999) 153–159.
52. N.F . Kassel and J.C. Torner, Size of intracranial aneurysms. *Neurosurg.* **12** (1983) 291–297.
53. C. Kim, J. Cervos-Navarro, H. Kikuchi, N. Hashimoto and F. Hazama, Alterations in cerebral vessels in experimental animals and their possible relationship to the development of aneurysms. *Surg. Neurol.* **38** (1992) 331–337.
54. M. Kojima, H. Handa, N. Hashimoto, C. Kim and F. Hazima, Early changes of experimentally induced cerebral aneurysms in rats: Scanning electron microscopy study. *Stroke* **17** (1986) 835–841.
55. T.M. Kosierkiewicz, S.M. Factor and D.W. Dickson, Immunocytochemical studies of atherosclerotic lesions of cerebral berry aneurysms. *J. Neuropath. Exp. Neurol.* **53** (1994) 399–406.
56. H. Kraus, *Thin Elastic Shells*, J. Wiley, New York (1967).
57. S.K. Kyriacou and J.D. Humphrey, Influence of size, shape and properties on the mechanics of axisymmetric saccular aneurysms. *J. Biomech.* **29** (1996) 1015–1022. Erratum **30** (1997) 761.
58. S.K. Kyriacou, C. Schwab and J.D. Humphrey, Finite element analysis of nonlinear orthotropic hyperelastic membranes. *Comp. Mech.* **18** (1996) 269–278.
59. S.K. Kyriacou, A. Shah and J.D. Humphrey, Inverse finite element characterization of nonlinear hyperelastic membranes. *J. Appl. Mech.* **64** (1997) 257–262.
60. E.R. Lang and M. Kidd, Electron microscopy of human cerebral aneurysms. *J. Neurosurg.* **22** (1965) 554–562.
61. B.L. Langille, Remodeling of developing and mature arteries: Endothelium, smooth muscle, and matrix. *J. Cardiovasc. Pharmacol.* **21** (1993) S11–S17.
62. Y. Lanir, Constitutive equations for fibrous connective tissues. *J. Biomechanics* **16** (1983) 1–12.

63. J.M. Lee and D.R. Boughner, Tissue mechanics of canine pericardium in different test environments. Evidence for time-dependent accommodation, absence of placticity, and new roles for collagen and elastin. *Circ. Res.* **49** (1981) 533–544.
64. K.O. Lim and D.R. Boughner, Low frequency dynamic viscoelastic properties of human mitral valve tissue. *Cardiovasc. Res.* **10** (1976) 459–465.
65. M. Low, K. Perktold and R. Raunig, Hemodynamics in rigid and distensible saccular aneurysms: A numerical study of pulsatile flow characteristics. *Biorheol.* **30** (1993) 287–298.
66. D.J. MacDonald, H.M. Finlay and P.B. Canham, Directional wall strength in saccular brain aneurysms from polarized light microscopy. *Ann. Biomed. Eng.* **28** (2000) 533–542.
67. F.B. Meyer, J. Huston and S.S. Riederer, Pulsatile increases in aneurysm size determined by cine phase-contrast MR angiography. *J. Neurosurg.* **78** (1993) 879–883.
68. W.R. Milnor, *Hemodynamics*, Williams and Wilkens, Baltimore (1989).
69. C. Mimata, M. Kitaoka, S. Nagahiro, K. Iyama, H. Hori, H. Yoshioka and Y. Ushio, Differential distribution and expressions of collagens in the cerebral aneurysmal wall. *Acta Neuropathol.* **94** (1997) 197–206.
70. G. Neil-Dwyer, J.R. Bartlett, A.C. Nicholls, P. Narcisi and F.M. Pope, Collagen deficiency and ruptured cerebral aneurysms. *J. Neurosurg.* **59** (1983) 16–20.
71. D.A. Nichols, F.B. Meyer, D.G. Piepgras and P.L. Smith, Endovascular treatment of intracranial aneurysms. *Mayo Clinic Proceedings* **69** (1994) 272–285.
72. S.H.M. Nyström, Development of intracranial aneurysms as revealed by electron microscopy. *J. Neurosurg.* **20** (1963) 329–337.
73. J.R. Ostergaard, Risk factors in intracranial saccular aneurysms. *Acta. Neurol. Scand.* **80** (1989) 81–98.
74. J.R. Ostergaard, E. Reske-Nielsen and H. Oxlund, Histological and morphometric observations on the reticular fibers in the arterial beds of patients with ruptured intracranial saccular aneurysms. *Neurosurg.* **20** (1987) 554–558.
75. L. Parlea, R. Fahrig, D.W. Holdsworth and S.P. Lownie, An analysis of the geometry of saccular intracranial aneurysms. *Am. J. Neuroradiol.* **20** (1999) 1079–1089.
76. W.R. Phillips, *Mineral Optics: Principles and Techniques,* W.H. Freeman, San Francisco (1971), pp. 171–190.
77. C. Richardson and O. Kofman, Cranial bruit with intracranial saccular aneurysms. *Trans. Am. Neurol. Assoc.* **76** (1951) 151–154.
78. J.M. Ryan and J.D. Humphrey, Finite element based predictions of preferred material symmetries in saccular aneurysms. *Ann. Biomed. Engr.* **27** (1999) 641–647.
79. A.L. Sahs, Observations on the pathology of saccular aneurysms. *J. Neurosurg.* **24** (1966) 792–806.
80. L.N. Sekhar and R.C. Heros, Origin, growth and rupture of saccular aneurysms: A review. *Neurosurg.* **8** (1981) 248–260.
81. L.N. Sekhar, R.P. Sclabassi, M. Sun, H.B. Blue and J.F. Wasserman, Intra-aneurysmal pressure measurements in experimental saccular aneurysms in dogs. *Stroke* **19** (1988) 353–356.
82. P. Seshaiyer, A.D. Shah, S.K. Kyriacou and J.D. Humphrey, Multiaxial mechanical behavior of human saccular aneurysms. *Comp. Meth. Biomech. Biomed. Engr.* **4** (2001) 281–290.
83. S. Scott, G.G. Ferguson and M.R. Roach, Comparison of the elastic properties of human intracranial arteries and aneurysms. *Can. J. Physiol. and Pharmacol.* **50** (1972) 328–332.
84. A.D. Shah, J.L. Harris, S.K. Kyriacou and J.D. Humphrey, Further roles of geometry and properties in saccular aneurysm mechanics. *Comp. Meth. Biomech. Biomed. Engr.* **1** (1997) 109–121.
85. A.D. Shah and J.D. Humphrey, Finite strain elastodynamics of saccular aneurysms. *J. Biomech.* **32** (1999) 593–599.
86. T.E. Simkins and W.E. Stehbens, Vibrational behavior of arterial aneurysms. *Lett. Appl. Engr. Sci.* **1** (1973) 85–100.

87. J.H.F. Smith, P.B. Canham and J. Starkey, Orientation of collagen in the tunica adventitia of the human cerebral artery measured with polarized light and the universal stage. *J. Ultrastruct. Res.* **77** (1981) 133–145.
88. J. Starkey, The analysis of three-dimensional orientation data. *Can. J. Earth Sci.* **30** (1993) 1355–1362.
89. W.E. Stehbens, Histopathology of cerebral aneurysms. *Arch. Neurol.* **8** (1963) 272–285.
90. W.E. Stehbens, Ultrastructure of aneurysms. *Arch. Neurol.* **32** (1975) 798–807.
91. W.E. Stehbens, The pathology of intracranial arterial aneurysms and their complications. In: J.L. Fox (ed.), *Intracranial Aneurysms*, Springer-Verlag, New York (1983).
92. W.E. Stehbens, Pathology and pathogenesis of intracranial berry aneurysms. *Neurol. Res.* **12** (1990) 29–34.
93. H.J. Steiger, R. Aaslid, S. Keller and H.J. Reulen, Strength, elasticity and viscoelastic properties of cerebral aneurysms. *Heart Vessels* **5** (1986) 41–46.
94. H.J. Steiger, D.W. Liepsch, A. Poll and H.J. Reulen, Hemodynamic stress in terminal saccular aneurysms: A laser doppler study. *Heart Vessels* **4** (1988) 162–169.
95. H.J. Steiger, Pathophysiology of development and rupture of cerebral aneurysms. *Acta Neurochir. Suppl.* **48** (1990) 1–57.
96. J. Suzuki and H. Ohara, Clinicopathological study of cerebral aneurysms. *J. Neurosurg.* **48** (1978) 505–514.
97. M. Toth, G.L. Nadasy, I. Nyary, T. Kerenyi, M. Orosz, G. Molnarka and E. Monos, Sterically inhomogeneous viscoelastic behavior of human saccular cerebral aneurysms. *J. Vasc. Res.* **35** (1998) 345–355.
98. H. Ujiie, K. Sato, H. Onda, A. Oikkawa, M. Kagawa, K. Atakakura and N. Kobayashi, Clinical analysis of incidentally discovered unruptured aneurysms. *Stroke* **24** (1993) 1850–1856.
99. H. Ujiie et al., Effects of size and shape (aspect ratio) on the hemodynamics of saccular aneurysms: A possible index for surgical treatment of intracranial aneurysms. *Neurosurg.* **45** (1999) 119–130.
100. R. Vilarta and B.C. Vidal, Anisotropic and biomechanical properties of tendons modified by exercise and denervation: Aggregation and macromolecular order in collagen bundles. *Matrix* **9** (1989) 55–61.
101. S.A. Wainwright, W.D. Biggs, J.D. Currey and J.M. Gosline, *Mechanical Design in Organisms*, Princeton University Press (1976), pp. 88–93.
102. D.O. Wiebers, J.P. Whisnant, T.M. Sundt and W.M. O'Fallon, The significance of unruptured intracranial aneurysms. *J. Neurosurg.* **66** (1987) 23–29.
103. D.O. Wiebers et al., Unruptured in tracranial aneurysms – risk of rupture and risks of surgical intervention. International study of unruptured intracranial aneurysms investigators. *N. Engl. J. Med.* **339** (1998) 1725–1733.
104. P. Whittaker, M.E. Schwab and P.B. Canham, The molecular organization of collagen in saccular aneurysms assessed by polarized light microscopy. *Conn. Tiss. Res.* **17** (1988) 43–54.
105. P. Whittaker, D.R. Boughner and R.A. Kloner, Role of collagen in acute myocardial infarct expansion. *Circulation* **84** (1991) 3123–3134.
106. M. Wolman and T.A. Gillman, A polarized light study of collagen in dermal wound healing. *Br. J. Exp. Path.* **53** (1972) 85–89.

Journal of Elasticity **61:** 83–111, 2000.

A Model of Arterial Adaptation to Alterations in Blood Flow

ALEXANDER RACHEV
Institute of Mechanics, Bulgarian Academy of Sciences, 1113 Sofia, Bulgaria.
E-mail: rachev@bgcict.acad.bg

Received 20 April 2000; in revised form 8 December 2000

Abstract. Mechanisms of arterial adaptation to changes in blood flow rates were tested by comparing the predictions of a proposed theoretical model with available experimental data. The artery was modeled as an elastic membrane made of a nonlinear, incompressible, elastic material. Stimulation of the vascular smooth muscle was modeled through the generation of an active component of circumferential stress. The muscular tone was modulated by flow-induced shear stress sensed by the arterial endothelium, and is responsible for the vasomotor adjustment of the deformed arterial diameter in response to changes in blood flow. This study addresses the hypothesis that the synthetic and proliferative activity of smooth muscle cells, leading to a change in arterial dimensions, is shear stress dependent and is associated with changes in the contractile state of the smooth muscle cells and changes in the circumferential wall stress. Remodeling to a step change in flow was formulated as an initial-value problem for a system of first order autonomous differential equations for the evolution of muscular tone and evolution of arterial geometry. The governing equations were solved numerically for model parameters identified from experimental data available in the literature. The model predictions for the time variation of the geometrical dimensions and their asymptotic values were found to be in qualitative agreement with available experimental data. Experiments for validating the introduced hypotheses and further generalizations of the model were discussed.

Mathematics Subject Classification(s) (2000): 74L15.

Key words: finite elastic deformations, flow-induced shear stress, circumferential strain and stress, arterial mechanics, vasomotor response, remodeling.

1. Introduction

Like all tissues whose physiological function is associated with exposure to mechanical forces, arteries are sensitive to changes in their mechanical environment, i.e. to arterial pressure and blood flow alterations. The complex of processes that causes a long-term transformation of an artery subjected to sustained changes in pressure and/or flow is called remodeling. Remodeling includes processes in which endothelial cells and smooth muscle cells are involved. Endothelial cells form a one cell-thick layer, which together with the underlying membrane is called the intima, and is in direct contact with flowing blood. Smooth muscle cells are one of the basic structural components of the middle layer called the media, the major load-bearing part of the arterial wall. Apart from smooth muscle cells, the media consists of collagen fibers, elastin, and a ground substance matrix.

At the micro-level, cellular and intracellular, remodeling manifests itself as growth, division, cell loss, change in size, shape and orientation of cells, cell migrations, as well as synthesis of collagen, and synthesis and degradation of extracellular matrix. These processes involve interaction of multiple ionic and enzymic pathways, whose precise nature remains unknown.

At the macro-level, remodeling appears as changes in the geometrical dimensions and mechanical response of the artery, which do not result from the deformation caused by the altered loads. Sometimes geometrical and mechanical alterations are accompanied by changes in wall structure and composition. Moreover, remodeling may lead to changes in the arterial response when arterial smooth muscle cells are stimulated to relax or contract. Because arteries are permanently subjected to loads, remodeling also affects the strain and stress fields in the arterial wall.

Compensatory enlargement of arteries in response to increased flow has been demonstrated in animal models and human vessels. Kamiya and Togawa [30] showed that an arterio-venous fistula between the canine common carotid artery and the external jugular vein causes increased flow rate in the artery proximal to the shunt, and thus creates a marked increase in the flow-induced shear stress at the intima. Over time, the deformed arterial radius increased and tended to restore the normal baseline levels of mean shear stress of about 1.5 Pa. Similar results have been observed in [6, 28, 32, 33]. Compensatory enlargement in response to increased flow was also recorded during normal development [18, 55] and hypertension [2]. More detailed analysis of the arterial response shows that it involves two successive phases. Firstly, an acute increase in arterial lumen occurs, resulting from the temporary dilatation of the artery. The process is mediated by the endothelium through the release of substances such as endothelium-derived relaxing factor (EDRF), the principal component of which, NO, causes relaxation of medial smooth muscle cells [29, 32, 33]. Vasomotor response is followed by long-term reconstruction of the media due to proliferation and migration of the smooth muscle cells in such a way that the undeformed lumen of the vessel increases. Arterial enlargement causes an increase in wall tension and thereby increases the average circumferential stress. A compensatory thickening of the arterial wall was experimentally observed, which seems to restore the normal values of the wall stress [25, 39, 43, 60].

Reduced flow elicits a different response. It again comprises two successive processes, but in general, remodeling does not follow a time course that is simply the reverse of that observed under increased flow conditions. At first, the decrease in wall shear stress sensed by the endothelium evokes a smooth muscle contraction and shrinkage of the vessel. Reductions in the release of EDRF or changes in concentration of specific vasoconstrictors are candidates as mediators of that process [34, 35]. Sustained decrease in flow provokes further processes, which occur mainly in the intima and result in wall thickening. Normally, remodeling is a self-limiting process and leads to a restoration of the baseline value of the wall shear stress. Smooth muscle cells migrate into the intima and produce matrix proteins, resulting in a structure that resembles the underlying media. This new tissue is referred to as fibrocellular intimal hypertrophy [25]. Sometimes the remodeling process may not lead to a configuration that maintains the normal level of wall shear stress. The proliferation process may continue and lead to the formation of a stenosis or obliteration of the lumen. The grown tissue is matrix-free, poorly organized and forms so-called intimal hyperplasia [25]. The exact mechanisms involved in the different modes of arterial wall remodeling in response to altered blood flow remain unclear.

Changes in flow also cause remodeling of endothelial cells in the intima. This is characterized by loss of cells due to decreased flow or proliferation due to increased flow. The net effect is an adaptation that maintains endothelial cell density despite the changes in arterial diameter resulting from remodeling of the media [33]. When an artery is denuded of intima, i.e. the endothelial cells are removed, a change in flow does not cause either acute vasomotor response or wall remodeling [32, 40]. Hence the shear deformation of the endothelial cells appears to be the first in the chain of events that cause a change in smooth muscle tone, resulting in further remodeling of the wall.

The mechanisms by which arteries adapt to chronic blood flow alterations differ in young and mature animals. It was found that the carotid arteries of adult rabbits subjected to reduced blood flow exhibited decreased internal diameter although no significant changes in vessel mass and wall constituents were observed [32–34]. Similarly, it has been shown that remodeling elicited by elevated blood flow ultimately produces a vessel with major properties similar to a control artery [19]. As for young animals, remodeling results not only in a change in geometrical dimensions but also affects wall structure and composition, disturbing the normal process of development and maturation [8, 17, 33, 63].

It is worth noting that other factors such as nervous stimuli and the local hormonal and metabolic environment might affect smooth muscle cell activity and wall remodeling. Recent investigations performed in cell and tissue culture systems make it possible to eliminate non-mechanical factors and independently evaluate the contribution of the mechanical environment. The results from these studies have shown that pressure and flow conditions are major determinants of the mechanical properties and geometrical dimensions of arteries [4, 41].

In summary, the vessel remodels to maintain baseline levels of certain mechanical characteristics such as flow-induced wall shear stress at the intima and wall tensile stress in the media. Therefore, remodeling represents a locally controlled adaptive response tending to cope with changes in the mechanical environment. Experimental evidence that the arterial response to changes in pressure and flow is a local phenomenon, which is described in terms of mechanical quantities such as strains and stresses, suggests the need to develop mathematical models based on continuum mechanics. The results of these models aim to predict geometrical and mechanical response in arteries following changes in their mechanical environment. The information is important because a vessel's dimensions and mechanical properties determine the major functional role of arteries namely, to transport blood to organs and tissues and to transform the highly pulsatile heart output into a flow of moderate fluctuations.

Comparison between theoretical predictions and experimental findings justifies the acceptance or rejection of introduced model hypotheses for mechanical quantities that drive and govern the adaptation process. If no appropriate experiments exist, modeling can suggest the kinds of new experiments that are needed and methodology for analyzing the data. It is expected that the results obtained from continuum mechanics models may advance the level of understanding of the role that mechanical factors play during normal arterial development and maturation, and might help to reveal the mechanical aspects of the genesis and progression of certain vascular pathologies. Additionally, results from model studies could promote the development of therapeutic interventions with the aim of restoring the mechanical loads on arteries to normal levels. Finally, knowledge of mechanisms underlying arterial wall remodeling might be used to control the mechanical environment in bioreactors for tissue engineered blood vessels.

Past theoretical studies have mainly focused on the arterial remodeling under hypertensive conditions. Fung [21] has proposed a stress-growth law, which relates growth rate of the mass of an artery to the actual stresses. A general three-dimensional theory for a finite stress-induced volumetric growth has been developed in [55] considering a solid as a collection of growing differential elements. Following this approach, theoretical studies on aortic growth during development and under variable pressure have been performed [56–59]. The authors have postulated that the transverse growth of smooth muscle fibers depends linearly on the passive fiber stress, while the longitudinal growth depends on both fiber stress and the shear stress on the endothelium. Another approach based on a global consideration of the arterial wall was proposed in [45, 48, 49]. The authors studied the dynamics of geometrical adaptation in response to sustained increase in blood pressure. It was assumed that during remodeling the geometry of the arterial cross-section and mechanical properties of arterial tissue change in a manner that restores the baseline values of the flow-induced shear stress at the intima, the normal stress distribution across the arterial wall, and the normal arterial compliance. In all models of arterial remodeling the active response of smooth muscle cells has been neglected.

As briefly explained above, arterial response to altered blood flow involves processes in which the smooth muscle cells play a key role. Therefore, relevant modeling should include both the short- and long-term contribution of the vascular smooth muscle. The aim of this study is to propose a mathematical model for both the acute vasomotor response and the long-term geometrical remodeling of arteries induced by sustained changes in blood flow. The model aims to give a probable interpretation of some experimental results available in the literature and to suggest new type of experiments. The study addresses the hypothesis that the synthetic and proliferative activity of smooth muscle cells, leading to a change in arterial dimensions, is shear stress dependent and is associated with changes in the contractile state of the smooth muscle and changes in the circumferential wall stress.

2. Methods

In the beginning a mathematical model of an artery subjected to physiological load is proposed. Afterwards, with account for some ionic processes in the stimulated smooth muscle cells, an evolution equation for a state parameter describing their contractile state is deduced. Finally, remodeling rate equations for the dynamics of the vessel's geometrical parameters are proposed following the approach used in [48, 49]. Remodeling in response to a step change in flow is formulated as an initial-value problem for a system of first-order autonomous differential equations coupled to the equations describing the deformation of the arterial wall.

2.1. MATHEMATICAL MODEL OF ARTERIAL WALL

An artery is considered to be a circular membrane made of nonlinear elastic and incompressible material. Based on experimental findings that in some cases an artery remodels without altering significantly its structure and composition [19, 33, 34], mechanical properties are considered to be constant during the remodeling process. The vessel is inflated by physiologic arterial pressure and extended longitudinally to its *in situ* length. When appropriately stimulated, vascular muscle cells are capable of contraction or relaxation, thereby changing the deformed arterial radius while keeping the loads on the vessel unchanged.

As has been demonstrated in several studies, [9, 61], when an arterial ring segment is cut radially, it springs open and takes on a form close to a circular sector. This fact indicates the existence of residual strains in the arterial wall, which arise from bending the artery from the opened-up configuration into the unloaded one. Analysis of stress distributions in the arterial wall on the basis of a thick-walled cylindrical model taking into account residual strains showed that residual strains significantly reduce the stress gradient across the arterial wall under physiological loads [9].

Because the arterial wall is assumed to be an elastic membrane, the circumferential and axial stresses are considered to be uniformly distributed across the wall thickness, while the radial stress is assumed zero everywhere. It follows from the overall equilibrium of the vessel in the radial and axial directions that, when all loads are removed and the smooth muscle cells are fully relaxed, the artery is in the zero-stress state. Hence, the membrane assumption excludes the existence of residual strains and stresses in the arterial wall of the unloaded arterial segment. Therefore, the state of no load when the smooth muscle is fully relaxed is accepted as a reference state to define the strain measures at any current deformed state.

Under applied loads the artery undergoes an axisymmetric finite deformation. The mean stretch ratios in the longitudinal and circumferential directions are

$$\lambda_1 = \frac{l}{L}, \qquad \lambda_2 = \frac{r}{R}, \tag{1}$$

where R and r are the mid-wall radii of the artery at the unloaded and deformed state respectively, and L and l are the corresponding axial lengths. Hereafter the subscripts 1 and 2 will be used to refer to the longitudinal and circumferential direction, respectively. The non-vanishing components of the Green strain tensor are

$$e_1 = \frac{1}{2}(\lambda_1^2 - 1), \qquad e_2 = \frac{1}{2}(\lambda_2^2 - 1). \tag{2}$$

Considering the vascular material as incompressible, the deformed thickness h of the artery is related to the thickness H at the zero-stress sate by the relation

$$h = \frac{H}{\lambda_1 \lambda_2}. \tag{3}$$

The deformed inner radius of the artery is

$$r_i = \lambda_2 R - \frac{H}{2\lambda_1 \lambda_2}. \tag{4}$$

A description of the stress state in the arterial wall requires modeling of the contribution of activated smooth muscle cells to load bearing. This study follows the approach used in [47]. The stresses per unit deformed area (Cauchy stress) are represented as a sum of two parts: (i) passive stresses, which are borne by the wall material at the state when the vascular muscle is fully relaxed; and (ii) active stresses developed by the smooth muscle when it is appropriately stimulated [14, 20].

Following the general theory of nonlinear elastic membranes [27], the passive longitudinal and passive circumferential stresses are given by

$$\sigma_{1\mathrm{p}} = \lambda_1^2 \frac{\partial W}{\partial e_1}, \qquad \sigma_{2\mathrm{p}} = \lambda_2^2 \frac{\partial W}{\partial e_2}, \tag{5}$$

where $W(e_1, e_2)$ is the strain energy function. It completely describes the passive mechanical properties of the wall material and is determined experimentally. The subscript "p" indicates the passive stress.

The magnitude of the active stress depends on several factors. First is the intrinsic capability of stimulated muscle cells to generate stress when the vessel is kept at constant diameter, so called isometric contraction. It is established that the contractile activity of smooth muscle cells occurs when the intercellular concentration of free calcium ions Ca^{++} exceeds a threshold concentration of 10^{-7} M. The concentration of Ca^{++} is modulated by the applied stimulation through processes, which are discussed in the next section. At the homeostatic state the relationship between the developed active stress under isometric conditions and free calcium concentration is close to sigmoidal with a pronounced linear portion over the physiological range of stresses [51]. On the other hand, the active stress developed at constant stimulation depends on the current configuration of contractile proteins within the muscle cells, which in turn depends on the deformed state of the arterial wall. There exists an optimum deformed diameter at which the active tension developed by the smooth muscle cells under isometric conditions has a maximum value, whereas below and above certain values of the diameter the muscle is not capable of developing active tension. The diameter-active tension relationship was experimentally determined for some arteries [10, 15]. It was found that under physiological conditions arteries normally operate on the ascending part of the curve representing this relationship. Finally, because the active stress is averaged across the wall, the magnitude of the active stress depends on the area ratio of the smooth muscle cells and the other structural components of the arterial cross-section. This ratio varies with species and location along the arterial tree, with a tendency to increase with increasing distance from the heart. In most large arteries smooth muscle cells mainly align in the circumferential direction and generate predominantly circumferential active stress [10, 15, 20].

Thus, the active circumferential stress is proposed in the following form

$$\sigma_{2a} = c\psi\lambda_2 F(\lambda_2), \tag{6}$$

where c is the calcium ion concentration, which exceeds the threshold concentration capable of initiating the active response; ψ is the ratio of the area occupied by the smooth muscle cells to the total area, both having a normal in the circumferential direction; $F(\lambda_2)$ is a function that accounts for the length-active tension relationship. For fixed values of c and ψ the function $F(\lambda_2)$ is proportional to the circumferential Lagrangian stress (defined per unit undeformed area). The factor λ_2 in the equation (6) accounts for the fact that the actual active stress (Cauchy stress), σ_{2a}, is defined per unit deformed area.

Equation (6) can be modified by normalization of function $F(\lambda_2)$ as follows

$$\sigma_{2a} = S\lambda_2 f(\lambda_2), \tag{7}$$

where

$$S = Mc\psi, \qquad f(\lambda_2) = \frac{F(\lambda_2)}{M}, \tag{8}$$

and $M = \max\ F(\lambda_2)$.

The parameter S is directly related to the ionic state of the muscle cells. It depends on the applied stimulus according to the active tension-dose relationship, specific for each type of stimulation. The value of S represents the maximum Lagrangian stress developed by the smooth muscle cells for given intensity of the muscular stimulation and at the optimal stretch ratio for which $f(\lambda_2) = 1$. Thus the parameter S, being proportional to the concentration of calcium ions Ca^{++} in the muscular cells, reflects the inherent capability of the smooth muscle to generate active tension. However, the magnitude of the active tension depends, according to equation (6), on the actual stretch ratio λ_2, which in turn is affected by the initial geometry of the artery and the applied load. The active circumferential stress σ_{2a} determines the apparent muscular tone developed by the muscle, while the value of parameter S gives the current level of contractile activity of smooth muscle cells developed in response to applied stimulation. It is evident from equation (7) that the magnitude of the active stress itself does not uniquely characterize the response of the smooth muscle to given excitation since identical values of σ_{2a} might be generated by different combinations of the muscle parameter S and deformation parameter λ_2.

The equation of overall equilibrium of the vessel in the radial direction relates the total mean circumferential stress to applied pressure, axial stretch ratio and initial dimensions as follows

$$\sigma_{2p}(\lambda_2) + S\lambda_2 f(\lambda_2) - P\left[\frac{\lambda_1\lambda_2^2 R}{H} - \frac{1}{2}\right] = 0, \tag{9}$$

where P is the applied arterial pressure.

At constant pressure and axial stretch ratio, equation (9) gives implicitly the relationship between the deformation parameter λ_2 and the muscle parameter S of an artery with fixed initial dimensions R and H. It follows that

$$\frac{d\lambda_2}{dS} = -\frac{\lambda_2 f(\lambda_2)}{m}, \tag{10}$$

where

$$m = \frac{d\sigma_{2p}}{d\lambda_2} + S\left[f(\lambda_2) + \lambda_2\frac{df(\lambda_2)}{d\lambda_2}\right] - \frac{2PR\lambda_1\lambda_2}{H}. \tag{11}$$

Similarly, at constant S and λ_1 equation (9) yields

$$\frac{d\lambda_2}{dP} = \frac{1}{m}\left[\frac{R\lambda_1\lambda_2^2}{H} - \frac{1}{2}\right]. \tag{12}$$

The derivative $d\lambda_2/dP$ must be positive. Otherwise the arterial diameter decreases when the transmural pressure increases and the vessel becomes unstable [44]. Hence $m > 0$ for any value of λ_2 and S. Then it follows from equation (10), that at isobaric contraction (under constant internal pressure), a larger value of the parameter S causes a smaller constricted arterial radius.

Finally, assuming the flow Q through the artery to be constant, the shear stress at the inner arterial surface is given by Poiseuille's formula

$$\tau = \frac{4\eta Q}{\pi r_i^3}, \tag{13}$$

where η is the blood viscosity.

2.2. EVOLUTION EQUATION FOR THE CONTRACTILE STATE OF VASCULAR SMOOTH MUSCLE

The artery is first considered under conditions at which the pressure, flow and muscular tone parameter have their baseline values P_0, Q_0, and S_0, respectively. The mid-wall radius and wall thickness in the unloaded state are denoted by R_0 and H_0. Hereafter, the subscript "zero" will be used to refer to the artery under normal flow conditions. The mean flow rate is changed in a step-wise manner from Q_0 to Q^* and kept constant, while the pressure remains constant at its baseline value. At the moment of the flow jump, the vessel maintains its inner radius at the control value r_{i0}. According to equation (13) the change in the flow causes an instantaneous jump in the shear stress at the arterial lumen from τ_0 to $\tau' = \tau_0(Q^*/Q_0)$. Over time the magnitude of the flow-induced shear stress varies as a result of changes in the inner radius caused by the vasomotor response and resultant geometrical remodeling.

As mentioned in the Introduction, the arterial response to increased or reduced flow begins with shear stress-dependent vasodilatation or vasoconstriction. To account for the effects of altered shear stress on the contractile state of smooth muscle cells an evolution equation for the muscle parameter S is proposed. As seen from equation (8), the time variation of S depends on variations of intercellular calcium concentration c and area ratio parameter ψ. It is evident that during the fast vasomotor response of a given artery to blood flow alterations, ψ does not change. Exploiting the experimental observations that for mature vessels the amount of major structural components of the wall material remains unchanged, it is assumed in this study that the parameter ψ remains constant during the whole process of remodeling as well. Thus, the equation that describes the dynamics of the muscle parameter S can be deduced from the processes that control the calcium concentration within the cells.

The basic mechanisms which lead to increase of Ca^{++} concentration are: (i) incoming flux of calcium ions due to concentration drop between the cell and its environment; and (ii) release of Ca^{++} from intercellular stores such as the sarcoplasmatic reticulum, where the ions are in the weakly bound form. The type and intensity of the applied stimulation affect both processes. On the other hand, there exist simultaneous processes, which cause a decrease in Ca^{++} concentration: active transport outside of cell by so-called calcium pumps, sodium-calcium exchange mechanisms, and reasorbtion of Ca^{++} in the sarcoplasmic reticulum. Following the phenomenological approach proposed in [46] and its modification given in [1], the rate of change of calcium concentration in the smooth muscle cells is described by the following balance equation

$$\frac{\mathrm{d}c}{\mathrm{d}t} = \varphi - \frac{1}{T}c. \tag{14}$$

The non-negative function φ represents the processes that lead to an increase in calcium concentration. The value of φ depends on the applied stimulus reflecting the coupling between stimulus and transmembrane flux through receptor-operated channels or release of calcium from intracellular stores. The second term on the right-hand side of equation (14) accounts for the decrease of calcium within cells, which is assumed to be proportional to the actual Ca^{++} concentration. Dependence of the pump rate upon intercellular calcium concentration has been reported in [50]. T is a positive constant that gives the decay rate of calcium concentration.

Using φ_0 and c_0 to represent the values that correspond to the steady state of Ca^{++} concentration under normal conditions, the following equation results

$$0 = \varphi_0 - \frac{1}{T} c_0. \tag{15}$$

Subtraction of equation (15) from (14) yields

$$\frac{dc}{dt} = \varphi - \varphi_0 - \frac{1}{T}(c - c_0). \tag{16}$$

Because the intima adapts to maintain a constant endothelial cell density, a change in arterial radius resulting from remodeling does not modulate the stimulus sensed by the endothelial cells and caused by altered shear stress. It is assumed as a first approximation that

$$\varphi - \varphi_0 = -\beta(\tau - \tau_0), \tag{17}$$

where β is a constant of proportionality and τ is the current value of the flow-induced shear stress, calculated from equation (13) for $Q = Q^*$. Equation (17) phenomenologically reflects the effects of altered shear stress at the endothelium on membrane permeability and internal calcium sources. It follows from equations (16) and (17) that an increase in shear stress over its baseline value causes a decrease in intercellular Ca^{++} concentration resulting in relaxation of the vascular smooth muscle. A decrease in shear stress over its baseline value has an opposite effect on Ca^{++} concentration and leads to contraction of the smooth muscle cells. These conclusions are in agreement with mechanisms of flow dependent constrictor and dilator vascular effects that relate changes in the shear stress at the inner surface to changes in intracellular Ca^{++}concentration in the vascular smooth muscle cells as proposed in [3]. Observations that shear stress-induced EDRF reduces calcium influx and thereby lowers intercellular calcium were given in [5].

Making use of equations (8), (16) and (17) and introducing dimensionless variables, the equation for the evolution of the muscle parameter S describing the ionic state of smooth muscle cells takes the following form

$$\frac{d\widetilde{S}}{dt} = \frac{1}{T_{S1}}\left[1 - \frac{\tau(\widetilde{H}, \widetilde{R}, \lambda_2)}{\tau_0}\right] - \frac{1}{T_{S2}}(\widetilde{S} - 1), \tag{18}$$

where $\widetilde{S} = S/S_0$, $\widetilde{H} = H/H_0$, and $\widetilde{R} = R/R_0$. $T_{S1} = S_0/M\beta\psi\tau_0$ and $T_{S2} = 1/T$ are time constants.

The muscle parameter S varies within the range $[0, S_{max}]$ according to the stimulation-active tension relationship. Hence the dimensionless parameter $\widetilde{S}$ has to satisfy the inequality $0 \leqslant \widetilde{S} \leqslant \widetilde{S}_{\max}$, where $\widetilde{S}_{\max} = S_{\max}/S_0$. The value $\widetilde{S} = 0$ corresponds to the state of maximal smooth muscle relaxation, and $\widetilde{S}_{\max}$ to the state of maximal contraction.

2.3. REMODELING RATE EQUATIONS FOR THE CASE OF INCREASED FLOW

Following the approach given in [48, 49], changes in the arterial dimensions resulting from flow-induced remodeling are monitored by time variations of arterial mid-wall radius and wall thickness in the state of no load. It has been experimentally established that geometrical remodeling in response to increased flow results predominately in an increase in arterial radius. It is postulated in this study that enlargement of the undeformed mid-wall radius is associated with an increase in the passive circumferential stress from its baseline value. The simplest analytical function that describes this relationship is a linear one. Thus, the following remodeling rate equation for mid-wall radius is proposed

$$\frac{\mathrm{d}\widetilde{R}}{\mathrm{d}t} = \frac{1}{T_R}\left[\frac{\sigma_{2\mathrm{p}}(\lambda_2)}{\sigma_{2\mathrm{p}0}} - 1\right], \tag{19}$$

where T_R is a time constant.

Generalizing experimental observations for growth and remodeling of tissues, Rodbard [54] has proposed that tension applied to relaxed muscle causes its longitudinal growth. Equation (19) is consistent with this statement since the circumferential stretch ratio and its corresponding passive stress increase during arterial vasodilatation occurring as a prime response to increased blood flow.

The transversal remodeling, which provides sufficient wall thickness to restore the normal value of the circumferential wall stress, is described by the following equation

$$\frac{\mathrm{d}\widetilde{H}}{\mathrm{d}t} = \frac{1}{T_{H1}}\left[\frac{\sigma_{2T}(\lambda_2, S)}{\sigma_{2T0}} - 1\right], \tag{20}$$

where $\sigma_{2T} = \sigma_{2\mathrm{p}} + \sigma_{2\mathrm{a}}$, is the total circumferential stress, and T_{H1}, is a time constant. Justifying this assumption are the experimental observations that despite a marked increase in arterial radius, which leads to an increase of medial wall tension, the arterial wall adapts to restore the baseline value of the circumferential stress. The assumption of wall thickening driven by the total mean circumferential stress has been used in [45, 48, 49] for modeling arterial remodeling under sustained hypertension and has yielded results which are in accordance with experimental observations. It is consistent with the phenomenological stress-growth law proposed by Fung [21] that increased stress can cause increase in tissue mass.

Equations (18)–(20) are coupled to the equation of equilibrium (9) and equations (4) and (13) for the flow-induced shear stress and have to be solved simultaneously. They represent a system of first-order non-linear autonomous equations subjected to side conditions. The initial conditions that correspond to the time $t = t_0$ of the onset of the step change in flow are

$$\widetilde{S}(t_0) = 1, \qquad \widetilde{R}(t_0) = 1, \qquad \widetilde{H}(t_0) = 1. \tag{21}$$

The formulated initial-value problem has a single solution. The new homeostatic state is given by the stationary solution of the governing equations after setting the time derivatives in equations (18)–(20) equal to zero. The values of the variables at that state, denoted by asterisks, are as follows

$$\begin{aligned} &R^* = R_0\sqrt[3]{\frac{Q^*}{Q_0}}, \qquad H^* = H_0\sqrt[3]{\frac{Q^*}{Q_0}}, \qquad r_i^* = r_{i0}\sqrt[3]{\frac{Q^*}{Q_0}}, \\ &\lambda_2^* = \lambda_{20}, \qquad \tau^* = \tau_0, \qquad \sigma_{2p}^* = \sigma_{2p0}, \qquad S^* = S_0. \end{aligned} \tag{22}$$

2.4. REMODELING RATE EQUATION FOR THE CASE OF REDUCED FLOW

Geometrical remodeling in response to reduced flow develops mainly as intimal thickening. It is postulated in this study that cellular and intracellular processes, which result in thickening of the arterial wall are associated with the increased contractile activity of the smooth muscle cells. The following evolution equation for the dimensionless wall thickness is proposed

$$\frac{\mathrm{d}\widetilde{H}}{\mathrm{d}t} = \frac{1}{T_{H2}}(\widetilde{S} - 1), \tag{23}$$

where T_{H2} is a time constant. Making use of equations (7), (10) and (11), it is easy to prove that under isobaric contraction

$$\frac{\mathrm{d}T_{2a}}{\mathrm{d}S} = \frac{f(\lambda_2)}{m}\left[\frac{\mathrm{d}\sigma_{2p}}{\mathrm{d}\lambda_2} + Sf(\lambda_2) - \frac{2\sigma_{2T} + P}{\lambda}\right], \tag{24}$$

where $T_{2a} = Sf(\lambda_2)$ is Lagrangian active stress. For deformations in the physiological range the right-hand side of equation (24) is always positive. Hence Lagrangian active stress, and therefore the active tension developed by the smooth muscle cells, increases when S increases. Allowing for this, equation (23) is consistent with the hypothesis of Rodbard [54] suggesting that increased contracting tension causes the muscle to thicken. The speculations in [54] refer to the case when the muscle contracts isometrically while equation (23) assumes an isobaric contraction. However, as it was shown in [13] the isometrically and isobarically contracted vessels tend to fall along a single pressure-strain curve, indicating equivalence of both the modes of contraction.

The evolution equation for the change in mid-wall radius is based on the experimental observations that mature vessels exhibit practically no change in vessel mass as a result of structural remodeling in response to reduced flow [32–34].

Moreover, adaptation is achieved without altering the vessel length [32]. Because the wall material is considered incompressible, a constancy of mass implies a constant arterial cross-sectional area at the state of no load. Considering this finding, it is assumed in this study that the area of the undeformed cross-section remains unchanged during the entire remodeling process. Taking into account the definitions of the dimensionless mid-wall radius and thickness, this condition yields

$$\widetilde{H}\widetilde{R} - 1 = 0. \tag{25}$$

Equations (18), (23) and (25) are again coupled to equations (9) and (13) and are subjected to the initial conditions (21). The system of governing equations has a stationary solution

$$\tau^* = \tau_0, \qquad S^* = S_0, \qquad r_{\mathrm{i}}^* = r_{\mathrm{i}0}\sqrt[3]{\frac{Q^*}{Q_0}}, \tag{26}$$

while the homeostatic geometrical parameters R^* and H^*, and the circumferential stretch ratio λ_2^* are calculated from the system arising from equations (4), (9) and (25).

$$\begin{gathered} \lambda_2^* R^* - \frac{H^*}{2\lambda_1\lambda_2^*} = r_{\mathrm{i}0}\sqrt[3]{\frac{Q^*}{Q_0}}, \qquad H^* R^* = H_0 R_0, \\ \sigma_{2\mathrm{p}}(\lambda_2^*) + S_0\lambda_2^* f(\lambda_2^*) - P\left[\frac{\lambda_2^{*2}\lambda_1 R^*}{H^*} - \frac{1}{2}\right] = 0. \end{gathered} \tag{27}$$

It is worth noting that both remodeling rate equations (20) and (23) describe an increase in wall thickness, but the mechanisms responsible for thickening are different and mutually exclusive. Equation (20) accounts for the smooth muscle proliferation caused by the increased total circumferential stress when the muscular tone is less than normal. In contrary, equation (23) yields the output of processes driven by the increased muscular contractility, accompanied by a reduced total stress when compared to the control value.

3. Results

To illustrate the results that follow from the proposed mathematical model it is necessary to specify all the model parameters for a real artery. Experimental data given in Cox [10, 11] for the passive and active response of a canine carotid artery kept at its *in situ* axial stretch were used. The initial dimensions of the artery were taken to be $R_0 = 1.5$ mm, $H_0 = 0.6$ mm. The *in situ* axial stretch ratio was accepted constant during remodeling $\lambda_1 = 1.5$. Experimental data of Cox represent pressure – normalized diameter relationships recorded at *in situ* axial stretch ratio. The shrinkage of the undeformed vessel diameter at zero pressure due to axial extension was taken to be 0.85 (Weizcaecker, 2000 private communication).

The experimental data were processed to determine the passive circumferential mean stress – mean strain constitutive equation. The analytical form of this relation was chosen in accordance to the general form of the strain energy function suitable to describe the passive mechanical properties of arterial tissue, [12, 22]

$$W = a_1 e_1^2 + a_2 e_2^2 + 2a_3 e_1 e_2 + c \exp\left(b_1 e_1^2 + b_2 e_2^2 + 2b_3 e_1 e_2\right), \tag{28}$$

where a_i, c, b_i $(i = 1, 2, 3)$ are material constants.

It follows from equation (5) that

$$\sigma_{2p} = 2\lambda_2^2 \left\lfloor a_2 e_2 + a_3 e_1 + c(b_2 e_2 + b_3 e_1) \exp\left(b_1 e_1^2 + b_2 e_2^2 + 2b_3 e_1 e_2\right)\right\rfloor. \tag{29}$$

The coupling between circumferential and axial strain depends on the constants a_3, b_1, and b_3, from which at least one must not to be zero. In this study the constitutive equation and the experimental data used for its identification refer only to the case when the axial strain has the fixed value corresponding to *in situ* axial extension. Utilizing the least square procedure for parameter identification, and setting $b_1 = b_3 = 0$, the material constants were determined as follows: $a_2 = 64.864$ kPa, $a_3 = 14,937$ kPa, $b_2 = 21.468$, $c = 4.736 \times 10^{-4}$ kPa.

The normalized function accounting for the active tension-stretch relationship (equation (6)) was found as follows

$$f(\lambda_2) = k_1(\lambda_2 - k_2), \quad 0.680 \leqslant \lambda_2 \leqslant 1.505, \tag{30}$$

where $k_1 = 1.213$, $k_2 = 0.680$. Function $f(\lambda_2)$ represents the ascending part of the normalized tension-stretch ratio curve along which the active response operates under normal physiological pressures. The mean arterial pressure is $P = 13.33$ kPa, and the mean flow under normal conditions was specified so that the corresponding shear stress at the endothelium would be $\tau_0 = 1.5$ Pa. The maximal Lagrangian active stress $S_0 = 15$ kPa was chosen to correspond to the basal muscular tone existing under normal conditions. S varies from $S = 0$ to $S_{\max} = 60$ kPa to span the range of active muscle stress from zero (fully relaxed muscle) to a maximal active Cauchy stress for canine carotid artery of the order 100 kPa [11]. Altered blood flow is simulated by a step change from Q_0 to $Q^* = kQ_0$, where k takes values of $k = 0.2$; 0.33; 0.5; 2; 3; and 5.

Unfortunately, quantitative experimental data for the time course of geometrical remodeling are not available in the literature except as a general description of the dynamics of the arterial response to altered blood flow [34]. Therefore, the model results obtained in the present study are tested for their ability to predict the trends of the arterial wall response to altered blood flow. Allowing for the fact that the vasomotor response is a much faster process compared to geometrical remodeling, the following values of the time constants in equations (18)–(20), (23) were chosen to give a reasonable time course for the fast vasomotor response and time course for the vessel's dimensions: $T_{S1} = 2$ min; $T_{S2} = 2.5$ min; $T_{H1} = 10\,000$ min; $T_R = 15\,000$ min; $T_{H2} = 15\,000$ min.

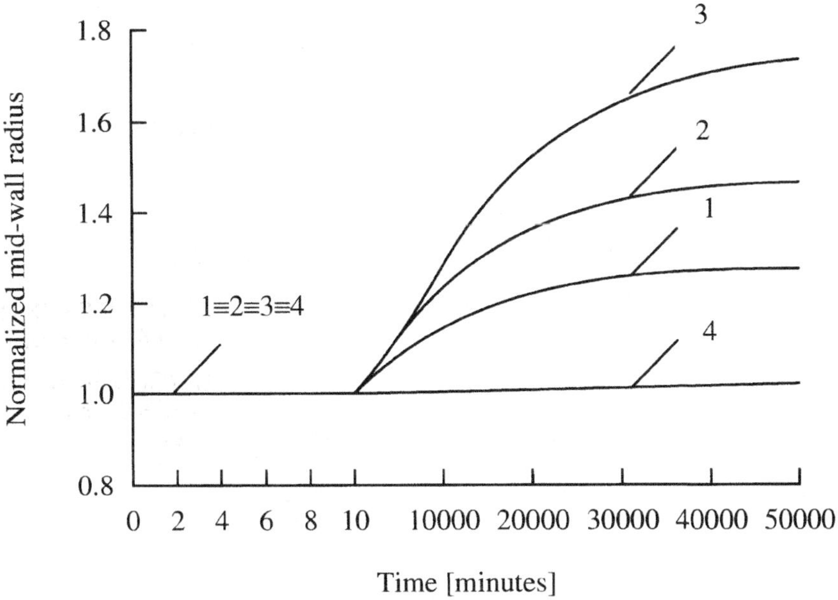

Figure 1. Time course of the normalized initial mid-wall radius $\widetilde{R} = R/R_0$ after increasing blood flow by factors of two (line 1), three (line 2) and five (line 3). R_0 is the mid-wall radius at no load conditions under normal flow. R is the current mid-wall radius at no load conditions under increased flow. Line 4 refers to the case of a twofold increase in flow and a tenfold increase in the time constant T_{S1}.

The governing equations of the process of arterial adaptation and the coupled equations for arterial wall deformation were solved numerically using Gear's methods, [24], subject to the equation of equilibrium (9) at each time step. Solutions are obtained for the time interval of 50 000 min (approximately 35 days) to cover a real time interval within which both the fast vasomotor response and geometrical remodeling take place [33].

Figures 1–3 (line 1) illustrate the time course of the dimensionless initial mid-wall radius $\widetilde{R}$, thickness $\widetilde{H}$, and the normalized deformed inner radius $\widetilde{r} = r_i/r_{i0}$ caused by 100% increase in blood flow. The time course of wall thickness is practically identical with those of mid-wall radius except for a small lag during the early phase of the geometrical remodeling. The deformed internal radius increases rapidly in the first several minutes and then relatively slowly attains the asymptotic value ($r_i^* = 1.26r_{i0}$), which restores the normal shear stress level (Figure 3, line 1). Variation of the muscle tone represented in terms of the smooth muscle parameter $\widetilde{S} = S/S_0$ and variation of stretch ratio λ_2 are shown in Figure 4 and Figure 5 (line 1). The non-monotonic pattern of variation of $\widetilde{S}$ (Figure 4, line 1) and stretch ratio λ_2 (Figure 5, line 1), as well as the constancy of the undeformed arterial thickness and mid-wall radius during the early period of flow alteration (Figures 1 and 2, line 1), show that the vessel's response is initially due to arterial dilatation. As geometrical remodeling proceeds, however, initial dimensions increase slowly and

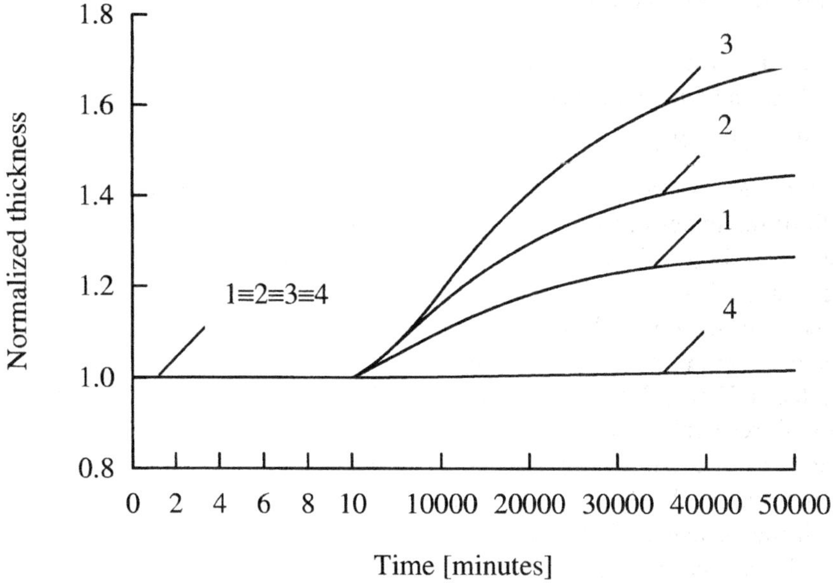

Figure 2. Time course of the normalized initial wall thickness $\widetilde{H} = H/H_0$ after increasing blood flow by factors of two (line 1), three (line 2) and five (line 3). H_0 is the thickness at no load conditions under normal flow. H is the wall thickness at no load conditions under increased flow. Line 4 refers to the case of a twofold increase in flow and a tenfold increase in the time constant T_{S1}.

monotonically to their adapted values. Simultaneously, the muscle cells restore their normal ionic state ($\widetilde{S} = 1$) as shown in Figure 4, and the circumferential stretch ratio returns to its baseline level ($\lambda_{20} = 1.141$) (Figure 5). Hence, the vessel restores both the passive and active stresses as they exist under normal flow conditions.

When blood flow was increased by factors of three and five, the time course of the changes in the geometrical parameters exhibits patterns similar to those for the case of 100% increase in flow (lines 2 and 3 in Figures 1–5). However, the artery attains a new steady state more slowly. Moreover, there exists a time period during which the smooth muscle cells become maximally relaxed and keep the artery completely dilated until the vessel enlarges. Because the loads on the artery remain unchanged, the time course of the circumferential stretch ratio depends on variation of the vessel's dimensions and contractile state of the vascular smooth muscle according to the equation of equilibrium (9). An increase in the initial mid-wall radius R causes increase of λ_2, while increase in initial thickness H and muscle parameter S have an opposite effect. During the period of acute vasomotor response the circumferential stretch increases solely due to relaxation of the smooth muscle cells, and is limited by the capability of the muscle to achieve maximal relaxation (lines 2 and 3 in Figure 5 in time interval 0–10 min). The limit value of λ_2 is a solution of the equilibrium equation (9) when S equals zero.

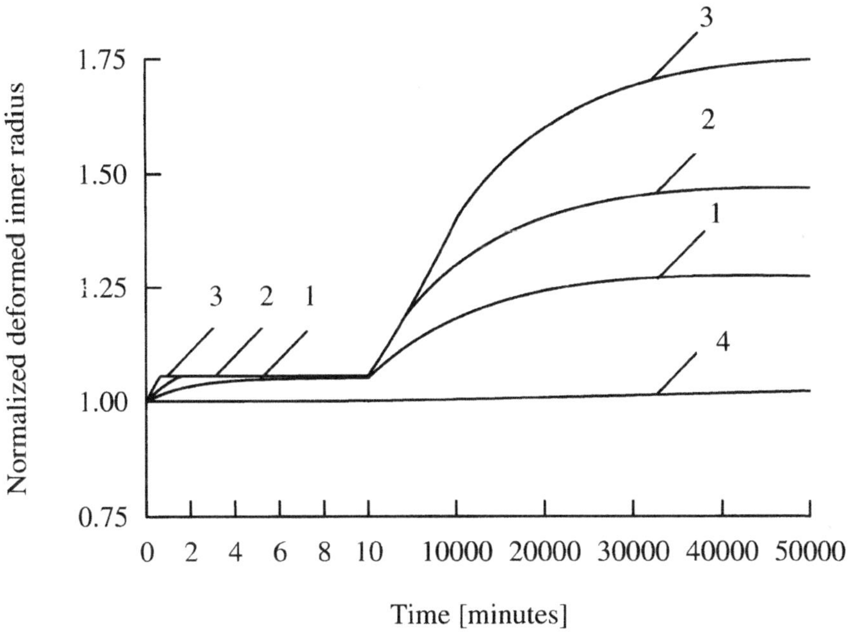

Figure 3. Time course of the normalized deformed inner radius $\widetilde{r}_i = r_i/r_{i0}$ after increasing blood flow by factors of two (line 1), three (line 2) and five (line 3). r_{i0} is the deformed inner radius under normal flow. r_i is the current deformed inner radius under incresed flow. Line 4 refers to the case of a twofold increase in flow and a tenfold increase in the time constant T_{S1}.

As seen in Figure 5, in the cases where the flow is increased by factors of three and five, the circumferential stretch ratio λ_2 increases until approximately 5000 min and 12 000 min, respectively. This is because the mid-wall radius increases faster than the wall thickness and the passive circumferential stress increases while the smooth muscle is still maximally relaxed. As the muscle tone begins to restore its baseline value, the vessel constricts and the circumferential stretch ratio gradually decreases to its baseline value.

The time variations of the geometrical and mechanical parameters following a reduction in flow to 50% of its control value are shown in Figures 6–10. The results again show a monotonic change in the wall dimensions and deformed inner radius (Figures 6–8, line 1). The primary response to reduced flow is vessel constriction (Figure 10, line 1), due to increased vasomotor tone (Figure 9, line 1). Geometrical remodeling again restores the control level ($\widetilde{S} = 1$) of the smooth muscle cell ionic state (Figure 9, line 1). In contrast to the case of arterial enlargement however, the circumferential stretch ratio reaches an asymptotic value ($\lambda_2^* = 1.069$), which is smaller than the control value ($\lambda_{20} = 1.141$). Hence the passive and active stresses in the adapted vessel become smaller than the corresponding control values. Results show that adaptation to decreased flow proceeds faster than response to 100% increase in flow, what is in agreement with experimental observations in [32, 33].

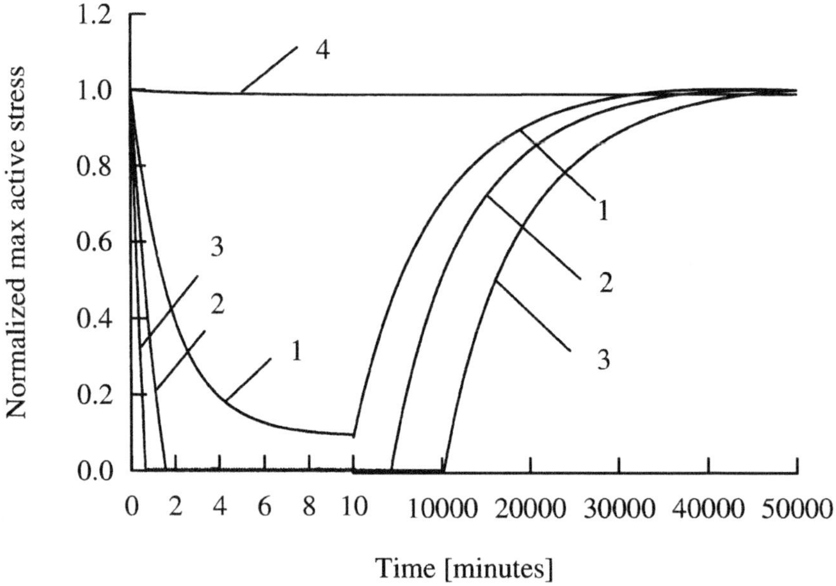

Figure 4. Time course of the normalized maximal active stress $\tilde{S} = S/S_0$ after increasing blood flow by factors of two (line 1), three (line 2) and five (line 3). S_0 is the maximal active stress under normal flow conditions. S is the current maximal active stress under increased flow. Line 4 refers to the case of a twofold increase in flow and a tenfold increase in the time constant T_{S1}.

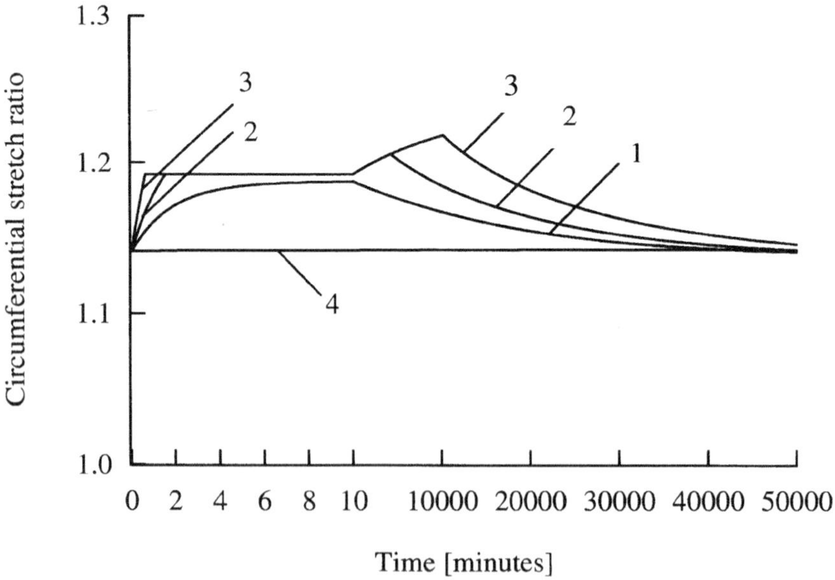

Figure 5. Time course of the circumferential stretch ratio λ_2 after increasing blood flow by factors of two (line 1), three (line 2) and five (line 3). Line 4 refers to the case of a twofold increase in flow and a tenfold increase in the time constant T_{S1}.

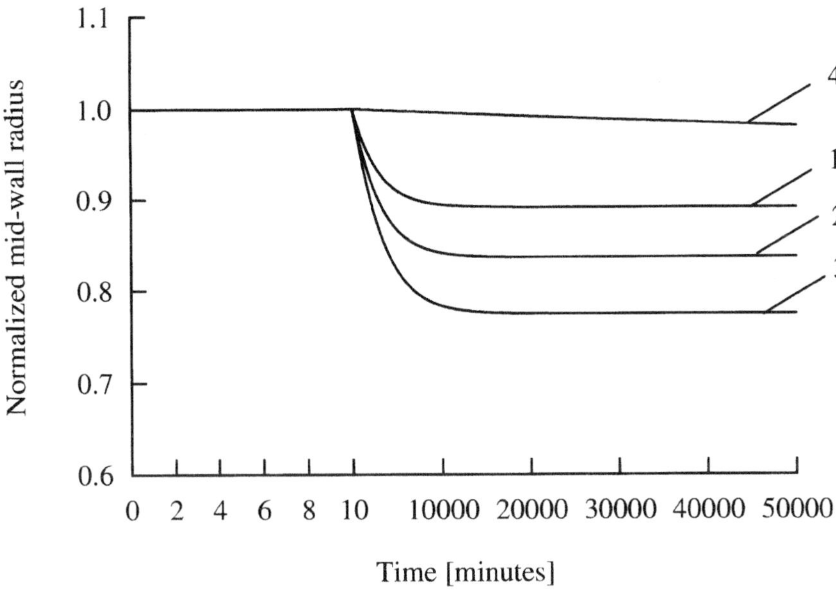

Figure 6. Time course of the normalized initial mid-wall radius $\widetilde{R} = R/R_0$ after reducing blood flow by factors of two (line 1), three (line 2) and five (line 3). R_0 is the mid-wall radius under no load conditions at normal flow. R is the current mid-wall radius under no load conditions at reduced flow. Line 4 refers to the case of a 50% reduction in flow and a tenfold increase in the time constant T_{S1}.

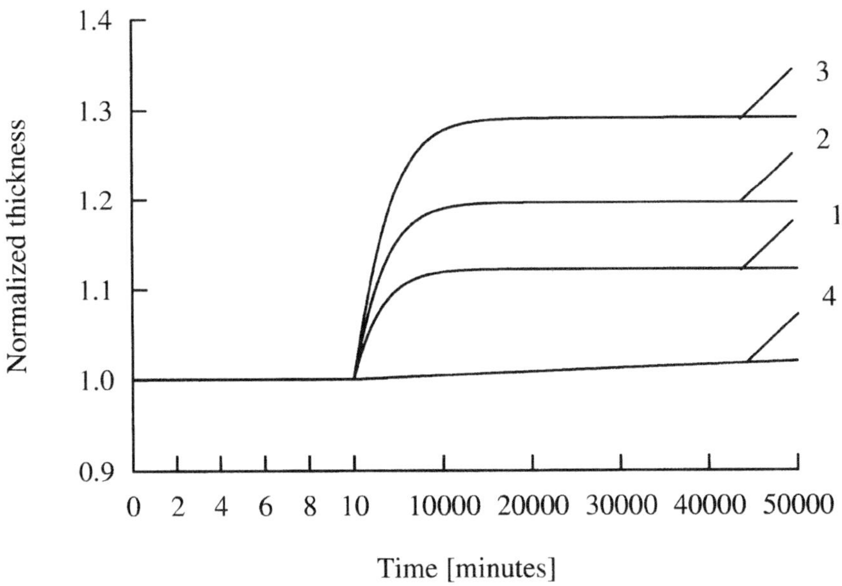

Figure 7. Time course of the normalized wall thickness $\widetilde{H} = H/H_0$ after reducing blood flow by factors of two (line 1), three (line 2) and five (line 3). H_0 is the wall thickness under no load conditions at normal flow. H is the current wall thickness under no load conditions at reduced flow. Line 4 refers to the case of a 50% reduction in flow and a tenfold increase in the time constant T_{S1}.

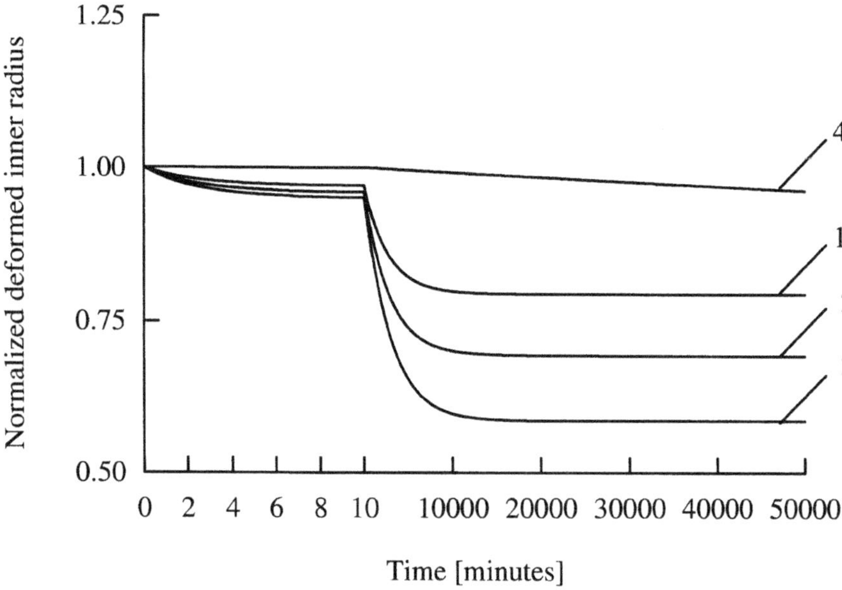

Figure 8. Time course of the normalized deformed inner radius $\tilde{r}_i = r_i/r_{i0}$, after reducing blood flow by factors of two (line 1), three (line 2) and five (line 3). r_{i0} is the deformed inner radius at normal flow. r_i is the current deformed inner radius at reduced flow. Line 4 refers to the case of a 50% reduction in flow and a tenfold increase in the time constant T_{S1}.

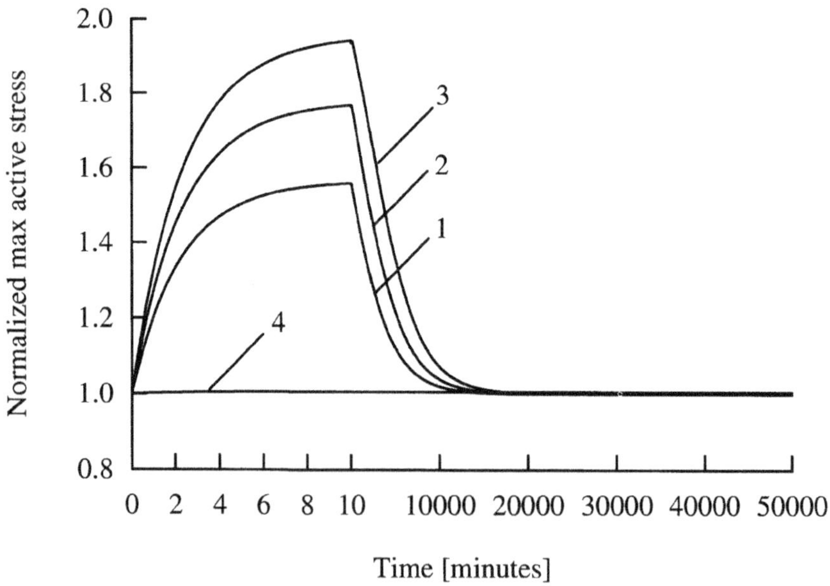

Figure 9. Time course of the normalized maximal active stress $\tilde{S} = S/S_0$ after reducing blood flow by factors of two (line 1), three (line 2) and five (line 3). S_0 is the maximal active stress under normal flow conditions. S is the current value of the maximal active stress under reduced flow conditions. Line 4 refers to the case of a 50% reduction in flow and a tenfold increase in the time constant T_{S1}.

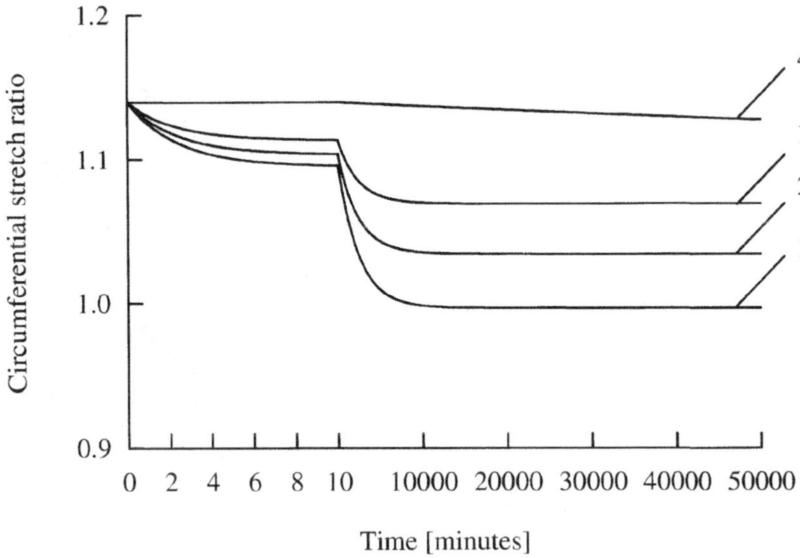

Figure 10. Time course of the circumferential stretch ratio λ_2 after reducing blood flow by factors of two (line 1), three (line 2) and five (line 3). Line 4 refers to the case of a 50% reduction in flow and a tenfold increase in the time constant T_{S1}.

Time course of geometrical and mechanical parameters have similar pattern when blood flow is threefold (curve 2) and fivefold (curve 3) as less than the control value.

According to the proposed model the arterial response includes several dynamical processes which develop simultaneously. Muscular tone, represented by the parameter $\widetilde{S}$, varies due to the deviation of the shear stress from its baseline value and mechanisms that tend to diminish the intercellular concentration of calcium ions. The dynamics are controlled by the time constants T_{S1} and T_{S2}. The parametric study shows that an increase in the time constant T_{S1} reduces the ability of the artery to produce vasomotor adjustment to the altered blood flow (lines 4 in Figures 1–10). Increasing the value of the time constant T_{S2} affects mainly the early stage of vasomotor response (not shown).

Geometrical remodeling caused by increased flow manifests itself as a temporal variation of the undeformed radius, controlled by the time constant T_R, and a temporal variation of wall thickness controlled by time constant T_{H1}. The dynamics of wall thickening in response to flow reduction are controlled by time constant T_{H2}. As a general tendency, a decreased rate of geometrical remodeling leads to an increase in the duration of the adaptation process and prolongs the period of disturbed muscular tone (results not shown).

4. Discussion

So far, a deviation of flow-induced shear stress from its baseline value has been recognized to be the single mechanical parameter that serves as a mediator between changes in blood flow and geometrical remodeling. An abrupt change in shear stress, however, could rapidly diminish almost completely, as a result of a vasomotor adjustment of the arterial lumen to altered flow conditions. It seems reasonable to seek interrelations between rate of change of arterial dimensions and other mechanical values that originate from flow changes and persistently remain altered while the artery remodels. The proposed remodeling rate equations (19), (20) and (23) are the results of such a strategy. An indirect argument in favor of the notion that there may be mechanical values that correlate with the geometrical outcomes of the remodeling process comes from observations that the magnitude of vessel remodeling is directly dependent on the duration of flow elevation [60]. Hence, despite the fact that flow-induced shear stress might be normalized as a result of acute vasomotor response, there exist other mechanical parameters that do not attain a steady state while remodeling is in progress.

The basic hypothesis in this study is that geometrical remodeling in response to altered blood flow is associated with sustained changes in the contractile state of the smooth muscle cells, which is modulated by the flow-induced shear stress sensed by the arterial endothelium. The contractile activity affects the stress and strains experienced by the smooth muscle cells and thereby affects their secretory function, as well as the pattern and the rate of cells growth and division. There are several experimental studies that support this hypothesis. The expression by vascular smooth muscle cells of a growth factor specific to the medial layer when they were exposed to mechanical strain under cell culture conditions was reported in [62]. Cell division and protein synthesis that has been elicited by chronic non-oscillatory wall tension was observed in segments of rabbit ear artery in [23].

The model describes the acute vasomotion and chronic geometrical remodeling that follow a change in blood flow. The theoretical results that the remodeling process is stable and results in restoration of the baseline values of the wall shear stress under normal conditions are in agreement with experimental observations [6, 25, 28, 30, 32, 33]. The model also predicts that the ionic state of vascular smooth muscle cells also reverts to the normal homeostatic state and the muscle restores its contractile activity to the control level. Recent findings of Hayashi (2000, private communication) support this model prediction. Hayashi showed that after 8 weeks there are no significant differences in the active circumferential stress calculated at the same circumferential strain among arteries exposed to low flow, high flow, and normal flow conditions.

The results obtained are in agreement with experimental observations for arterial enlargement and thickening in response to increased blood flow [25, 39, 43, 60]. Thus, this study supports the hypothesis that arterial enlargement due to circumferential proliferation of smooth muscle cells could be considered to be driven by the increased passive circumferential stress during isobaric dilatation. A concomitant increase in the total circumferential stress causes an increase in wall thickness, tending to maintain the strain and stress state as it exists under normal conditions. Because of the moderate basal tone, which large arteries exhibit, the fast vasodilator response to a more pronounced increase in flow cannot itself restore the flow shear stress. The model predicts that this makes the artery remain longer in the state of maximum dilatation and therefore increases the duration of the geometrical adaptation (lines 2 and 3, Figures 1–5). To explain the slower adaptation to elevated flow compared to that due to reduced flow, this scenario was assumed by Langille in [33].

The model predictions for narrowing of arteries in response to reduced flow are in agreement with experimental findings available in the literature [32–35], and support the hypothesis adopted here of a link between an increase in wall thickness and increased ionic state in the smooth muscle cells. In contrast to the case of increased flow, the model predicts that, in the newly developed homeostatic state, the circumferential strain and stress are smaller than control values. Thus, under subnormal flow conditions, wall remodeling restores the ionic state of the smooth muscle cells but tolerates reduced wall stress.

It is possible that the proposed relations expressed mathematically as remodeling rate equations (19), (20) and (23) do not reflect a causal link between geometrical and mechanical parameters. The parameters involved in these equations might only be independent and concomitant products of a remodeling process. However, if the remodeling rate equations reflect a reliable correlation between rates of change of the current arterial geometry and stress or strain measures, the latter might be accepted as the "driving stimuli" of remodeling.

The pulsatile nature of the cardiac output imposes a cyclic pressure and strain on the arterial wall. Arteries undergo small pulsatile circumferential deformations relative to the deformed state caused by the mean pressure and *in situ* axial extension. It is useful to analyze the short- and long-term effects of changes in blood flow on the pulsatile strain. The cyclic strain in response to a pulse pressure ΔP is determined by the arterial compliance defined as $C = \Delta r_\mathrm{i}/r_\mathrm{i}\Delta P$, where r_i is the inner radius at mean pressure and *in situ* length, and Δr_i is the change in radius [38]. The compliance characterizes the overall mechanical response of an artery to a physiological variation of pressure around its mean value. It follows from the equilibrium equation, after neglecting terms of smaller order, that the compliance is given by the formula $C = r/E_\mathrm{inc}h$, where $E_\mathrm{inc} = \lambda_2^2 \mathrm{d}\sigma_2/\mathrm{d}e_2$ is the incremental elastic modulus of arterial material calculated at the deformed state caused by the mean pressure; r and h are the corresponding deformed mid-wall radius and wall thickness.

The fast vasomotor response that follows a change in blood flow alters both the deformed vessel's dimensions and E_inc, and thereby affects the arterial compliance. As shown in [16], isobaric constriction increases arterial compliance at most pressures, while isobaric dilatation has an opposite effect. Altered compliance affects the magnitude of the pulsatile strain, which the arterial wall experiences due to pulsatile pressure. Hence, as a result of the acute short-term response, smooth muscle cells alter their contractile state and are exposed to abnormal pulsatile stretching. It has been experimentally established that cyclic overstretching, as well as loss of stretch experienced by the smooth muscle cells, provokes their synthetic and proliferative phenotype [37]. On the other hand, the endothelial cells, which after fast vasomotor adjustment may not be exposed to abnormal mean shear stress, become subjected to abnormal cyclic strains and abnormal cyclic shear stress due to changes in arterial compliance. This could trigger a cascade of enzymic and ionic processes that lead to upregulation of the growth and division of the underlying smooth muscle cells and ultimately, a change in arterial geometry.

Allowing for the findings in [16], the acute vasomotor constriction in response to reduced blood flow leads to an increase in the pulsatile stretching imposed on the arterial wall by the pulsatile pressure. It was found that increased cyclic stretching of arterial smooth muscle cells *in vitro* induces an increase in the synthesis of collagen and extracellular matrix [36]. A change from a contractile into a synthetic phenotype characterized by increased proliferation and protein secretion was also observed in cultured vascular smooth muscle cells undergoing cyclic mechanical strains [7, 42, 52]. On the other hand, vasomotor dilatation, which follows an increase in flow, causes a decrease in the compliance of most conduit arteries [16]. This means that the smooth muscle cells are exposed to less cycling stretch, which in turn elicits production of extracellular matrix, leading to wall thickening. A similar proliferative response was discussed in [31] to explain the high rates of intimal hyperplasia in low compliance PTFE grafts.

In the long-term the circumferential strain, as well as the ratio of mid-wall radius to wall thickness reach their baseline values after an increase in blood flow (equations (22)). Because the mechanical properties are considered unchanged, the value of the incremental modulus E_{inc} is also reestablished. Hence, the normal value of arterial compliance is restored and the magnitude of the pulsatile stretch returns to normal. When blood flow is reduced the model predicts that the adapted artery becomes thicker and smaller and the circumferential strain is smaller than the control value. Due to the mechanical non-linearity of the vascular material the incremental modulus decreases. The geometrical change (higher h/R ratio) and the change in the incremental mechanical response (lower E_{inc}) have opposite effects on the magnitude of the compliance. One might speculate that the net effect of remodeling is that arterial compliance becomes close to its baseline value. Therefore, it seems reasonable to hypothesize that under both increased and reduced flow conditions remodeling adapts not only the mean shear stress at the intimal layer, but also the cyclic strain to which smooth muscle and endothelial cells are exposed. Flow-induced and endothelium-triggered changes in muscular tone might therefore be responsible for geometrical remodeling through its effects on the magnitude of cyclic arterial wall strain.

If the interdependence between the shear stress and muscular tone and/or the interdependence between the muscular tone and processes responsible for geometrical remodeling are impaired, the arterial response to altered blood flow does not follow its normal pattern. Shear deformation of the endothelial cells appears to be the first in the chain of events that result in smooth muscle relaxation or contraction. The ability to sense the deviation of shear stress from its normal level might be impaired due to endothelial dysfunction or decreased deformability of the endothelial cells. Theoretical predictions of the model for the case of large values of the time constant T_{S1}, simulating very slow or even missing sensitivity to flow-induced shear stress, show that the ionic state in the smooth muscle cells remains practically unchanged (lines 4 in Figures 4 and 9). Therefore, neither fast vasomotor adjustment of the arterial radius nor geometrical remodeling becomes apparent (lines 4 in Figures 1–3 and 5–7). The conclusion that wall adaptation is endothelium dependent is in agreement with experimental observations for missing flow-induced response in arteries denuded of endothelium [32, 40].

The potential remodeling mechanisms identified in this study should be tested experimentally. The results suggest two possible approaches. The first involves monitoring the time course of mechanical parameters such as passive and active stress, and circumferential strains calculated from data recorded under conditions of flow-induced remodeling. Interpretation of this information within the framework of the proposed model might answer the question whether the introduced "driving parameters" in the evolution equations indeed accompany the geometrical changes caused by the altered flow. On the other hand, it is possible to impose cyclic pressure on arterial segments maintained under culture conditions while controlling the contractile state of the vascular smooth muscle. Thus, it will be possible to mimic the mechanical environment according to the mechanisms hypothesized in present theoretical study. Comparison of recorded geometrical changes, if they occur, to the pattern of the geometrical changes resulting from control experiments of flow-induced remodeling can show whether the relations given by the proposed evolution equations in fact reflect causal relationships.

The limitations in applicability of proposed model come from the assumptions used in its construction. Because only lengths and loads can be directly measured, calculation of stress, strains and parameters of the active response requires relevant modeling of the deformation process and constitutive formulation of the mechanical properties of arterial tissue with account for smooth muscle cell contraction and relaxation. According to this model, the stimulation of smooth muscle cells results in the addition of an active stress to the passive stress borne by the other structural components. The model can predict some of the behaviors demonstrated for large conduit arteries [47]. However, some experimental observations of smaller arteries cannot be interpreted in the framework of this constitutive formulation. As has been assumed in [64], stimulation might affect the interrelations between smooth muscle cells and extracellular structural components and thus alters the mechanical properties of arterial tissue termed in this study as passive ones. In this case the effects of muscle contraction/relaxation on arterial compliance might also be different.

The thin-wall membrane model used in this study disregards the deferential growth across the wall and the contribution of residual strains. Further generalizations of the model should consider arteries as thick-walled tubes and the effects of residual stress should be taken into account. It is also necessary to verify experimentally the assumption that the cross sectional area of the arterial wall remains constant throughout the remodeling process induced by a chronic reduction in blood flow, or whether it returns to its original value at the end of the process. Modified equations for smooth muscle cells contraction and additional evolution equations are needed for modeling flow-induced remodeling of arteries during development and maturation, which account for changes in geometry, composition and mechanical properties. In general, there is a need for additional data on the time variation of geometrical and mechanical parameters during remodeling.

In conclusion, this study proposes a relatively simple mathematical model in terms of continuum mechanics. The model does not account implicitly for the series of interrelated electrical, chemical, mechanical and biological processes involved in smooth muscle contraction and relaxation, regulation of vascular cell mitosis and apoptosis rates, cell migration, control of matrix synthesis and degradation, and regulation of matrix reorganization. This study focuses on three main events that occur after a change in blood flow: (i) a change in the ionic state of smooth muscle cells caused by the altered shear stress sensed by the endothelium; (ii) a change in the stress state in the arterial wall due to altered muscular tone; and (iii) a change in the geometrical dimensions of the arterial cross-section due to remodeling. The model suggests plausible hypotheses for arterial wall remodeling and predicts the main features of arterial response to changes in blood flow. It is believed that the present study gives insight into the processes of wall remodeling and might provoke future experimental and theoretical investigations.

Acknowledgment

The author would like to thank Dr. E. Manoach for his help in writing the computer program for the numerical study.

References

1. H. Achakri, A. Rachev, N. Stergiopulos and J.-J. Meister, A theoretical investigation of low frequency diameter oscillations of muscular arteries. *Ann. Biomedical Eng.* **22** (1994) 253–263.
2. G.L. Baumbach and D.D. Heistad, Remodeling of cerebral arterioles in chronic hypertension. *Hypertension* **13** (1989) 968–972.
3. J.A. Beven and I. Laher, Pressure and flow-dependent vascular tone. *FASEB J.* **5** (1991) 2267–2273.
4. N. Brady, R. Merval, J. Benessiano, J.-L. Samuel and A. Tedgul, Differential effects of pressure and flow on DNA and protein synthesis and on fibrinectin expression by arteries in a novel organ culture system. *Circ. Res.* **77** (1995) 684–694.
5. J.E. Brayden, Membrane hyperpolarization is a mechanism of endothelium-dependent cerebral vasodilatation. *Am. J. Physiol.* **259** (1990) 668–673.
6. R.D. Brownlee and B.L. Langille, Arterial adaptation to altered blood flow. *Canad. J. Physiol. Parmacol.* **69** (1991) 978–983.

7. J. Chamley-Campbell, G.R. Campbell and R. Ross, The smooth muscle cell in culture. *Physiol. Rev.* **59** (1979) 1–61.
8. A. Cho, L. Mitchell, D. Koopmans and B.L. Langille, Effects of changes in blood flow rate on cell death and cell proliferation in carotid arteries of immature rabbits. *Circ. Res.* **81**(3) (1997) 328–337.
9. C.J. Chuong and Y.C. Fung, On residual stress in arteries. *ASME J. Biomech. Eng.* **108** (1986) 189–192.
10. R.H. Cox, Comparision of carotid artery mechanics in the rat, rabbit, and dog. *Am. J. Physiol.* **234** (1978) 280–288.
11. R.H. Cox, Regional variation of series elasticity in canine arterial smooth muscles. *Am. J. Physiol.* **234** (1978) 542–551.
12. H. Demiray, H.W. Weizacker and K. Pascale, A mechanical model for passive behavior of rats carotid artery. In: *Biomedizinische Technik*, Band 31, Heft 3 (1986), pp. 46–52.
13. P.B. Dobrin, Isometric and isobaric contraction of carotid arterial smooth muscle. *Am. J. Physiol.* **225** (1973) 659–663.
14. P.B. Dobrin, Influence of initial length on length-tension relationship of vascular smooth muscle. *Am. J. Physiol.* **225** (1973) 664–670.
15. P.B. Dobrin, Vascular mechanics. In: D.F. Bohr, A.P. Somlyo and H.V. Sparks Jr. (eds), *Handbook of Physiology. The Cardiovascular System. Vascular Smooth Muscle*, Am. Physiol. Soc., Bethesda (1983), Chapter 3, pp. 65–104.
16. P.B. Dobrin and A.A. Rovick, Influence of vascular smooth muscle on contractile mechanisms and elasticity of arteries. *Am. J. Physiol.* **217** (1969) 1644–1651.
17. I. Di Stefano, D.R. Koopmans and B.L. Langille, Modulation of arterial growth of the rabbit carotid artery associated with experimental elevation of blood flow. *J. Vasc. Res.* **1** (1998) 1–7.
18. A. Ben Driss, J. Benessiano, P. Poitevin, B.I. Levy and J.B. Michel, Arterial expansive remodeling induced by high flow rates. *Am. J. Physiol.* **272** (1997) 851–858.
19. W. Fath, H.M. Burkhart, S.C. Miller, M.C. Dalshing and J.L. Unthank, Wall remodeling after wall shear rate normalization in rat mesenteric colaterals. *J. Vas. Res.* **35** (1998) 257–264.
20. Y.C. Fung, *Biodynamics: Circulation*, Springer-Verlag, New York (1984).
21. Y.C. Fung, *Biomechanics: Motion, Flow, Stress, and Growth*, Springer-Verlag, New York (1990).
22. Y.C. Fung and P. Patitucci, Pseudoelasticity of arteries and choice of its mathematical expressions. *Am. J. Physiol.* **35** (1979) 620–631.
23. W.R. Hume, Proline and thymidineuptake in rabbit ear artery segments in vitro increased by chronic tangential load. *Hypertension* **2** (1980) 738–743.
24. C.W. Gear, *Numerical Initial Value Problems in Ordinary Differential Equations*, Prentice-Hall, Inc., Englewood Cliffs, NJ (1971).
25. S. Glagov, C.K. Zarnis, N. Masawa, C.P. Xu, H. Bassiouny and D.P. Giddens, Mechanical functional role of non-atherosclerotic intimal thickening. *Frontiers Med. Biol. Engng.* **1** (1993) 37–43.
26. B.S. Gow, The influence of vascular smooth muscle on the viscoelastic properties of blood vessels. In: D. Bergel (ed.), *Cardiovascular Fluid Dynamics,* Academic Press (1972), pp. 65–110.
27. A.E. Green and J.E. Adkins, *Large Elastic Deformations and Non-Linear Continuum Mechanics*, Clarendon Press, Oxford (1960).
28. J.R. Guyton and C.J. Hartley, Flow restriction of one carotid artery in juvenile rats inhibits growth of arterial diameter. *Am. J. Physiol.* **248** (1985) 540–546.
29. L. Kaiser, S.S. Hull Jr. and H.V. Sparks Jr., Methylene blue and ETYA block flow-dependent dilation in canine femoral artery. *Am. J. Physiol.* **250** (1986) 974–981.
30. A. Kamiya and T. Togawa, Adaptive regulation of wall shear stress to flow change in canine carotid artery. *Am. J. Physiol.* **239** (1980) 14–21.

31. D.N. Ku and R.C. Allen, Vascular grafts. In: J.D. Bronzino (ed.), *The Biomedical Engineering Handbook*, CRC Press (1995), pp. 1871–1878.
32. B.L. Langille, Remodeling of developing and mature arteries: Endothelium, smooth muscle and matrix. *J. Cardiovasc. Pharmacol.* **21**(Suppl. 1) (1993) 11–17.
33. B.L. Langille, Blood flow-induced remodeling of the artery wall. In: J.A. Bevan, G. Kaley and G. Rubanyi (eds), *Flow-Dependent Regulation of Vascular Function*, Oxford University Press, New York (1995), pp. 227–299.
34. B.L. Langille, M.P., Bendeck and W. Keeley, Adaptations of carotid arteries of young and mature rabbits to reduced carotid blood flow. *Am. J. Physiol.* **256** (1989) 931–939.
35. B.L. Langille and F. O'Donnell, Reductions in arterial diameter produced by chronic decrease in blood flow are endothelium-dependent. *Science Wash. DC* **231** (1986) 405–407.
36. D.Y.M. Leung, S.G. Glagov and M.B. Mathews, Cycling stretching stimulates synthesis of components by arterial smooth muscle cells in vivo. *Science* **116** (1976) 455–477.
37. S. Lehoux and A. Tedgui, Signal transduction of mechanical stresses in the vascular wall. *Hypertension* **32** (1998) 338–345.
38. Sir J. Lighthill, *Mathematical Biofliuddynamics*, Society for Industrial and Applied Mathematics, Philadelphia, PA (1975).
39. H. Masuda, H. Bassiouny, S. Glagov and C.K. Zarins, Artery wall restructuring in response to increased flow. *Surg. Forum* **40** (1989) 285–286.
40. H. Masuda, K. Kawamura, T. Sugiyama and A. Kamiya, Effects of endothelial denudationin flow-induced arterial dilatation. *Frontiers. Med. Biol. Engng.* **1** (1993) 57–62.
41. T. Matsumoto, E. Okumura, Y. Miura and M. Sato, Mechanical and dimensional adaptation on rabbit carotid artery cultured in vitro. *Mech. Biol. Eng. Comput.* **37** (1999) 252–256.
42. G.K. Owens, Role of mechanical strain in regulation of differentiation of vascular muscle cells. *Circ. Res.* **79** (1996) 1054–1055.
43. F. Pourageaud and J.G. De Mey, Structural properties of rat mesenteric small arteries after 4-wk exposure to elevated or reduced blood flow. *Am. J. Physiol.* **273** (1997) 1699–1706.
44. C.M. Quick, H.L. Baldick, N. Safabakhsh, T.J. Lenihan, K.-J.J. Li, H.W. Weizsacker and A. Noordergraaf, Unstable radii in muscular blood vessels. *Am. J. Physiol.* **271** (1996) 2669–2676.
45. A. Rachev, Theoretical study of the effect of stress-dependent remodeling on arterial geometry under hypertensive conditions. *J. Biomechanics* **8** (1997) 819–827.
46. A. Rachev, Ts. Ivanov and K. Boev, A model for contraction of smooth muscle. *Mech. Comp. Mat.* **2** (1980) 259–263 (translated from Russian).
47. A. Rachev and K. Hayashi, Theoretical study of the effects of vascular smooth muscle contraction on strain and stress distributions in arteries. *Ann. Biomedical Eng.* **27** (1999) 459–468.
48. A. Rachev, N. Stergiopulos and J.-J. Meister, Theoretical study of dynamics of arterial wall remodeling in response to changes in blood pressures. *J. Biomechanic* **5** (1996) 635–642.
49. A. Rachev, N. Stergiopulos and J.-J. Meister, A model for geometrical and mechanical adaptation of arterics to sustained hypertension. *ASME J. Biomech. Eng.* **120** (1998) 9–17.
50. H. Rausmussen and P. Barrett, Calcium messenger system: An integrated view. *Physiol. Rev.* **64** (1984) 938–984.
51. C.M. Rembold and R.A. Murphy, Latch-bridge model in smooth muscle: $[Ca2+]_I$ can quantitatively predict stress. *Am. J. Physiol.* **259** (1990) C251–C259.
52. P. Reusch, H. Wagdy, R. Reusch, E. Wilson and H. Ives, Mechanical strain increases smooth muscle and decreases nonmuscle myosin expression in rat vascular smooth muscle cells. *Circ. Res.* **79** (1996) 1046–1053.
53. W. Risau, Vasculogennesis, angiogenesis, and endothelial cell differentiation during embryonic development. In: R.N. Feinberg, G.K. Sherer and R. Auerach (eds), *The Development of the Vascular System*, Karger, Basel (1991), pp. 58–68.

54. S. Rodbard, Negative feedback mechanisms in the architecture and function of the connective and cardiovascular tissue. *Perspectives in Biology and Medicine* **13** (1970) 507–527.
55. E.K. Rodriguez, A. Hoger and A.D. McCulloch, Stress-dependent finite growth in soft elastic tissues. *J. Biomechanics* **27** (1994) 455–467.
56. L.A. Taber, A theoretical study of aortic growth. In: *Advances in Bioengineering*, BED-Vol. 28, ASME (1994), pp. 1–2.
57. L.A. Taber, Biomechanical growth laws for muscle tissue. *J. Theor. Biol.* **193**(2) (1998) 201–213.
58. L.A. Taber, A model for aortic growth based on fluid shear and fiber stresses. *ASME J. Biomech. Eng.* **120** (1998) 348–354.
59. L.A. Taber and D.W. Eggers, Theoretical study of stress-modulated growth in the aorta. *J. Theoret. Biol.* **180** (1996) 343–357.
60. A. Tulis, J.L. Unthank and R.L. Prewitt, Flow-induced arterial remodeling in rat mesenteric vasculature. *Am. J. Physiol.* **274** (1998) 874–882.
61. R.N. Vaishnav and J. Vossoughi, Estimation of residual strains in aortic segments. In: C.W. Hall (ed.), *Biomedical Engineering Recent Developments*, Pergamon Press, New York (1983), pp. 330–333.
62. E. Wilson, Q. Mai, K. Sudhir, R.H. Weiss and H.E. Ives, Mechanical strain induces growth of vascular smooth muscle cells via autocrine action of PDFG. *J. Cell Biol.* **123** (1993) 741–747.
63. L.C. Wong and B.L. Langille, Developmental remodeling of the internal elastic lamina of rabbit arteries: Effect of blood flow. *Circ. Res.* **78**(5) (1996) 799–805.
64. P.J. Zeller and T.C. Skalak, Contribution of individual structural components in determining the zero-stress state in small arteries. *J. Vasc. Res.* **35** (1998) 8–17.

Journal of Elasticity **61:** 113–141, 2000.

Computational Mechanics of the Heart

From Tissue Structure to Ventricular Function

M. P. NASH
University Laboratory of Physiology, University of Oxford, Parks Road, Oxford OX1 3PT, U.K.
E-mail: martyn.nash@physiol.ox.ac.uk

P. J. HUNTER
Department of Engineering Science, The University of Auckland, Auckland, New Zealand.
E-mail: p.hunter@auckland.ac.nz

Received 4 May 2000; in revised form 30 October 2000

Abstract. Finite elasticity theory combined with finite element analysis provides the framework for analysing ventricular mechanics during the filling phase of the cardiac cycle, when cardiac cells are not actively contracting. The orthotropic properties of the passive tissue are described here by a "pole–zero" constitutive law, whose parameters are derived in part from a model of the underlying distributions of collagen fibres. These distributions are based on our observations of the fibrous-sheet laminar architecture of myocardial tissue. We illustrate the use of high order (cubic Hermite) basis functions in solving the Galerkin finite element stress equilibrium equations based on this orthotropic constitutive law and for incorporating the observed regional distributions of fibre and sheet orientations. Pressure–volume relations and 3D principal strains predicted by the model are compared with experimental observations. A model of active tissue properties, based on isolated muscle experiments, is also introduced in order to predict transmural distributions of 3D principal strains at the end of the contraction phase of the cardiac cycle. We end by offering a critique of the current model of ventricular mechanics and propose new challenges for future modellers.

Mathematics Subject Classifications (2000): 74B20, 74L15.

Key words: finite elastic deformation, cardiac mechanics, orthotropic constitutive relations, fibrous-sheet tissue structure.

Abbreviations: FE – finite element; FEM – finite element method; LA – left atrium; LV – left ventricle; LVEDP – left ventricular end-diastolic pressure; RA – right atrium; RV – right ventricle

1. Introduction

The heart is an electro-mechanical pump with remarkable properties, but one whose failure is the leading non-accidental cause of death in developed countries. The mechanical function of the heart depends crucially on its material properties. Understanding these properties in relation to the structure of the tissue, and how changes in the tissue structure thereby affect the pumping function of the intact heart, is a central goal of cardiac mechanics research. The purpose of this paper is to review our current knowledge of myocardial structure and the models used to describe it, and to show how the material constitutive laws are used to understand the mechanics of the heart via finite element analysis of the right and left ventricles.

Analyzing the behaviour of any soft biological tissue, at the isolated tissue level and the intact organ level, presents many fascinating challenges to both the experimental materials scientist and the computational modeller. Soft tissues are invariably anisotropic and inhomogeneous, exhibit nonlinear and time-dependent properties, and can seldom be treated within just a mechanics framework since they are often electrically active, contain blood vessels and grow! The complex anatomy of a soft tissue organ also presents a computational challenge, because there are no axes of symmetry to exploit and boundary loads are invariably spatially nonuniform and time-varying. Of course, this complexity is precisely what makes the study of soft tissue so interesting. Fortunately, we have sufficient computational resource to solve the relevant boundary value problems and sufficient graphics capability on a desktop computer to display the results of 3D computational modelling.

One of the most comprehesive reviews of early cardiac mechanics modelling is provided by [45], which summarises the evolution from the late nineteenth century thin-walled analytical models of ventricular wall stress [43] to the biophysically realistic computational models during the 1970's. Key developments during this period include:

(i) the necessity to incorporate finite deformation analysis as opposed to the small-strain (a.k.a. classical or linear) elasticity theory (first illustrated by [26]);
(ii) the use of the finite element method (FEM) to efficiently and accurately describe the complex geometry of the ventricles (one of the earliest finite element (FE) models of left ventricle (LV) mechanics was formulated by [9]);
(iii) the significance of material anisotropy and heterogeneity on predicted wall stress [15]; and
(iv) the use of nonlinear stress-strain relations to predict more realistic wall deformations [16].

Subsequent reviews of [12, 24] describe how recent models of cardiac mechanics have integrated the developments of the earlier studies with detailed experimental measurements of anatomical geometry, tissue microstructure and myocardial material properties for both resting and active tissue.

The plan for this review of cardiac mechanics modelling is as follows. We first describe the biological structure – the anatomy of the heart and its microstructural architecture. We then present a mathematical framework for describing the finite elastic deformation of inhomogeneous, incompressible, anisotropic soft tissue, and how cubic Hermite finite element basis functions can be used to approximate the geometric fields describing the deformations that occur as the heart beats. A Galerkin finite element approximation of the stress equilibrium equations is presented and appropriate constitutive laws are formulated together with their underlying microstructural basis. Finally, a finite element model of the left and right ventricular myocardium, with a fibrous-sheet structure fitted from canine heart measurements, is loaded as it would be during the filling phase of the cardiac cycle and at the end of the contraction phase to demonstrate the importance of the anisotropic and inhomogeneous material properties described earlier.

2. The Anatomy and Tissue Structure of the Heart

The primary function of the heart is to pump blood throughout the body, delivering nutrients and removing wastes from each organ. The heart has four chambers: the *left* and *right ventricles* and the *left* and *right atria* (refer to Figure 1). The thinner-walled atria act as large low pressure blood reservoirs for the ventricles which are the predominant pumping chambers. Atrial myocardium is separated from that of the ventricles by the *basal skeleton*, which is a fibrous framework formed by the rings of four valves and surrounding connective tissue. The two *atrioventricular valves* connect the atria to their respective ventricles. The *mitral valve* consists of two leaflets (bicuspid) and prevents back flow from the LV to the left atrium (LA). Similarly the *tricuspid valve* controls the passage of blood from the right atrium (RA) to the right ventricle (RV). The remaining two *semilunar valves* join the outflow tracts of each ventricle with the great arteries into which the ventricle ejects blood. The *pulmonary valve* ensures forward flow of blood from the RV into the *pulmonary artery* and the *aortic valve* ensures blood flow from the LV into the *aorta*.

The LV is a thick-walled muscular chamber that pumps blood at physiologically high pressures (up to approximately 15 kPa or 110 mmHg during the normal heart cycle) to distal locations throughout the body. The cavity of the LV resembles a truncated ellipsoid in which both the inflow and outflow tracts are adjacent. In contrast, the RV pumps blood at comparatively low pressures (about one seventh the pressure of the LV) and wraps around the LV in a crescent-like fashion so that

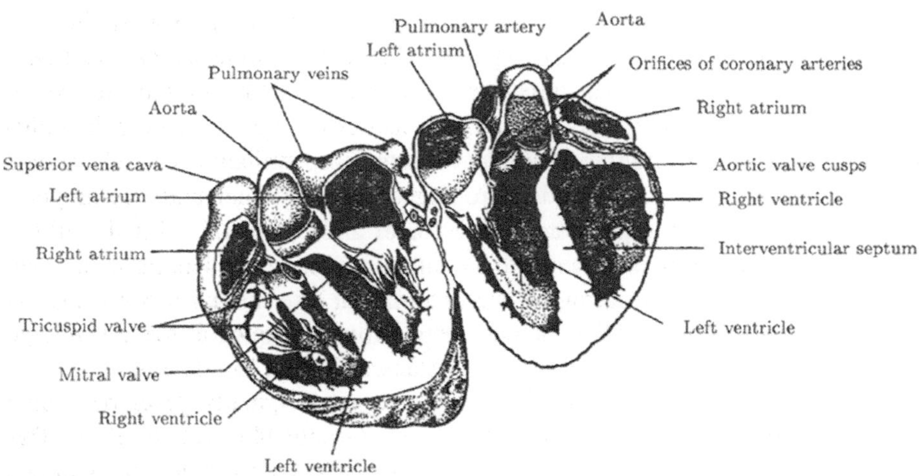

Figure 1. Longitudinal cross-section of the heart. From [18].

its cavity forms a shallow U-shape. The ventricles are separated by the *interventricular septum*, which normally functions as part of the LV and moves toward the LV free wall during *systole* (the contraction phase of the cardiac cycle).

Cardiac muscle cells or *myocytes* are typically cylindrical with lengths that range from 80 to 100 μm and diameters ranging from 10 to 20 μm. The fundamental contractile unit is the *sarcomere*, which is about 2 μm in length. The sarcomere spans between adjacent *Z-lines* along the longitudinal axis of the cell and contains the contractile apparatus. Cells consist of about 40–50 sarcomeres in series and branch and interconnect end-to-end through *intercalated disk junctions*. The branching angle is usually acute so that adjacent cells run almost parallel to one another. In this way, the contractile apparatus between cells is aligned for efficient mechanical function. Intercalated disks contain *nexi* or *gap junctions*, which provide electrical continuity between cells. Consequently, the electrical impulses propagate more rapidly along rather than across the axes of the muscle fibres.

Muscle cells within the cardiac walls generate contractile forces upon electrical depolarisation. Each cell is excited in turn as a wave of electrical activation propagates throughout the myocardium. Activation normally begins spontaneously in pacemaker cells of the *sinoatrial node*, which lies between the superior vena cava and the right atrium (see Figure 1). The first structures to be depolarised during the normal cardiac cycle are the atria. As the wave of electrical activation propagates throughout the atria, they contract and pump blood into their respective ventricles, which marks the final stage of ventricular filling. The ventricular myocardium is normally electrically isolated from the atria except for a group of slow conducting cells known as the *atrioventricular (AV) node*. This provides enough time for the atrial blood to be pumped into the ventricles prior to ventricular contraction.

The activation wave reaches the ventricular myocardium via the *AV bundle* (also known as the *common bundle*, or *bundle of His*), which bifurcates into *right* and *left bundle branches* at the top of the interventricular septum. Each branch passes down the septum and curls around into the apical portions of its respective ventricle. At this point the bundles divide into networks of fast conducting *Purkinje fibres*, which spread over and deliver the electrical impulse to the inner or *endocardial* portions of the ventricular myocardium. The activation wave generally proceeds from endocardial layers to the *epicardial* or outer portions of the ventricular myocardium. As the wave of excitation propagates, individual myocardial cells are sequentially stimulated to generate tensile forces and contract. At a macroscopic level, this causes the ventricles to contract and pump blood to the body.

Electrical activation initiates cell contraction, but also mechanical contraction affects the propagating electrical wave – especially during the reentrant arrhythmias that precede ventricular fibrillation. Changes in cell length and the cross-sectional area of cells have a direct effect on propagation pathways and also change the conductance of ion channels, thereby altering membrane potential. Some ion channels are in fact activated by supra-threshold levels of stretch (so-called *stretch activated channels* or SACs) [18].

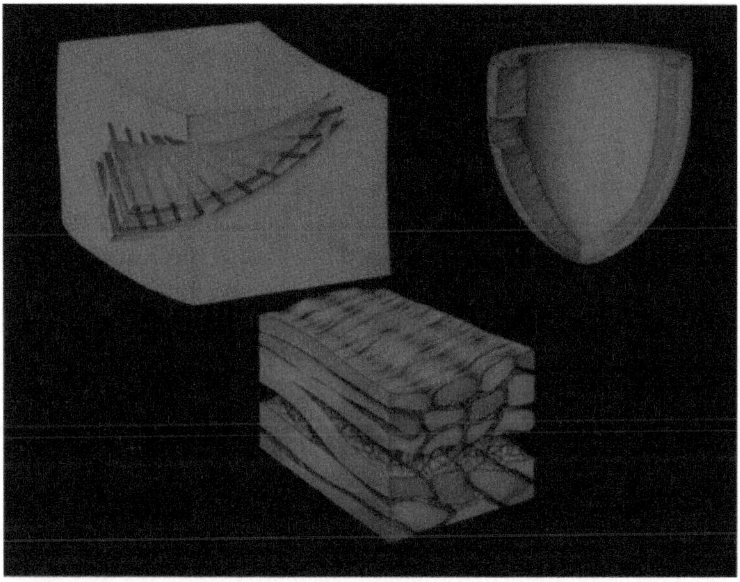

Figure 2. Schematic of cardiac microstructure showing fibre orientation and branching sheet structures. A transmural block of tissue (top left) from the wall of the LV (top right) contains *myocardial sheets* (yellow) comprised of myocytes, whose axes (red) define the *fibre direction*. These sheets interconnect (bottom) and are bound together by endomysial collagen surrounding the sheets and perimysial collagen between the sheets. From [19] by permission of the *American Journal of Physiology*.

Recent anatomical studies [19, 28] have shown that cardiac tissue is a composite of discrete layers of myocardial muscle fibres tightly bound by *endomysial collagen*, as illustrated in Figure 2. These *myocardial laminae* or *sheets* are loosely coupled by *perimysial collagen* and have the ability to slide over each other with relative ease [20]. Laminae are on average four to six cells thick and continuously branch in each direction throughout the ventricular walls. Their orientation is generally normal to the ventricular surfaces, except in the subendocardial and subepicardial regions, where they appear to become more aligned with the wall surfaces.

The *pericardium* is a fibrous sac that encompasses the entire heart to resist rapid increases in cardiac size. The inner wall of this sac is called the *parietal pericardium* and is continuous with the *epicardium* or *visceral pericardium* (the layer of connective tissue on the outer ventricular surface) at the base of the heart, where the great vessels enter and leave. A small amount of fluid within the pericardial sac provides lubrication for the continuous movement of the heart. The pericardium affects myocardial deformation and will need to be considered in future models of ventricular mechanics (see later).

The heart is nourished by blood flow through a network of vessels known as the *coronary vascular system*. These blood vessels enter the ventricular wall from the outside and bifurcate through about seven generations of *arteries* and *arterioles* before merging into a *capillary cascade*, where gas exchange with the tissue takes place. At this level, a cross-section of the muscle reveals a one-to-one ratio of capillaries to myocytes. The capillaries collect into *venules* and then *veins*, which eventually empty the de-oxygenated blood into the RA. Coronary flow is strongly influenced by myocardial mechanics. During systolic contraction, for example, arterial flow in the subendocardium is briefly reversed as the ventricles are compressed by the surrounding active muscle. Similarly, the mechanical deformation of the myocardial tissue is influenced by coronary perfusion in at least two respects: the spatially and temporally changing coronary blood pressure influences the tissue hydrostatic pressure; and the shift of fluid volume, especially during systole, introduces an effective compressibility into the otherwise incompressible tissue.

3. The Theory of Finite Elastic Deformations

Cardiac cells change length by over 20% during a normal heart beat, so mechanical analysis must be based on finite deformation elasticity theory. We first describe the kinematics of large deformation, then stress equilibrium and the finite element basis functions that extend the analysis to complex 3D ventricular shapes.

3.1. KINEMATICS

$\mathbf{x} = (x_1, x_2, x_3)$ gives the present position in rectangular cartesian coordinates of a material particle that occupied the place $\mathbf{X} = (X_1, X_2, X_3)$ in the reference state. In standard finite deformation theory, (X_1, X_2, X_3) are considered as material coordinates and a *deformation gradient tensor* $\mathbf{F}$ is defined, which carries the line segment $\mathrm{d}\mathbf{X}$ into $\mathrm{d}\mathbf{x} = \mathbf{F}\,\mathrm{d}\mathbf{X}$, or in component form, $\mathrm{d}x^i = F^i_M\,\mathrm{d}X^M$, where

$$F^i_M = \frac{\partial x_i}{\partial X_M}. \tag{1}$$

Polar decomposition, $\mathbf{F} = \mathbf{RU}$, splits $\mathbf{F}$ into the product of an orthogonal rotation tensor $\mathbf{R}$ and a symmetric positive definite stretch tensor $\mathbf{U}$ which contains a complete description of the material strain, independent of any rigid body motion (see [1]).

For inhomogeneous, anisotropic materials the orientation of the material axes may vary with location and so it is no longer convenient to identify the material axes in the undeformed body with the reference coordinates. Instead, a new material coordinate system (ν_1, ν_2, ν_3) is introduced, which is aligned with certain structural features of the material. For myocardium, a natural set of material axes are formed by identifying ν_1 with the muscle fibre direction, ν_2 with the sheet direction and ν_3 with the sheet-normal direction (refer to Figure 3). It is convenient to choose the base vectors for the ν_α-coordinate system to be orthogonal in the reference state. However, the ensuing deformation means that they are not orthogonal, in general, in the deformed configuration.

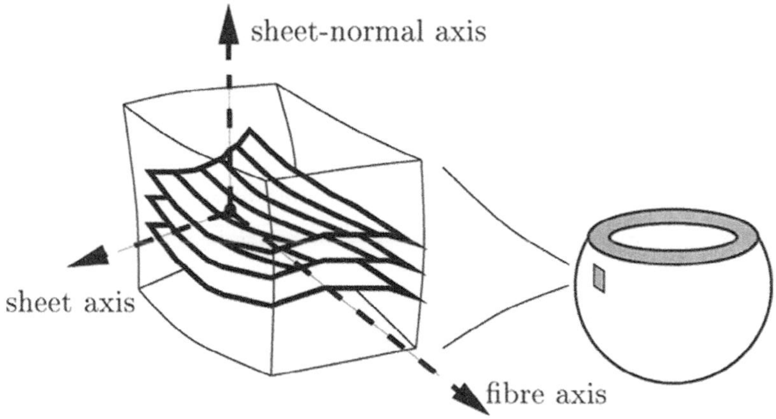

Figure 3. Microstructural material axes for myocardial tissue. The first of these directions is referred to as the *fibre axis* and coincides with the muscle fibre direction at each point. The *sheet axis* is defined to lie in the plane of the muscle layer and is perpendicular to the fibre direction. The third axis is defined to be orthogonal to the first two and is referred to as the *sheet-normal axis* as it is perpendicular to the muscle layer.

$\mathbf{A}_{\alpha}^{(\nu)}$, $\mathbf{A}_{(\nu)}^{\alpha}$ and $\mathbf{a}_{\alpha}^{(\nu)}$, $\mathbf{a}_{(\nu)}^{\alpha}$ denote the covariant and contravariant base vectors in the undeformed (upper-case) and deformed (lower-case) configurations, respectively. The corresponding metric tensors are denoted by $A_{\alpha\beta}^{(\nu)}$, $A_{(\nu)}^{\alpha\beta}$ and $a_{\alpha\beta}^{(\nu)}$, $a_{(\nu)}^{\alpha\beta}$. The undeformed covariant base vectors $\mathbf{A}_{\alpha}^{(\nu)}$ can be defined to be unit vectors by choosing the ν_α-coordinates to be a measure of physical arc-length in the undeformed state. The covariant base vectors and metric tensors for the ν_α-coordinate system are:

$$\begin{aligned} \mathbf{A}_{\alpha}^{(\nu)} &= \frac{\partial X_k}{\partial \nu_\alpha}\mathbf{g}_k^{(x)}, & \mathbf{a}_{\alpha}^{(\nu)} &= \frac{\partial x_k}{\partial \nu_\alpha}\mathbf{g}_k^{(x)}, \\ A_{\alpha\beta}^{(\nu)} &= \mathbf{A}_{\alpha}^{(\nu)}\cdot\mathbf{A}_{\beta}^{(\nu)}, & a_{\alpha\beta}^{(\nu)} &= \mathbf{a}_{\alpha}^{(\nu)}\cdot\mathbf{a}_{\beta}^{(\nu)}, \end{aligned} \tag{2}$$

where $\mathbf{g}_k^{(x)}$ are the base vectors of the rectangular Cartesian reference axes.

The Green strain tensor with respect to fibre-sheet material coordinates is then

$$E_{\alpha\beta} = \frac{1}{2}\left(a_{\alpha\beta}^{(\nu)} - A_{\alpha\beta}^{(\nu)}\right). \tag{3}$$

3.2. STRESS EQUILIBRIUM AND THE PRINCIPLE OF VIRTUAL WORK

Stress equilibrium is expressed via the following equation derived from the principle of virtual work:

$$\begin{aligned} &\int_{V_0} T^{\alpha\beta} F_\beta^j \delta v_{j,\alpha}\, \mathrm{d}V_0 \\ &\quad = \int_{V_0} \rho_0\left(b^j - f^j\right)\delta v_j\, \mathrm{d}V_0 + \int_{S_2} p_{(\mathrm{appl})}\frac{g_{(\xi)}^{3M}}{\sqrt{g_{(\xi)}^{33}}}\frac{\partial x_j}{\partial \xi_M}\delta v_j\, \mathrm{d}S, \end{aligned} \tag{4}$$

where $T^{\alpha\beta}$ and $F^j_\beta (= \partial x_j/\partial \nu_\beta)$ are respectively the second Piola–Kirchhoff stresses and deformation gradients expressed relative to the fibre-sheet material coordinates, $\delta \mathbf{v} = \delta v_j \mathbf{i}_j$ are virtual displacements expressed relative to the reference coordinate system (see [23]), ρ_0 is the tissue density, b^j and f^j are respectively the components of the body force and acceleration vectors, $p_{(\mathrm{appl})}$ is the pressure (i.e., physical stress) applied to the surface S_2 with normal direction ξ_3, and $g^{MN}_{(\xi)}$ are contravariant metric tensors for the ξ_M coordinate system (described below), with covariant base vectors and metric tensors for the undeformed and deformed states defined by:

$$\begin{aligned} \mathbf{G}^{(\xi)}_M &= \frac{\partial X_k}{\partial \xi_M}\mathbf{g}^{(x)}_k, & \mathbf{g}^{(\xi)}_M &= \frac{\partial x_k}{\partial \xi_M}\mathbf{g}^{(x)}_k, \\ G^{(\xi)}_{MN} &= \mathbf{G}^{(\xi)}_M \cdot \mathbf{G}^{(\xi)}_N = \frac{\partial X_k}{\partial \xi_M}\frac{\partial X_k}{\partial \xi_N}, & g^{(\xi)}_{MN} &= \mathbf{g}^{(\xi)}_M \cdot \mathbf{g}^{(\xi)}_N = \frac{\partial x_k}{\partial \xi_M}\frac{\partial x_k}{\partial \xi_N}. \end{aligned} \tag{5}$$

Equation (4) is the starting point for the analysis of a body undergoing large elastic deformations. For further details see [3, 4].

3.3. RESIDUAL STRAIN

Mathematical models of large deformation mechanics compute material strains (and hence stresses) with respect to a well defined reference configuration, for which the strain components are usually assumed to be zero. However, if no such state exists for an intact specimen such as the heart, the closest approximation is the no-load state, in which some structures may be pre-stretched or residually stressed. The residual strains within the specimen may be approximated by introducing the concept of a “growth tensor” [32].

4. The Finite Element Method for Finite Elasticity

The complex anatomy of the heart greatly influences its mechanical function and little insight is gained by treating it as a simple geometric shape, such as a cylinder or sphere. Consequently, numerical analysis is required to solve the above stress equilibrium equations and the FEM provides the most convenient framework.

4.1. BASIS FUNCTIONS

For FE analysis of finite deformation elasticity problems, it is convenient to choose geometric variables as the dependent variables and interpolate the geometric coordinates (x_1, x_2, x_3) defined at the finite element nodes. Element coordinates (ξ_1, ξ_2, ξ_3) are then normalised material coordinates (see Figure 4), which move with the deforming body and provide the parameterisation of the undeformed and deformed geometric variables:

$$X_i = \Psi_n(\xi_1, \xi_2, \xi_3) X^n_i, \tag{6}$$

$$x_i = \Psi_n(\xi_1, \xi_2, \xi_3) x^n_i, \tag{7}$$

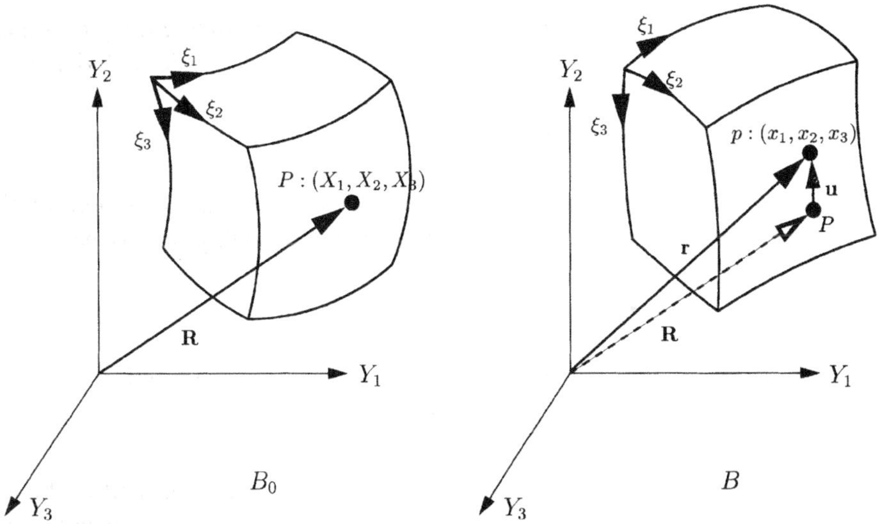

Figure 4. The finite element material coordinate system (ξ_1, ξ_2, ξ_3) is used to identify a material point P in the reference state, which moves through displacement $\mathbf{u}$ to a new spatial location p in the deformed state.

where $\Psi_n(\xi_1, \xi_2, \xi_3)$ are the chosen 3D basis functions (see below) and X_i^n and x_i^n are the element nodal values (and derivatives) of the ith geometric coordinate in the undeformed and deformed states, respectively.

Since strain (and hence stress) depends on the gradients of these geometric variables, it is convenient to choose cubic Hermite finite element basis functions to ensure derivative continuity across element boundaries. For further information see, for example, [48]. The 1D cubic Hermite interpolation for a field variable u, with nodal parameters u_1 and u_2, and derivatives $(\mathrm{d}u/\mathrm{d}\xi)_1$ and $(\mathrm{d}u/\mathrm{d}\xi)_2$, is given by:

$$u(\xi) = \Psi_1^0(\xi)u_1 + \Psi_1^1(\xi)\left(\frac{\mathrm{d}u}{\mathrm{d}\xi}\right)_1 + \Psi_2^0(\xi)u_2 + \Psi_2^1(\xi)\left(\frac{\mathrm{d}u}{\mathrm{d}\xi}\right)_2, \tag{8}$$

where the basis functions (shown in Figure 5) are:

$$\begin{aligned} \Psi_1^0(\xi) &= 1 - 3\xi^2 + 2\xi^3, & \Psi_2^0(\xi) &= \xi^2(3 - 2\xi), \\ \Psi_1^1(\xi) &= \xi(\xi - 1)^2, & \Psi_2^1(\xi) &= \xi^2(\xi - 1). \end{aligned} \tag{9}$$

Instead of using the nodal derivative $(\mathrm{d}u/\mathrm{d}\xi)_n$ that depends on the local element ξ-coordinate in the two adjacent elements (which may have different physical lengths), it is more useful to define a global node derivative $(\mathrm{d}u/\mathrm{d}s)_N$, where s is the arc-length and N is the global node number. The ξ-coordinate derivative is then:

$$\left(\frac{\mathrm{d}u}{\mathrm{d}\xi}\right)_n = \left(\frac{\mathrm{d}u}{\mathrm{d}s}\right)_N \left(\frac{\mathrm{d}s}{\mathrm{d}\xi}\right)_n, \tag{10}$$

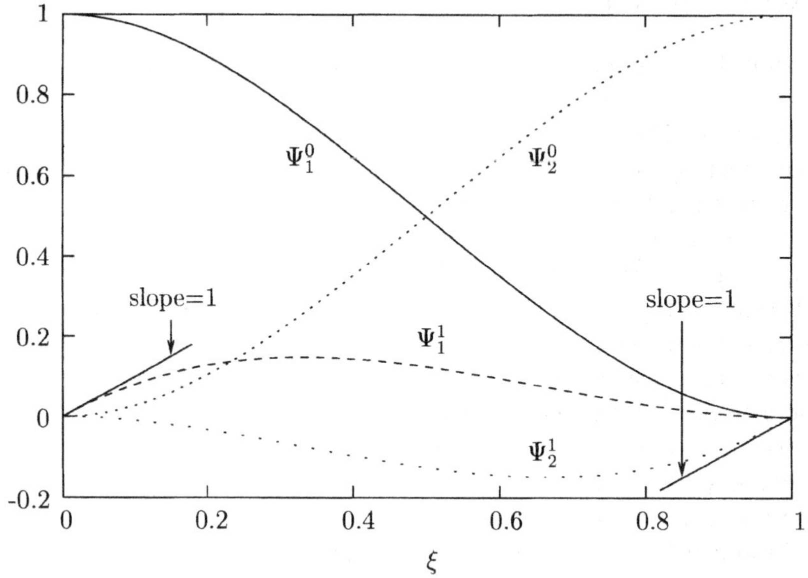

Figure 5. Cubic Hermite basis functions.

where $(\mathrm{d}s/\mathrm{d}\xi)_n$ is an *element scale factor* which scales the arc-length derivative of global node N to the ξ-coordinate derivative of element node n. The result is that $(\mathrm{d}u/\mathrm{d}s)$ is implicitly constrained to be continuous across element boundaries rather than $(\mathrm{d}u/\mathrm{d}\xi)$. The construction of tricubic Hermite basis functions is a straightforward extension of the 1D basis functions and results in 8 parameters per node for each field variable u.

4.2. GALERKIN EQUILIBRIUM EQUATIONS

Using the interpolation functions Ψ_n we can define virtual displacement fields δv_j as:

$$\delta v_j = \Psi_n(\xi_1, \xi_2, \xi_3)\delta v_j^n, \tag{11}$$

where δv_j^n are arbitrary virtual nodal displacements. Substituting equation (11) into the equilibrium equations (equation (4)) and setting the coefficient of each arbitrary displacement component δv_j^n to zero, gives:

$$\int_{V_0} T^{\alpha\beta} F_\beta^j \Psi_{n,\alpha} \,\mathrm{d}V_0 = \int_{V_0} \rho_0 (b^j - f^j) \Psi_n \,\mathrm{d}V_0 + \int_{S_2} p_{(\mathrm{appl})} \frac{g_{(\xi)}^{3M}}{\sqrt{g_{(\xi)}^{33}}} \frac{\partial x_j}{\partial \xi_M} \Psi_n \,\mathrm{d}S. \tag{12}$$

To evaluate the integrals in equation (12), they must first be transformed from the reference coordinate space to the ξ_M-coordinate space using the appropriate

Jacobian:

$$\iiint_{V_0} T^{\alpha\beta} F^j_\beta \Psi_{n,\alpha} \sqrt{G^{(\xi)}}\, \mathrm{d}\xi_3\, \mathrm{d}\xi_2\, \mathrm{d}\xi_1 = \iiint_{V_0} \rho_0 \big(b^j - f^j\big) \Psi_n \sqrt{G^{(\xi)}}\, \mathrm{d}\xi_3\, \mathrm{d}\xi_2\, \mathrm{d}\xi_1 + \iint_{S_2} p_{(\mathrm{appl})} g^{3M}_{(\xi)} \frac{\partial x_j}{\partial \xi_M} \Psi_n \sqrt{g^{(\xi)}}\, \mathrm{d}\xi_2\, \mathrm{d}\xi_1, \tag{13}$$

where $\sqrt{G^{(\xi)}} = \sqrt{\det\{G^{(\xi)}_{ij}\}}$ and $\sqrt{g^{(\xi)}} = \sqrt{\det\{g^{(\xi)}_{ij}\}}$ are the Jacobians of the 3D coordinate transformation with respect to the undeformed and deformed configurations, respectively. Note that the surface integral is transformed by substituting $J_{2D}\, \mathrm{d}\xi_2\, \mathrm{d}\xi_1$ for $\mathrm{d}S$, where the 2D Jacobian with respect to deformed coordinates is given by $J_{2D} = \sqrt{g^{(\xi)} g^{33}_{(\xi)}}$ [29].

The 3D integrals in equation (13) are evaluated over the undeformed volume and the 2D integral is computed over the portion of the deformed surface (denoted S_2) for which external pressure loads are applied. These integrals are replaced by a sum of integrals over the collection of element domains which constitute the FE model. Element integrals are evaluated numerically using Gaussian quadrature. Components of the second Piola–Kirchhoff stress tensor $T^{\alpha\beta}$ are evaluated at each Gauss point using the constitutive equations below.

4.3. GALERKIN INCOMPRESSIBILITY CONSTRAINT

For incompressible materials, an additional scalar hydrostatic pressure field is introduced into the constitutive equations (see below). The extra constraint necessary to determine the parameters of the hydrostatic pressure field arise from the requirement that the third strain invariant $I_3 = 1$ for incompressible materials.

For a Galerkin formulation, the form of the incompressibility constraints is given by:

$$\iiint_{V_e} \big(\sqrt{I_3} - 1\big) \Psi^p \sqrt{G^{(\xi)}}\, \mathrm{d}\xi_3\, \mathrm{d}\xi_2\, \mathrm{d}\xi_1 = 0, \tag{14}$$

where V_e denotes the domain of the element and Ψ^p are the basis functions used to approximate the 3D hydrostatic pressure field. Note that the undeformed 3D Jacobian $\sqrt{G^{(\xi)}}$ is introduced since the integrals are evaluated with respect to the undeformed configuration.

5. Myocardial Material Properties

Heart muscle contains connective tissue and cells, surrounded by fluid-filled extracellular space. Both components consist primarily of water. The nonlinear viscoelastic and poroelastic nature of myocardial tissue has been modelled [14, 44], but this aspect of the material properties is neglected here, where for simplicity we treat the myocardium as an incompressible, elastic solid.

5.1. CONSTITUTIVE RELATIONS

For an incompressible material components of the second Piola–Kirchhoff stress tensor are given by the derivatives of the strain energy function $W(\mathbf{E})$ with respect to the components of $\mathbf{E}$ and a *hydrostatic pressure* (denoted by p), which does not contribute to the deformation, and hence strain energy, of the material [23]:

$$T^{\alpha\beta} = \frac{1}{2}\left(\frac{\partial W}{\partial E_{\alpha\beta}} + \frac{\partial W}{\partial E_{\beta\alpha}}\right) - p a^{\alpha\beta}_{(\nu)}. \tag{15}$$

The parameters of such a constitutive law can be obtained directly from experiment without reference to the underlying tissue structure. But an approach which incorporates parameters that directly reflect mechanical or structural properties of the material yields a more useful constitutive relation. For example, observed spatial variation in collagen distributions can be related to material constitutive parameters. Section 5.2 details the development of a microstructurally-based constitutive law for passive heart tissue.

Biaxial tension tests on thin sections of ventricular myocardium [6, 10, 33, 46] reveal highly nonlinear, anisotropic stress-strain behaviour (typical of most soft biological tissues). The stress-strain properties along each of the microstructurally relevant directions are quite different, reflecting in part the organisation of collagen relative to these three axes. Figure 6 schematically summarises typical stress-strain behaviour of myocardium when stretched along each of the three microstructural axes [12]. The most striking difference between each of the three axes is the limiting strain for an elastic response. When the tissue is stretched along the fibre direction the limiting extension ratio* is about 1.3, whereas the limiting extension

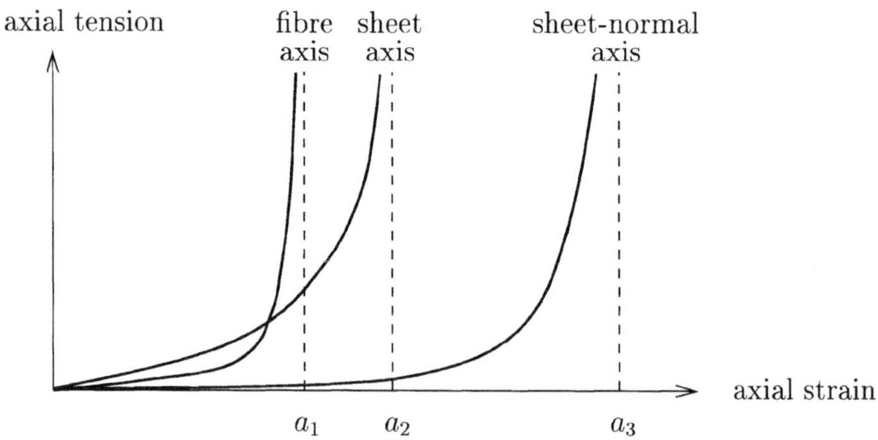

Figure 6. Typical nonlinear stress-strain properties of ventricular myocardium. The parameters a_1, a_2 and a_3 represent the limiting strains for elastic deformations along the fibre, sheet and sheet-normal axes, respectively. Note the highly nonlinear behaviour as the elastic limits are approached.

* Relative to a resting sarcomere length of approximately 1.95 μm for the unloaded muscle.

ratio for the sheet axis is approximately 1.5. Below an extension ratio of 1.5 very little tension is developed in the direction of the sheet-normal, but tension increases rapidly above this and irreversible damage occurs when this extension ratio exceeds about 1.7 [12].

Variations in the axial limiting strains can be explained by the organisation of the extracellular connective tissue matrix. The high fibre stiffness is most probably due to intracellular titin protein together with the tightly bound endomysial collagen coils that surround individual myocytes [31]. As the tissue is stretched along the fibre axis, these coils straighten and it is the taut length of the collagen that determines the limiting fibre strain [21, 22]. In contrast, the relatively low sheet-normal stiffness is most likely to be due to the sparse array of perimysial collagen links in the cleavage planes between myocardial sheets [19]. Based on this information, a fully 3D orthotropic constitutive law, which incorporates material properties that can be directly estimated from the tissue, is described next.

5.2. THE "POLE–ZERO" CONSTITUTIVE LAW FOR MYOCARDIUM

Stress-strain behaviour along one axis is very nearly independent of the degree of lateral stretch [33]. This means that the contribution to the total strain energy of the stretch along one of the material axes is nearly independent of the contribution from the other two axes – there is, of course, a small degree of cross-axis coupling via the hydrostatic pressure, but this coupling does not occur in a biaxial tension test experiment because the hydrostatic pressure is zero (since the out-of-plane stress is zero). For this reason, the strain energy function is separated into individual expressions in terms of the stretch along each of the material axes. It is also evident from the biaxial tests that the axial stress is very low for small axial strains, but increases rapidly as the strain approaches the limiting strain for that axis. These characteristics, microstructural observations and biaxial test results have been encapsulated in the *pole–zero strain energy function* for myocardium given by:

$$
\begin{aligned}
W = {} & k_{11}\frac{E_{11}^2}{|a_{11}-E_{11}|^{b_{11}}} + k_{22}\frac{E_{22}^2}{|a_{22}-E_{22}|^{b_{22}}} + k_{33}\frac{E_{33}^2}{|a_{33}-E_{33}|^{b_{33}}} \\
& + k_{12}\frac{E_{12}^2}{|a_{12}-E_{12}|^{b_{12}}} + k_{13}\frac{E_{13}^2}{|a_{13}-E_{13}|^{b_{13}}} + k_{23}\frac{E_{23}^2}{|a_{23}-E_{23}|^{b_{23}}},
\end{aligned}
\tag{16}
$$

where the constitutive parameters (a's, b's and k's) have the following interpretations: The *limiting strains* or *poles*, denoted $a_{\alpha\beta}$, are physical properties of the tissue that may be measured directly from microstructural observations. In particular, [22] used elastica theory on the collagen helices surrounding myofibres to determine the yield strain (pole) of $a_{11} = 0.52$ along the fibre axis. Alternatively, these yield strains can be estimated by fitting the model directly to experimental stress-strain data as is done, for example, by [33]. The second group of parameters, denoted $b_{\alpha\beta}$, are related to the curvature of the uniaxial stress-strain relationships for each mode of deformation and have been estimated using the biaxial tension test results of [33]. Lastly, the $k_{\alpha\beta}$ parameters weight the contribution of the corresponding mode of deformation to the total strain energy of the material. Estimation of these coefficients is discussed below.

The constitutive parameters of equation (16) are naturally split into six groups, one for each mode of deformation. These groups correspond to the six independent components of Green's strain tensor. The first three terms in equation (16) refer to the three *axial* modes of deformation (fibre, sheet and sheet-normal, denoted 11, 22 and 33, respectively). The parameters associated with these terms have been estimated using a combination of microstructural observations, biaxial tension test results and non-invasive magnetic resonance imaging data, and are listed in Table I. The remaining terms relate to modes of *shear* deformation between the microstructural axes (fibre/sheet, fibre/sheet-normal and sheet/sheet-normal, denoted by subscripts 12, 13 and 23, respectively).

Equation (16) can be regarded as a first order approximation of a power series in the pole–zero terms. A more comprehensive description would include cross-product terms reflecting coupling between different modes of axial and shear deformation and may be warranted following further experimental testing (it cannot be justified on the basis of the biaxial experiments).

It is reasonable to assume that the cellular structures responsible for resisting the shearing deformations are exactly those structures responsible for limiting axial deformations. These load-bearing connections are simply the collagen struts that link individual cardiac fibres and sheets. The important implication of this assumption is that the parameters of the shear terms in equation (16) are strongly correlated to the parameters of the axial terms. The fibre distribution model described below has been used to help understand and quantify some of these parameter correlations.

Table I. Material properties of myocardium for the pole–zero constitutive law.

Type	Axial properties		Shear properties	
Coefficients	k_{11}	1.937	k_{12}	1.0
	k_{22}	0.028	k_{13}	1.0
	k_{33}	0.310	k_{23}	1.0
Poles	a_{11}	0.523	a_{12}	0.731
	a_{22}	0.681	a_{13}	0.731
	a_{33}	1.037	a_{23}	0.886
Curvatures	b_{11}	1.351	b_{12}	2.0
	b_{22}	5.991	b_{13}	2.0
	b_{33}	0.398	b_{23}	2.0

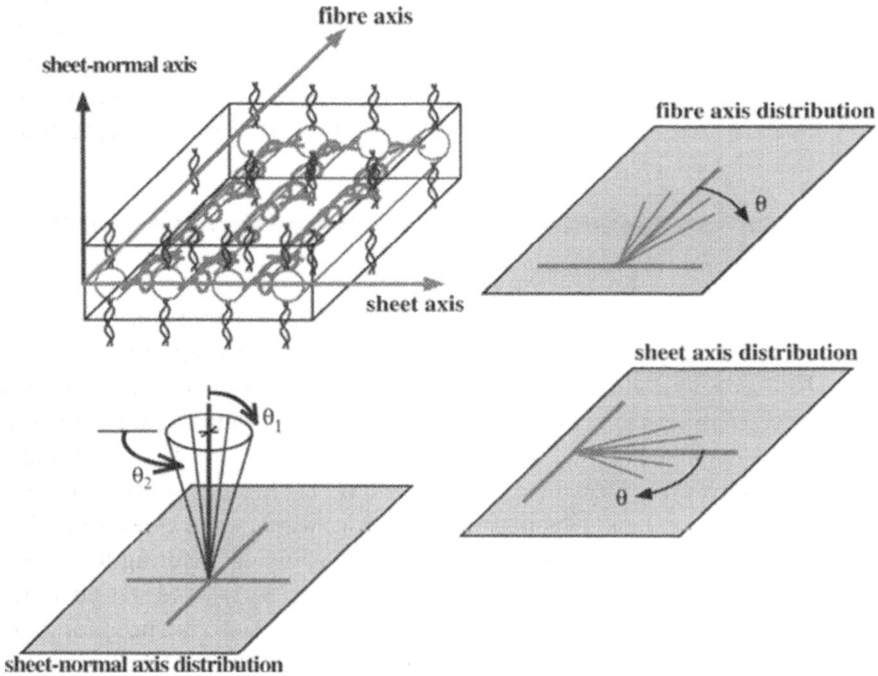

Figure 7. The fibre distribution model: orientation of each fibre family about its mean orientation.

Experiments involving shear deformations of cardiac tissue are currently being performed to further validate these correlations [7, 8] and further comments on measuring these shearing characteristics are given in Section 8.

5.3. A FIBRE DISTRIBUTION MODEL FOR CARDIAC TISSUE

A biophysical model of cardiac muscle elasticity is presented here to help understand the correlation between the axial and shear parameters of the pole–zero constitutive law for myocardium [12]. The main assumption of this fibre distribution model is that three families of fibrous connective tissue (mainly collagen) are responsible for storage of the total strain energy of the myocardium. This implies that some axial and shear deformations must be strongly correlated since they involve the same underlying collagen microstructure. The fibre orientations within each family are assumed to be normally distributed about a mean direction, which is aligned with one of the microstructural material axes (see Figure 7). Note that in the following description the term 'fibre' refers to a collagen connection within a fibre family and not a cardiac muscle fibre. The latter will be referred to as a 'myocyte'.

The first fibre family consists of the large coiled perimysial fibres that are closely associated with the myocytes [21, 31]. The mean direction of this family is

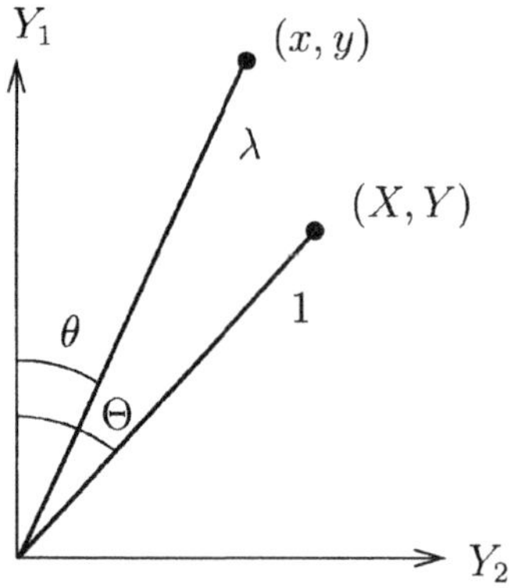

Figure 8. Kinematic analysis of a typical deforming fibre.

coincident with the longitudinal axis of the local myocytes and individual collagen fibres are assumed to lie in the plane of the myocardial sheet. The second family has a mean orientation centred about the sheet axis (which also lies in the sheet plane, but is perpendicular to the myocyte axis) and consists of tightly bound endomysial collagen [2]. The third family of fibres is assumed to have an axisymmetric distribution about a mean direction aligned with the local sheet-normal axis. This family consists of the sparse array of perimysial collagen struts that connect the myocardial sheets.

The variation of connective tissue fibre orientations about their mean directions is assumed to be different for each family of fibres. These variations are defined by standard deviations that describe the distribution of each family of fibres, as illustrated in Figure 7. The first standard deviation defines the variation of the large coiled perimysial collagen fibres about the mean myocyte axis and is therefore relatively small. The second standard deviation defines the variation of the direction of in-sheet endomysial collagen about the local sheet axis and is greater than the first. Two further standard deviations define the axisymmetric variation of the inter-sheet collagen strut direction.

To evaluate the contribution that one particular fibre of a family makes to the total strain energy, consider a unit length fibre in the reference state at an angle of Θ to the Y_1-axis, as illustrated in Figure 8. The Y_1 and Y_2 axes are not material axes (they do not change with material deformation), but rather are local *orthogonal* reference axes with the Y_1 coordinate defined to be aligned with one of the microstructural material axes. In the deformed state, the fibre has length λ and is oriented at an angle of θ to the Y_1-axis. In the undeformed state $X = \cos\Theta$ and $Y = \sin\Theta$, and in the deformed state $x = \lambda\cos\theta$ and $y = \lambda\sin\theta$.

Consider now a particular state of strain in the tissue, characterised by extension ratios λ_1 and λ_2 along the local reference axes Y_1 and Y_2, respectively. These extension ratios can be expressed using:

$$\lambda_1 = \frac{x}{X} = \lambda\frac{\cos\theta}{\cos\Theta}, \qquad \lambda_2 = \frac{y}{Y} = \lambda\frac{\sin\theta}{\sin\Theta}. \tag{17}$$

An expression for the angle of the deformed fibre, θ, is obtained by dividing the extension ratios as follows:

$$\frac{\lambda_2}{\lambda_1} = \frac{\tan\theta}{\tan\Theta} \quad \text{or} \quad \theta = \tan^{-1}\left(\frac{\lambda_2}{\lambda_1}\tan\Theta\right). \tag{18}$$

Thus, given λ_1 and λ_2, a fibre at initial position Θ is rotated to an angle θ and stretched by an extension ratio λ as follows:

$$\lambda = \begin{cases} \lambda_1 \dfrac{\cos\Theta}{\cos\theta}, & \text{when } \theta < \dfrac{\pi}{2}, \\ \lambda_2 \sin\Theta, & \text{when } \theta = \dfrac{\pi}{2}. \end{cases} \tag{19}$$

The fibre strain is calculated from the extension ratio using

$$E_{\mathrm{f}} = \frac{1}{2}(\lambda^2 - 1). \tag{20}$$

The next step is to assume that the fibre orientations for each of the families are normally distributed about their mean directions. In this way, for example, the family of fibres associated with the myocyte axis may be approximated using the Gaussian probability distribution:

$$p_1(\Theta) = \frac{1}{\sqrt{2\pi}\sigma_1} \exp\left[-\frac{1}{2}\frac{\Theta^2}{\sigma_1^2}\right], \tag{21}$$

where σ_1 is the standard deviation. Note that the mean of this distribution is aligned with the local myocyte axis in the reference state and that Θ quantifies the *difference* between the direction of a particular fibre and the mean fibre direction.

The total strain energy (due to the deformation) stored in the family of fibres associated with the myocyte axis may be computed by summing up the individual strain energies of all fibres in the family. Equation (22) expresses this sum as the integral over all possible fibres since the probability distribution function varies continuously with the undeformed position Θ. In this expression, k_1, a_1 and b_1 are properties of the family of fibres associated with the myocyte axis. The dependence of the fibre strain E_{f} on Θ is defined using equations (18)–(20)

$$W_1 = \int_{-\pi/2}^{\pi/2} p_1(\Theta) \frac{k_1 E_{\mathrm{f}}^2}{(a_1 - E_{\mathrm{f}})^{b_1}} \, \mathrm{d}\Theta. \tag{22}$$

In a similar manner, the total strain energy stored in the family of fibres associated with the myocardial sheet axis may be calculated using

$$W_2 = \int_{-\pi/2}^{\pi/2} p_2(\Theta) \frac{k_2 E_{\mathrm{s}}^2}{(a_2 - E_{\mathrm{s}})^{b_2}} \, \mathrm{d}\Theta, \tag{23}$$

where $p_2(\Theta)$ is the Gaussian probability distribution function for the family of fibres associated with the sheet axis, k_2, a_2 and b_2 are properties of this family,

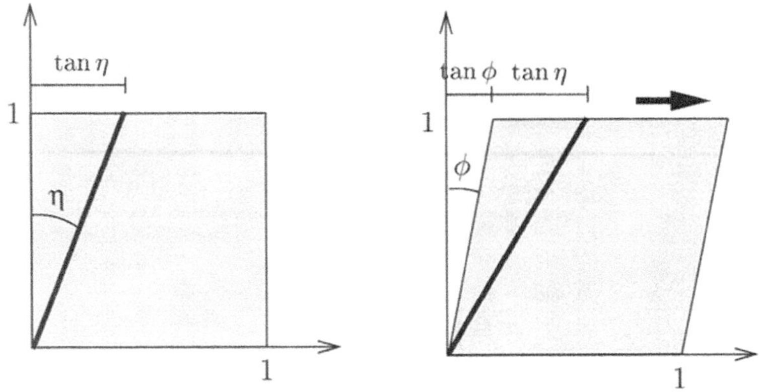

Figure 9. Kinematic analysis of a fibre during simple shear.

and E_s is the material strain along the sheet axis. Note that E_s and $p_2(\Theta)$ may be evaluated using expressions similar to equations (20) and (21), respectively.

The strain energy for the third family of sheet-normal fibres is calculated using equation (24), and the probability distribution function for this family is expressed in equation (25) in terms of the two standard deviations that describe an axisymmetric variation of inter-sheet collagen fibre orientations.

$$W_3 = \int_{\Theta_4=0}^{2\pi} \int_{\Theta_3=0}^{\pi/2} p_3(\Theta_3, \Theta_4) \frac{k_3 E_n^2}{(a_3 - E_n)^{b_3}} \, \mathrm{d}\Theta_3 \, \mathrm{d}\Theta_4 \tag{24}$$

with

$$p_3(\Theta_3, \Theta_4) = \frac{1}{2\pi \sigma_3 \sigma_4} \exp\left[-\frac{1}{2}\left(\frac{\Theta_3^2}{\sigma_3^2} + \frac{\Theta_4^2}{\sigma_4^2}\right)\right], \tag{25}$$

where E_n is the material strain along the family of collagen fibres associated with the sheet-normal axis and k_3, a_3 and b_3 are properties of this family. Finally, it is assumed that the combined strain energies from each of the three families sum to yield the total strain energy in the tissue.

For present purposes, the fibre distribution model is used to express the limiting strains for shear (namely, a_{12}, a_{13}, and a_{23} in equation (16)) as a function of the axial poles, since it was assumed that the same underlying distributions of collagen connections determine both the tensile and shear characteristics of the tissue. This relationship is derived by considering the kinematics of a typical fibre during a simple shear deformation as shown in Figure 9.

The bold line segment in Figure 9 represents a particular connective tissue fibre oriented at angle η to the mean direction for its family in the reference state. This undeformed fibre has length $\sqrt{1 + \tan^2\eta} = \sec\eta$. During the deformation the fibre moves through a shear angle of ϕ and due to the simple kinematics of the deformation, the deformed fibre length is $\sqrt{1 + (\tan\eta + \tan\phi)^2}$. The extension ratio (deformed length divided by undeformed length) of the fibre is defined in equation (26) as a function of the undeformed fibre angle η and the shear angle ϕ

$$\lambda_\eta = \sqrt{1 + \tan^2\phi \cos^2 \eta + \tan\phi \sin 2\eta}. \tag{26}$$

For a given shear angle ϕ, the fibre angle η^* that produces maximum stretch is found from equation (26) by solving $\partial\lambda_\eta/\partial\eta = 0$ for η^*, resulting in

$$\eta^* = \frac{1}{2}\tan^{-1}(2\cot\phi). \tag{27}$$

As ϕ increases from 0° to 90°, η^* decreases from 45° to 0°. The extension ratio of the fibre with maximum stretch is determined by substituting equation (27) back into equation (26). The maximum extension ratio for a given shear angle is

$$\lambda_{\max} = \frac{1}{2}\left(\kappa + \sqrt{4 + \kappa^2}\right), \tag{28}$$

where $\kappa = \tan\phi$. If this particular fibre yields when its Green's strain reaches the limit $a = \frac{1}{2}(\lambda_{\max}^2 - 1)$, then using some considerable manipulation (see [27]) the maximum possible *elastic* shear strain is

$$\kappa = \frac{2a}{\sqrt{1 + 2a}}. \tag{29}$$

The key point here is that the shear poles of equation (16) (namely, a_{12}, a_{13} and a_{23}) may be directly determined from the limiting strains of the fibre families. For example, consider simple shearing deformations within the plane of the myocardial sheet, referred to here as the $(1, 2)$-plane. The yield strain for a simple shear of the $(1, 2)$-plane in the direction of the myocyte axis (a 2–1 shear) is limited by the *sheet axis pole position* a_{22} since the collagen connections associated with the sheet axis family are put into tension. On the other hand, a 1–2 shear is limited by the *fibre axis pole position* a_{11} since the collagen fibres aligned with the myocyte axis sustain the load. Thus for a general shear of the $(1, 2)$-plane, a reasonable approximation to the limiting shear strain a_{12} may be determined by substituting the minimum of a_{11} and a_{22} into equation (29), which monotonically increases with a. The pole position for the in-plane $(1, 2)$ shear is defined using:

$$a_{12} = \begin{cases} \dfrac{2a_{22}}{\sqrt{1 + 2a_{22}}} & \text{if } a_{22} \leqslant a_{11}, \\[2ex] \dfrac{2a_{11}}{\sqrt{1 + 2a_{11}}} & \text{if } a_{22} > a_{11}. \end{cases} \tag{30}$$

Pole positions for the other shear terms may be determined in an analogous manner.

It remains then to estimate the coefficients and curvature parameters for the shear terms in equation (16). This can be achieved by applying the fibre distribution model to a range of kinematically simple experiments, which involve various axial and shear deformations. The relative contributions to the strain energy of the tissue can then be used to estimate these unknown parameters (see [27]). Table I lists the material properties used for the modelling in the next section.

6. Passive Inflation during Ventricular Diastole

As an example of applying the constitutive laws discussed above in a finite element model of the intact myocardium, the model is solved here for passive inflation by pressure boundary conditions applied to the left and right ventricular endocardial surfaces of the computational model, illustrated in Figure 10.

Passive ventricular function has been quantified using a variety of global mechanical indices, including the diastolic cavity pressures and volumes, longitudinal elongation, and axial twist of the base relative to the apex. Regional function has been quantified using diastolic strain distributions referred to cardiac wall and fibre coordinate systems (see [38]).

6.1. DIASTOLIC CAVITY PRESSURE AND VOLUME VARIATIONS

The global properties of the diastolic LV have been commonly characterised by the cavity pressure–volume relation. Many studies have quantified this relationship using various techniques in isolated, supported and *in-vivo* hearts. Three canine studies have been selected to assess the accuracy of the pressure–volume relationship predicted by the ventricular mechanics model. For comparison purposes, the ventricular volume has been transformed into a percentage volume change relative to the volume of the unloaded cavity, for which the cavity pressure is zero.

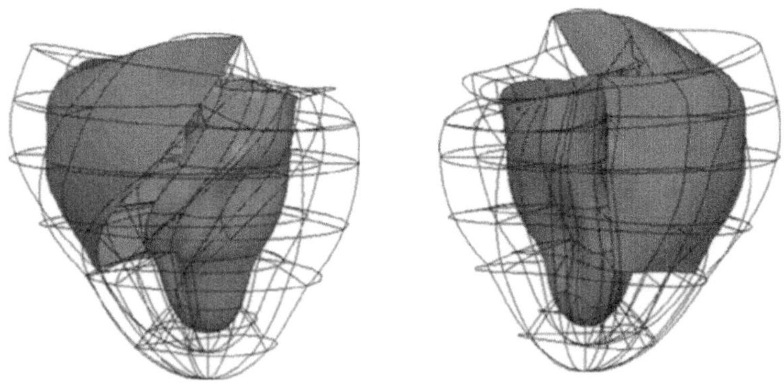

(a) Anterior view. (b) Posterior view.

Figure 10. Computational model of the cardiac ventricles showing LV and RV endocardial surfaces (shaded) and boundaries of the FE mesh, which consists of 120 tricubic Hermite elements.

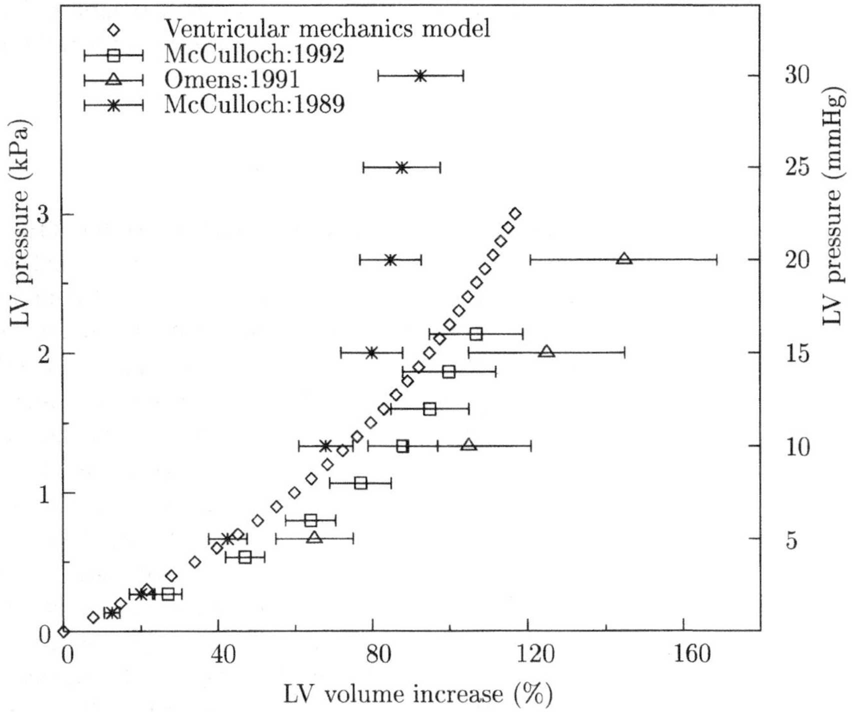

Figure 11. Diastolic pressure–volume relations for the canine LV. The ventricular mechanics model (◊) predicts realistic diastolic stiffening of the LV compared to three experimental studies, all from potassium arrested dog hearts subject to static pressure loading. Error bars show standard deviations.

The ventricular mechanics model was passively inflated to LV and RV pressures of 3.0 kPa (22.5 mmHg) and 0.6 kPa (4.5 mmHg), respectively, using thirty equal load steps. The LV volume was computed at each step, for which the model had converged to an error tolerance of 10^{-3}. Volume changes were normalised by the unloaded LV volume of 32 ml. The predicted normalised passive pressure–volume relationship is represented by diamonds (◊) in Figure 11. It is clear from this comparison that the ventricular mechanics model predicted sufficiently realistic diastolic pressure–volume characteristics.

In seven isolated potassium-arrested canine LVs, [30] quantified 3D principal strains in the equatorial region of the anterior midwall using biplane radiography of three transmural columns of radiopaque beads. Mean strains were reported at four normalised volume changes, which corresponded to approximately 5 ml increments in LV volume. The corresponding LV pressures were estimated from [30] and the authors commented that the angle between the radial axis and the principal axis of greatest thinning (E_3) was $<1°$. Observations from this study are represented by box symbols (□) in Figure 12.

In comparison, the ventricular mechanics model (◊ in Figure 12) predicts realistic wall thinning (E_3) and minimum in-plane (E_2) strains at the midwall of

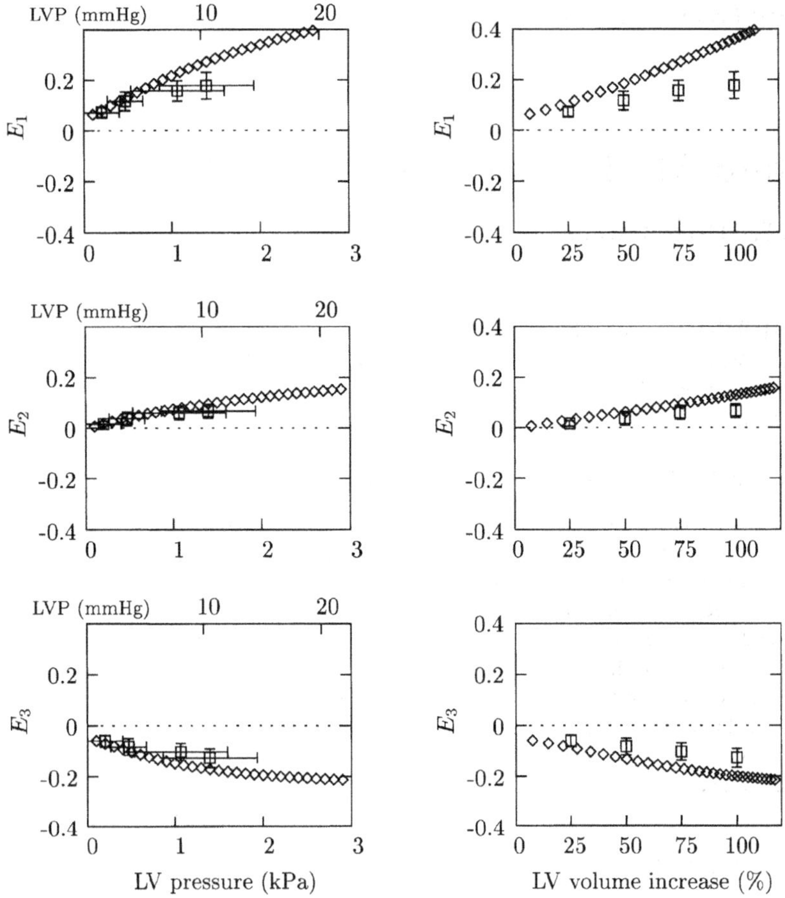

Figure 12. 3D principal strains (E_1, E_2 and E_3) versus LV pressure (LVP) and volume at the midwall of the anterior equatorial region during diastole. Ventricular mechanics model predictions (◇) are compared to the experimentally observed midwall principal strains (mean ± SD, $n = 7$) of [30]. See text for explanation.

the anterior LV. However, the predicted maximum principal strain (E_1) illustrates more compliance of the ventricular mechanics model as opposed to the isolated potassium-arrested hearts. More realistic predictions would possibly be achieved if the model had accounted for the heterogeneous material properties of ventricular myocardium.

7. End-Systolic Principal Strains

7.1. ACTIVE CONTRACTION OF MYOCARDIUM

When stimulated, cardiac muscle fibres generate contractile forces. For present purposes, it is assumed that cardiac muscle fibres only generate force in the direction of their longitudinal axes (although models that include active force development tranverse to the mean fibre axis can lead to more accurate 3D systolic shear strain predictions [38], but this remains to be verified experimentally). This assumption means that just one term must be added to the passive 3D stress tensor of equation (15) to model the active behaviour of myocardium. The additional active stress acts in the muscle fibre direction (aligned with the ν_1-coordinate), and so if the stress tensor is expressed with respect to the microstructural material axes, it is in fact only the T^{11} component that is modified thus:

$$T^{\alpha\beta} = \frac{1}{2}\left(\frac{\partial W}{\partial E_{\alpha\beta}} + \frac{\partial W}{\partial E_{\beta\alpha}}\right) - pa^{\alpha\beta}_{(\nu)} + Ta^{11}\delta^{\alpha}_{1}\delta^{\beta}_{1}, \tag{31}$$

where $T = T(t, \lambda_{11}, [\mathrm{Ca}^{2+}]_i)$ is the active tension generated by a fibre at time t. For the current modelling, active tension is defined to depend on the time varying muscle fibre extension ratio, $\lambda_{11} = \sqrt{2E_{11} + 1}$, and the concentration of free intracellular calcium, $[\mathrm{Ca}^{2+}]_i$ which is taken to characterise the *level of activation* of a cardiac muscle cell. It is assumed here that the transverse and shear strains have no effect on the active tension generated by the fibres. The steady state mechanical properties of active myocardium are summarised below at a constant level of activation. For more detailed descriptions refer to [11].

The majority of experimental studies that have quantified end-systolic distributions of strain have used the end-diastolic configuration as the reference state. Three such studies have been selected to assess the accuracy of end-systolic principal strain distributions predicted by the ventricular mechanics model. In these studies, E_1 represents maximum shortening and E_3 represents maximum stretch, which is generally associated with wall thickening. ϕ_1 represents the angle to the axis of principal shortening in the plane of the ventricular wall, where positive angles signifies anticlockwise rotations from the circumferential direction.

[42] imaged columns of radiopaque markers implanted in the anterior equatorial LV free wall of seven open-chest dogs. End-systolic 3D principal strains were computed with respect to the end-diastolic reference state. The peak systolic and end-diastolic LV pressures were 16.6 ± 2.5 kPa (125 ± 19 mmHg) and 0.6 ± 0.2 kPa (4.7 ± 1.5 mmHg), respectively. Observations from this study are represented by crosses ($\times$) in Figure 13.

Using similar methods, [39] measured transmural distributions of end-systolic 3D principal strains and directions referred to the end-diastolic state in the anterior equatorial LV free wall of seven open-chest dogs. The peak systolic LV pressure was 16.1 ± 2.9 kPa (121 ± 22 mmHg) and left ventricular end-diastolic pressure (LVEDP) was 0.3 ± 0.2 kPa (2.3 ± 1.5 mmHg). Observations from this study are represented by triangles ($\triangle$) in Figure 13.

[25] analysed the experimental results from a set of bead studies by [41] using non-homogeneous strain analysis. End-systolic 3D principal strains referred to the end-diastolic state were computed for the anterior LV free wall of six open-chest dogs. The peak systolic LV pressure was 15.5 ± 2.8 kPa (116 ± 21 mmHg)

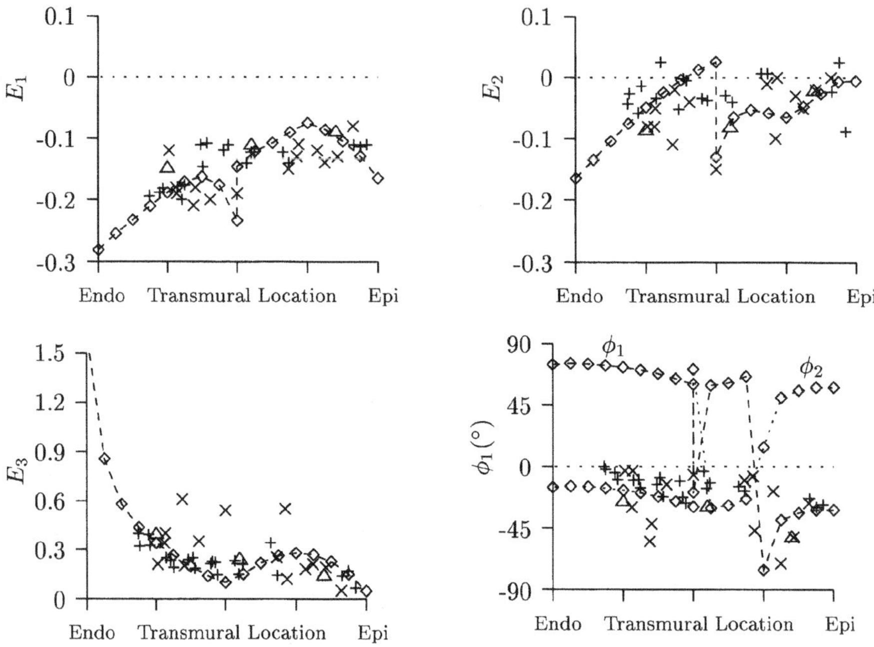

Figure 13. Transmural distributions of 3D principal strain at end-systole (referred to the end-diastolic state) for the equatorial region of the anterior wall. Ventricular mechanics model predictions (◊) are compared to experimental observations from [39] (△, $n = 7$), [42] (×, $n = 7$) and [41] (+, $n = 5$). The predicted in-plane angle to the second principal strain (ϕ_2) is shown for comparison purposes (see text for details).

and LVEDP was 0.7 ± 0.3 kPa (5 ± 2 mmHg). Observations from this study are represented by plus symbols (+) in Figure 13.

End-systolic strains for the anterior equatorial LV wall computed using the ventricular mechanics model (represented by diamonds (◊) in Figure 13) were referred to the predicted end-diastolic state for comparison with the experimental studies. Reasonable predictions were produced for the maximum shortening (E_1) and thickening (E_3) strains, although the large subendocardial thickening was unrealistic. There are two aspects of myocardial mechanics that are not currently included in the model, which would reduce the unreasonably large subendocardial thickening:

(1) the higher density of collagen fibres in the subendocardium, which would provide a greater constraint on wall thickening; and
(2) the compressibility of the subendocardium due to the movement of blood from subendocardium to subepicardium during systole.

Discontinuities in the distributions of E_1 and E_2 were due to the element interface in the LV midwall and may have revealed a need to either refine the model transmurally or increase the order of the interpolation scheme within the wall plane.

The predicted subepicardial in-plane angle to maximum shortening (ϕ_1) was realistic, but for deeper wall locations there are fundamental differences between the predicted and reported first principal angles. At approximately 25% of the wall thickness below the epicardium, E_1 and E_2 reach similar magnitudes and there is a marked change in the predicted principal angle ϕ_1 (this is not surprising, since the principal strains and axes are arbitrarily ranked in order of strain magnitude). Interestingly, for the inner three-quarters of the wall the predicted in-plane angle to the second principal strain (ϕ_2) shows reasonable agreement with the reported ϕ_1 measurements. This is not the case at the midwall, however, where the magnitudes of E_1 and E_2 are again very similar and the principal axes seem to switch orientations. For further discussion of the role of the fibre-sheet architecture in myocardial shear deformation and ventricular wall thickening, see [20].

8. Critique of the Ventricular Mechanics Model

In this section we critique the current model and indicate where we believe further developments could improve the model and hence our understanding of ventricular mechanics.

1. *Anatomical model.* The model does not account for myocardial fibres that lie oblique to the ventricular wall surfaces, since the imbrication angle is assumed to be zero at all locations. The computational framework has been developed to incorporate an imbrication angle into the anatomical description, but the ventricular mechanics model awaits further experimental work to characterise the spatial distribution of imbrication angles. This places doubt on model predictions at locations near the apex, where imbrication angles have been reported to descend steeply into the ventricular wall [34]. However, it is unclear at present whether a more realistic representation of the fibre angle distribution in the apical region will greatly influence the global mechanical behaviour of the ventricular myocardium.
2. *Finite element model.* The chosen combination of high order interpolation functions and FE mesh resolution produced sufficiently accurate strain distributions at all locations, except those near the apex. Apical elements of the present ventricular mechanics model must be refined in the longitudinal direction if these locations are of particular interest. Effects of apical mesh refinement have yet to be quantified, but it is unlikely to significantly alter the global ventricular mechanics.
3. *Passive elastic material response.* Several issues must be addressed regarding the passive constitutive law for myocardium (see Section 5.2). The pole–zero law is primarily based on *in-vitro* biaxial tension tests of thin sections of ventricular myocardium. The pole–zero axial weighting coefficients $k_{\alpha\alpha}$, derived from these experiments, may not be appropriate for *in-vivo* mechanics of the ventricles. These constitutive properties may be estimated using *in-vivo* recordings of ventricular deformation and cavity pressures.

   Further research is required to characterise the shear material response of ventricular myocardium. To reconcile the shear response in terms of tissue structure, Section 5.3 introduced a fibre distribution model for cardiac tissue. Further work is required to interpret shear deformations using this model and to use it to estimate the shear weighting coefficients of the pole–zero law. The shearing behaviour of small cubes of myocardial tissue have recently been measured under various degrees of tissue compression [7, 8] and work is currently underway to fit these results with the pole–zero model and, in particular, to determine whether the nonlinear coupling terms proposed as an extension to the power series in equation (16) are needed.

Finally, the pole–zero law does not provide a suitable description of the compressive response of cardiac tissue. Further experimental studies are required to more accurately model compressive myocardial response.

4. *Regional variation of material properties.* For present purposes, the model incorporated homogeneous material properties throughout the myocardium. Recent microstructural observations suggest that this is an oversimplification of ventricular wall properties. For example, the extent of branching between myocardial sheets changes across the ventricular wall [19]. This implies that the mechanical stiffness along the sheet-normal axis also varies transmurally. In addition, the transmural variation of collagen density has been measured for rat hearts [47], but the appropriate relationship between collagen density and mechanical stiffness remains to be established. Once quantified experimentally, spatially varying material properties could be readily incorporated into the ventricular mechanics model using standard FEM fitting and interpolation techniques. A likely benefit of incorporating stiffer subendocardial tissue properties (resulting from the higher than average density of collagen fibres in this region) may be to reduce the excessive subendocardial wall thickening currently predicted by the model.
5. *3D distributions of residual strain.* The present model accounts for the transmural residual fibre strain distribution, but does not incorporate other components of residual strain. [5] recently quantified 3D distributions of residual strain with respect to the anatomical fibre coordinates. This information could be readily incorporated into the ventricular mechanics model using the growth tensor described by [32].
6. *Myocardial fluid shifts and viscoelasticity.* It is well established that the ventricles stiffen with increased intracoronary blood volume [40]. [13] used an axisymmetric poroelastic model of the LV to illustrate the increase in predicted intramyocardial pressure when redistribution of intracoronary blood was suppressed. Furthermore, [44] used a model that included extracellular fluid flow, which was concluded to account for the insensitivity of experimentally observed passive stress-strain curves to loading rate and of stress-relaxation curves to the amount of stretch. To account for regional myocardial compressibility via changes in vascular volume, we have formulated a simple model of myocardial fluid shift [27]. However, the present study made no provision for the poroelastic or viscoelastic nature of ventricular myocardium, but such con-

siderations would be essential to study the time-varying mechanical response of the ventricles.

7. *Boundary constraints.* A model of the pericardial sac needs to be incorporated into the ventricular mechanics model to restrict filling [36, 37]. Moreover, it is clear that the atria, basal skeleton and chordae tendineae play important roles in restricting the motion of the adjacent portions of ventricular myocardium [35].
8. *Active material response.* For present purposes, a simple steady-state model of the active material response of ventricular myocardium was sufficient to quasi-statically simulate the systolic phase of the heart cycle. To account for the time-varying nature of ventricular activation on heart mechanics, a dynamic model of the active myocardial material response must be incorporated. The computational framework has been formulated to use the time dependent fading memory model for force generation of cardiac muscle fibres [11].

Acknowledgement

Our thanks to David Nickerson for his help in preparing this manuscript.

References

1. R.J. Atkin and N. Fox, *An Introduction to the Theory of Elasticity*. Longman, London (1980).
2. J.B. Caulfield and T.K. Borg, The collagen network of the heart. *Lab. Invest.* **40**(3) (1979) 364–372.
3. K.D. Costa, P.J. Hunter, J.M. Rogers, J.M. Guccione, L.K. Waldman and A.D. McCulloch, A three-dimensional finite element method for large elastic deformations of ventricular myocardium: Part I – Cylindrical and spherical polar coordinates. *ASME J. Biomech. Engrg.* **118**(4) (1996) 452–463.
4. K.D. Costa, P.J. Hunter, J.M. Rogers, J.M. Guccione, L.K. Waldman and A.D. McCulloch, A three-dimensional finite element method for large elastic deformations of ventricular myocardium: Part II – Prolate spherical coordinates. *ASME J. Biomech. Engrg.* **118**(4) (1996) 464–472.
5. K.D. Costa, K. May-Newman, D. Farr, W.G. O'Dell, A.D. McCulloch and J.H. Omens, Three-dimensional residual strain in midanterior canine left ventricle. *Amer. J. Physiol.* **273** (1997) H1968–H1976.
6. L.L. Demer and F.C.P. Yin, Passive biaxial mechanical properties of isolated canine myocardium. *J. Physiol. London* **339** (1983) 615–630.
7. S. Dokos, B.H. Smaill, A.A. Young and I.J. Le Grice, Anisotropic shear properties of passive ventricular myocardium. *Amer. J. Physiol.* (2001) (submitted).
8. S. Dokos, A.A. Young, B.H. Smaill and I.J. Le Grice, A triaxial-measurement shear-test device for soft biological tissues. *ASME J. Biomech. Engrg.* (2001) (in press).
9. P. Gould, D. Ghista, L. Brombolich and I. Mirsky, *In vivo* stresses in the human left ventricular wall: Analysis accounting for the irregular 3-dimensional geometry and comparison with idealised geometry analyses. *J. Biomech.* **5** (1972) 521–539.
10. J.D. Humphrey, R.K. Strumpf and F.C.P. Yin, Determination of a constitutive relation for passive myocardium: II. Parameter estimation. *ASME J. Biomech. Engrg.* **112** (1990) 340–346.
11. P.J. Hunter, A.D. McCulloch and H.E.D.J. ter Keurs, Modelling the mechanical properties of cardiac muscle. *Prog. Biophys. Molec. Biol.* **69** (1998) 289–331.

12. P.J. Hunter, M.P. Nash and G.B. Sands, Computational electromechanics of the heart. In: A.V. Panfilov and A.V. Holden (eds), *Computational Biology of the Heart*. Wiley, West Sussex, England (1997) Chapter 12, pp. 345–407.
13. J.M. Huyghe, T. Arts, D.H. van Campen and R.S. Reneman, Porous medium finite element model of the beating left ventricle. *Amer. J. Physiol.* **262** (Heart Circ. Physiol. 31) (1992) H1256–H1267.
14. J.M. Huyghe, D.H. van Campen, T. Arts and R.M. Heethaar, A two-phase finite element model of the diastolic left ventricle. *J. Biomech.* **24**(7) (1991) 527–538.
15. R.F. Janz and A.F. Grimm, Finite-element model for the mechanical behaviour of the left ventricle: Prediction of deformation in the potassium-arrested rat heart. *Circ. Res.* **30** (1972) 244–252.
16. R.F. Janz and A.F. Grimm, Deformation of the diastolic left ventricle. I. Nonlinear elastic effects. *Biophys. J.* **13** (1973) 689–704.
17. A.M. Katz, *Physiology of the Heart*, 2nd ed. Raven Press, New York (1992).
18. P. Kohl, P. Hunter and D. Noble, Stretch-induced changes in heart rate and rhythm: Clinical observations, experiments and mathematical models. *Prog. Biophys. Molec. Biol.* **71**(1) (1999) 91–138.
19. I.J. Le Grice, B.H. Smaill, L.Z. Chai, S.G. Edgar, J.B. Gavin and P.J. Hunter, Laminar structure of the heart: Ventricular myocyte arrangement and connective tissue architecture in the dog. *Amer. J. Physiol.* **269**(38) (1995) H571–H582.
20. I.J. Le Grice, Y. Takayama and J.W. Covell, Transverse shear along myocardial cleavage planes provides a mechanism for normal systolic wall thickening. *Circ. Res.* **77** (1995) 182–193.
21. D.A. MacKenna, J.H. Omens and J.W. Covell, Left ventricular perimysial collagen fibers uncoil rather than stretch during diastolic filling. *Basic Res. Cardiol.* **91**(2) (1996) 111–122.
22. D.A. MacKenna, J.H. Omens, A.D. McCulloch and J.W. Covell, Contributions of collagen matrix to passive left ventricular mechanics in isolated rat hearts. *Amer. J. Physiol.* **266** (1994) H1007–H1018.
23. L.E. Malvern, *Introduction to the Mechanics of a Continuous Medium*. Prentice-Hall, Englewood Cliffs, NJ (1969).
24. A.D. McCulloch, Cardiac biomechanics. In: J.D. Bronzino (ed.), *The Biomedical Engineering Handbook*. CRC Press, Boca Raton, FL (1995) Chapter 31, pp. 418–439.
25. A.D. McCulloch and J.H. Omens, Non-homogeneous analysis of three-dimensional transmural finite deformation in canine ventricular myocardium. *J. Biomech.* **24**(7) (1991) 539–548.
26. I. Mirsky, Ventricular and arterial wall stresses based on large deformation analyses. *Biophys. J.* **13**(11) (1973) 1141–1159.
27. M.P. Nash, Mechanics and material properties of the heart using an anatomically accurate mathematical model. PhD Thesis, The University of Auckland, New Zealand (1998).
28. P.M.F. Nielsen, I.J. Le Grice, B.H. Smaill and P.J. Hunter, Mathematical model of geometry and fibrous structure of the heart. *Amer. J. Physiol.* **260** (Heart Circ. Physiol. 29) (1991) H1365–H1378.
29. J.T. Oden, *Finite Elements of Nonlinear Continua*. McGraw-Hill, New York (1972).
30. J.H. Omens, K.D. May and A.D. McCulloch, Transmural distribution of three-dimensional strain in the isolated arrested canine left ventricle. *Am. J. Physiol.* **261** (Heart Circ. Physiol. 30) (1991) H918–H928.
31. T.F. Robinson, M.A. Geraci, E.H. Sonnenblick and S.M. Factor, Coiled perimysial fibers of papillary muscle in rat heart: Morphology, distribution and changes in configuration. *Circ. Res.* **63** (1988) 577–592.
32. E.K. Rodriguez, A. Hoger and A.D. McCulloch, Stress-dependent finite growth in soft elastic tissues. *J. Biomech.* **27**(4) (1994) 455–467.
33. B.H. Smaill and P.J. Hunter, Structure and function of the diastolic heart: Material properties of passive myocardium. In: L. Glass, P.J. Hunter and A.D. McCulloch (eds), *Theory of Heart:*

Biomechanics, Biophysics, and Nonlinear Dynamics of Cardiac Function. Springer, New York (1991) pp. 1–29.

34. D.D. Streeter, Jr., H.M. Spotnitz, D.P. Patel, J. Ross, Jr. and E.H. Sonnenblick, Fibre orientation in the canine left ventricle during diastole and systole. *Circ. Res.* **24** (1969) 339–347.
35. Y. Takayama, I.J. Le Grice, J.W. Holmes and J.W. Covell, Effects of chordal uncoupling on deformation in the papillary muscle. *FASEB J.* **8**(5) (1994) A591.
36. J.V. Tyberg and E.R. Smith, Ventricular diastole and the role of the pericardium. *Herz* **15**(6) (1990) 354–361.
37. G.S. Tyson, Jr., G.W. Maier, C.O. Olsen, J.W. Davis and J.S. Rankin, Pericardial influences on ventricular filling in the conscious dog: An analysis based on pericardial pressure. *Circ. Res.* **54** (1984) 173–184.
38. T.P. Usyk, R. Mazhari and A.D. McCulloch, Effect of laminar orthotropic myofiber architecture on regional stress and strain in the canine left ventricle. *J. Elasticity* **61** (2000) 143–164.
39. F.J. Villarreal, W.Y.W. Lew, L.K. Waldman and J.W. Covell, Transmural myocardial deformation in the ischemic canine left ventricle. *Circ. Res.* **68**(2) (1991) 368–381.
40. W.M. Vogel, C.S. Apstein, L.L. Briggs, W.H. Gaasch and J. Ahn, Acute alterations in left ventricular diastolic chamber stiffness: Role of the erectile effect of coronary arterial pressure and flow in normal and damaged hearts. *Circ. Res.* **51**(4) (1982) 465–478.
41. L.K. Waldman, Y.C. Fung and J.W. Covell, Transmural myocardial deformation in the canine left ventricle: Normal *in vivo* three-dimensional finite strains. *Circ. Res.* **57**(1) (1985) 152–163.
42. L.K. Waldman, D. Nossan, F. Villareal and J.W. Covell, Relation between transmural deformation and local myofiber direction in the canine left ventricle. *Circ. Res.* **63** (1988) 550–562.
43. R.H. Woods, A few applications of a physical theorem to membranes in the human body in a state of tension. *J. Anat. Physiol.* **26** (1892) 362–370.
44. M. Yang and L.A. Taber, The possible role of poroelasticity in the apparent viscoelastic behaviour of passive cardiac muscle. *J. Biomech.* **24**(7) (1991) 587–597.
45. F.C.P. Yin, Ventricular wall stress. *Circ. Res.* **49**(4) (1981) 829–842.
46. F.C.P. Yin, R.K. Strumpf, P.H. Chew and S.L. Zeger, Quantification of the mechanical properties of noncontracting canine myocardium under simultaneous biaxial loading. *J. Biomech.* **20** (1987) 577–589.
47. A.A. Young, I.J. Le Grice, M.A. Young and B.H. Smaill, Extended confocal microscopy of myocardial laminae and collagen network. *J. Microsc.* **192**(2) (1998) 139–150.
48. O.C. Zienkiewicz and R.L. Taylor, *The Finite Element Method. I. Basic Formulation and Linear Problems*, 4th ed. McGraw-Hill, Berkshire, UK (1994).

Journal of Elasticity **61:** 143–164, 2000.

Effect of Laminar Orthotropic Myofiber Architecture on Regional Stress and Strain in the Canine Left Ventricle

T.P. USYK, R. MAZHARI and A.D. MCCULLOCH*
Department of Bioengineering, The Whitaker Institute for Biomedical Engineering, University of California, San Diego, La Jolla, Calif., U.S.A. E-mail: amcculloch@ucsd.edu

Received 20 April 2000; in revised form 12 December 2000

Abstract. Recent morphological studies have demonstrated a laminar (sheet) organization of ventricular myofibers. Multiaxial measurements of orthotropic myocardial constitutive properties have not been reported, but regional distributions of three-dimensional diastolic and systolic strains relative to fiber and sheet axes have recently been measured in the dog heart by Takayama et al. [30]. A three-dimensional finite-deformation, finite element model was used to investigate the effects of material orthotropy on regional mechanics in the canine left ventricular wall at end-diastole and end-systole. The prolate spheroidal model incorporated measured transmural distributions of fiber and sheet angles at the base and apex. Compared with transverse isotropy, the orthotropic model of passive myocardial properties yielded improved agreement with measured end-diastolic strains when: (1) normal stiffness transverse to the muscle fibers was increased tangent to the sheets and decreased normal to them; (2) shear coefficients were increased within sheet planes and decreased transverse to them. For end-systole, orthotropic passive properties had little effect, but three-dimensional systolic shear strain distributions were more accurately predicted by a model in which significant active systolic stresses were developed in directions transverse to the mean fiber axis as well as axial to them. Thus the ventricular laminar architecture may give rise to anisotropic material properties transverse to the fibers with greater resting stiffness within than between myocardial laminae. There is also evidence that intact ventricular muscle develops significant transverse stress during systole, though it remains to be seen if active stress is also orthotropic with respect to the laminar architecture.

Mathematics Subject Classifications (2000): 74B20, 74L15.

Key words: myocardium, myofibers, constitutive equation, and orthotropy.

1. Introduction

Stress and strain distributions in the left ventricular wall are needed for understanding regional ventricular function, and they can have a significant effect on processes such as hypertrophy, remodeling and regional coronary blood flow.

Ventricular geometry has been studied in most quantitative detail in the dog heart [20]. Geometric models have been very useful in the analysis, especially the use of confocal and nonconfocal ellipses of revolution to describe the epicardial

---

* UCSD, Department of Bioengineering 9500 Gilman Drive La Jolla, CA 92093-0412, U.S.A.

and endocardial surfaces of the left ventricular walls. The canine left ventricle is reasonably modeled by a thick ellipsoid of revolution truncated at the base. Using a truncated ellipsoidal model, left ventricular geometry in the dog can be defined by the major and minor radii of two surfaces, the left ventricular endocardium, and a surface defining the free wall epicardium and the septal endocardium of the right ventricle. The cardiac ventricles have a complex three-dimensional muscle fiber architecture [28]. In the human or dog left ventricle, the muscle fiber angle typically varies continuously from about -60° (i.e. 60° clockwise from the circumferential axis) at the epicardium to about $+70^\circ$ at the endocardium. The fibrous architecture of the myocardium has motivated models of myocardial material symmetry as transversely isotropic. The transverse laminae are the first structural evidence for material orthotropy and have motivated the development of a models describing the variation of fiber, sheet and sheet-normal axes throughout the ventricular wall [15].

Regional distributions of three-dimensional deformation in resting ventricular myocardium have been measured from the displacements of markers that are either implanted in the ventricular wall [18, 24] or generated using noninvasive magnetic resonance tagging techniques [22, 23]. However, the direct measurement of local forces or stresses in the intact heart wall has not been reliable [8].

Previous investigations have used various models to compute transmural distributions of three-dimensional strain and stress. Guccione et al. [5, 6] and Humphrey and Yin [9] used cylindrical models, Guccione et al. [7] and Kantor et al. [12, 13] used axisymmetric models, and Costa et al. [3] used a fully nonsymmetric three-dimensional model with a realistic left ventricular geometry and muscle fiber orientations. All of these models used transversely isotropic constitutive laws consistent with biaxial tests [21, 32], which show that resting myocardial stiffness in the muscle fiber direction tends to be greater than in crossfiber directions.

However, recent studies [15, 16] have demonstrated that the well recognized cleavage planes [27] observed in transverse histological sections reflect a laminar organization of myofibers. Branching "sheets" consist of tightly coupled layers of myofibers about four cells thick that appear surrounded and loosely interconnected by perimysial fibrillar collagen. These observations suggest that transverse myocardial stiffness may be lower normal to the sheet plane than within it. It has also been proposed that this organization may permit significant shearing between adjacent laminae during systole [4, 15].

Although no multiaxial experimental testing of orthotropic material properties have yet been reported for myocardium, a recent experimental study measured regional distributions of three-dimensional diastolic and systolic strains relative to local fiber and sheet in the canine left ventricle [30].

In three-dimensional models of systolic contraction, the usual assumption has been that active systolic stress development is uniaxial and directed only along the mean myofiber axis. However a recent biaxial test study by Lin and Yin [17] showed that tetanized rabbit myocardium develops substantial transverse stiffness during tonic activation that can be 50% or more of fiber stiffness. Possible mechanisms for this include transverse force development at the crossbridge [33] and the dispersion of myofiber orientations about the mean [14].

Although muscle fiber and sheet architecture themselves generate significant mechanical heterogeneity, another typical assumption is that intrinsic material properties can be approximated as homogeneous when expressed with respect to regionally varying structural axes defined by the fiber and sheet structure. But there is some reason to question this assumption, too. The degree of branching and connective tissue interconnection is not transmurally uniform [15]. And coiled perimysial collagen fibers are thought run continuously from base to apex; if so, their density should be significantly greater at the apex. Finally, at the single cell level, evidence of electrophysiological and mechanical heterogeneity has been reported. The transient outward current of mid-myocardial M-cells is significantly different from that of epicardial cells in several species, and this is thought to underlying the normal morphology of the T wave in the electrocardiogram. Finally, a recent study of single cell mechanics in isolated rat and ferret cardiac myocytes demonstrated transmural differences in resting and active axial stiffness [1].

In this article we present a new model in which regional muscle fiber and sheet orientations have been included. We compared the results of the axisymmetric, orthotropic model of left ventricular filling and ejection with three-dimensional regional strains measured in the anesthetized dog [30] to test the following structurally based hypotheses: (1) transverse resting stiffness is lower in the sheet-normal direction than tangent to the sheet plane [15]; (2) interlaminar shear stiffnesses in planes normal to the sheets are lower than shear stiffnesses in the fiber-sheet plane because interlaminar shearing can be large during systole [15, 16]; (3) significant transverse active stresses are developed during systole in the intact myocardium [17].

2. Methods

The undeformed geometry of the body was described using a curvilinear system of world coordinates θ_A with covariant base vectors $G_A^{(\theta)}$, while θ_α and $g_A^{(\theta)}$ were used for the deformed geometry [2]. The finite element implementation used normalized element coordinates ξ_k with base vectors $G_k^{(\xi)}$ (Figure 1). The fibrous structure of the anisotropic myocardium was defined using locally orthonormal body coordinates X_i in which X_f is aligned with the local muscle fiber axis and lies in the epicardial tangent (ξ_1, ξ_2)-coordinate plane. X_s lies in the sheet plane and X_n is orthogonal to the sheet plane. These two coordinate systems are related through a transformation matrix (Appendix A). The Galerkin finite element equations for three-dimensional finite elasticity (virtual work formulation) were derived in prolate spheroidal coordinates as described by Costa et al. [2, 3].

Three-dimensional elements in prolate spheroidal coordinates with bicubic Hermite interpolation in the transmural and longitudinal coordinates and linear La-

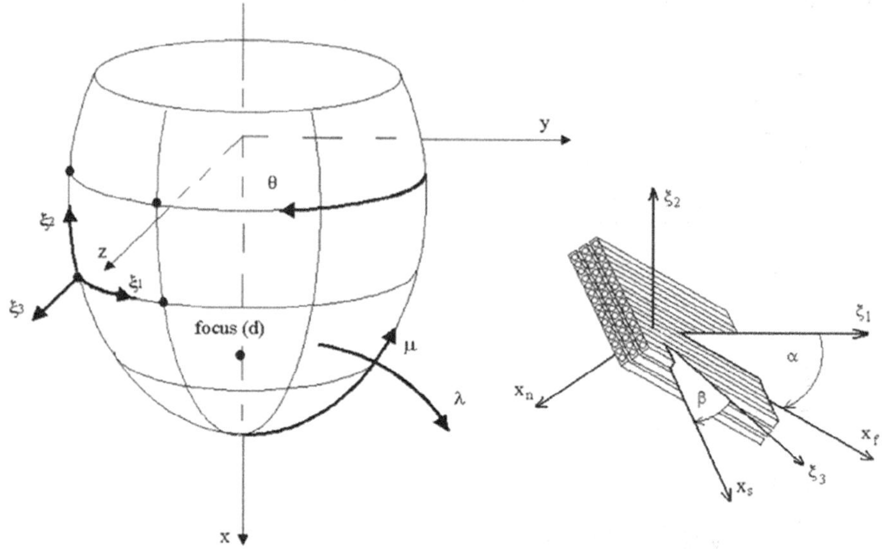

Figure 1. Prolate spheroidal coordinates $\{\theta_A\} = \{\Lambda, M, \Theta\}$ used to model a thick-walled confocal ellipsoidal body. The geometry can be modeled by curvilinear finite elements with normalized coordinates $\{\xi_1, \xi_2, \xi_3\}$ oriented in the $\{-\Theta, M, \Lambda\}$ directions, respectively. Fiber-sheet coordinates, defined by fiber angle (α) and sheet angle (β), comprised the fiber axis (X_f), sheet axis perpendicular to X_f within the sheet plane (X_s), and sheet-normal axis (X_n).

grange interpolation in the circumferential coordinate direction were used to describe the ventricular geometry. The nonlinear finite element equations were integrated using Gaussian quadrature and solved by a Newton-iterative method.

The geometry of the canine left ventricle was represented by a truncated ellipsoid of revolution. The focus, inner and outer surfaces dimensions of the prolate spheroid were calculated from experimental data collected from different dog hearts [4, 18, 30]. Prolate spheroidal and global Cartesian coordinates (see Figure 1) are related by:

$$Y_1 = d \cdot \cosh\Lambda \cos M; \qquad Y_2 = d \cdot \sinh\Lambda \sin M \cos\Theta;$$
$$Y_3 = d \cdot \sinh\Lambda \sin M \sin\Theta, \tag{1}$$

where d is focal length of the prolate spheroidal coordinate system.

Using experimental data [4, 18, 30], the left ventricular model was built with the following parameters: the preload pressure in the reference state for strain measurement was 3 mmHg; end-diastolic pressures were 8 mmHg, 13 mmHg and 17 mmHg; initial cavity volume V_0 was 13 ml; wall thickness in the equatorial region h_1 was 12 mm; wall thickness in the apical region h_2 was 10 mm. Using these parameters, we obtained: $a = 40.4$ mm (apex to base distance), $r_1 = 9.8$ mm, $r_2 = 21.7$ mm (inner and outer radii at the equator respectively). The focal length of the prolate spheropidal coordinate system d was 25.2 mm. The model of the

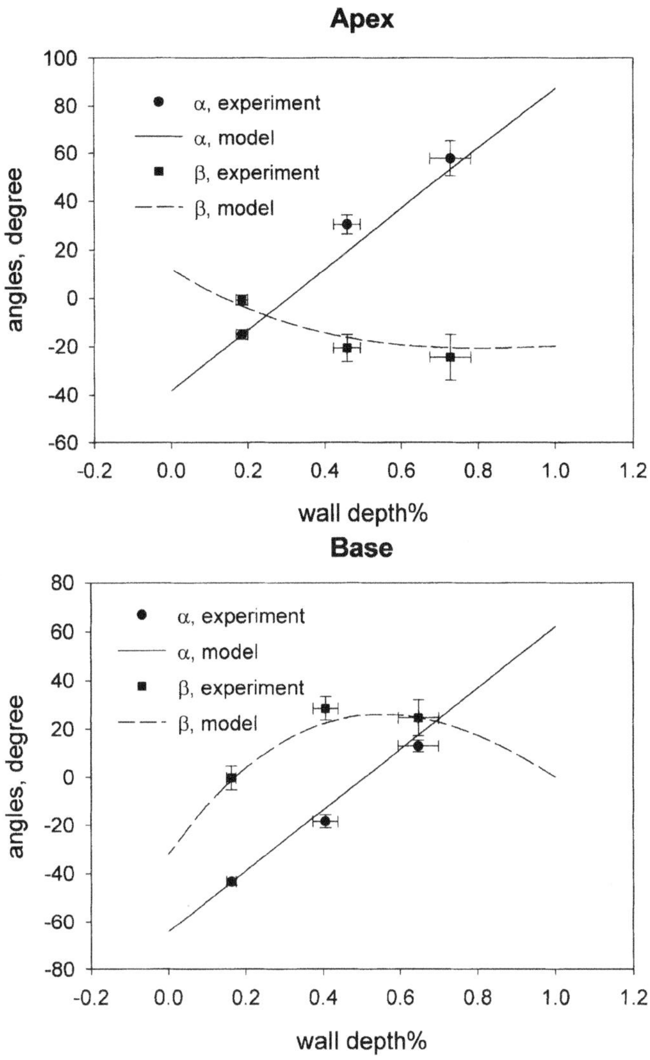

Figure 2. Dependence fiber and sheet angles with wall depth at apical and basal sites, and transmural distributions of experimental mean fiber and sheet angles at apical and basal sites.

left ventricle during of systolic ejection had an end-diastolic pressure of 13 mmHg (the reference configuration for the model of systolic ejection), and an end-systolic pressure of 117 mmHg.

Morphological studies in the dog have shown that circumferential fibers are located near the midwall at the base but are substantially closer to the epicardium at the apex [4]. The fiber angle was $-60°$ to $-35°$ at the epicardium and increased from base to apex. Fiber angle α varied linearly in the model from the endocardium ($\Lambda = 0.38$) to the epicardium ($\Lambda = 0.78$). Figure 2 shows the variation of α in the apical and basal sites as a function of wall depth. The fiber angle varied linearly from base to apex. The fiber angle in the endocardium varied from 60° at the base to 83° at the apex. The root-mean-squared error for α about the experimental values was $8.8 \pm 5.3°$.

Sheet orientation, described by the angle β, tended to be negative in the apical region and positive in the basal region. The variation of β in the left ventricular wall at the apical and basal sites is shown in Figure 2. In the model, the sheet angle varied linearly from base to apex, consistent with experimental measurements [4, 30] in the dog. The root-mean-squared error for β in these experimental measurements was $11.3 \pm 7.4°$.

The epicardial finite element nodes at the base had Θ and Λ fixed to simulate constraints of the relatively stiff mitral valve annulus [5] and the longitudinal coordinate M was fixed at all basal ($M = 120°$) and apical ($M = 5°$) nodes. The symmetric high-order finite element model required 3 elements and 8 nodes for convergence of the strain energy to within 0.5%.

Previous model investigations of passive left ventricular mechanics [2, 3, 7, 9, 11] all assumed that myocardium is transversely isotropic. In the present analysis, the stress-strain relationship of the passive left ventricle was defined by the following exponential strain-energy function W, which treats myocardium as nonlinear, orthotropic and nearly incompressible.

$$\begin{aligned} W &= C(\mathrm{e}^Q - 1)/2 + C_{\mathrm{compr}}(J \cdot \ln J - J + 1); \\ Q &= b_{ff}E_{ff}^2 + b_{ss}E_{ss}^2 + b_{nn}E_{nn}^2 + b_{fs}(E_{fs}^2 + E_{sf}^2) \\ &\quad + b_{fn}(E_{fn}^2 + E_{nf}^2) + b_{ns}(E_{ns}^2 + E_{sn}^2); \end{aligned} \tag{2}$$

where E_{ij} are components of Green's strain tensor $\mathbf{E}$ in an orthogonal coordinate system having fiber, sheet and sheet-normal (f, s, n) axes respectively; J is the determinant of the stretch tensor $\mathbf{U}$. The following material constants: $C = 0.88$ kPa; $b_{ff} = 18.5$; $b_{ss} = 3.58$; $b_{nn} = 3.58$; $b_{fs} = 2.8$; $b_{fn} = 2.8$; $b_{ns} = 2.8$; $C_{\mathrm{compr}} = 100$ kPa were chosen as a starting point, consistent with previous transversely isotropic analyses [7] of canine myocardium. The last term of equation (2) describes nearly incompressible behavior [25]. We found that this form gives more accurate numerical results than previous incompressible formulation [31] without significantly affecting myocardial volume.

Systolic contraction was modeled by defining the Cauchy stress tensor as the sum of the passive three-dimensional stress tensor $\mathbf{T}^{(\mathrm{p})}$ derived from the strain energy function and an active stress tensor $\mathbf{T}^{(\mathrm{a})}$:

$$\mathbf{T} = \mathbf{T}^{(\mathrm{p})} + \mathbf{T}^{(\mathrm{a})}. \tag{3}$$

The components $T_{ij}^{(\mathrm{a})}$ of the active stress tensor in prolate spheroidal coordinates were derived from the diagonal stress tensor $\mathbf{T}_{\mathrm{active}}$ referred to local fiber-sheet coordinates (X_f, X_s, X_n) using a rotation matrix $\mathbf{q}$, which defines the relation between the prolate spheroidal coordinate system and the local fiber-sheet coordinate system using deformed fiber and sheet angles:

$$\mathbf{T}^{(\mathrm{a})} = \mathbf{q}^{\mathrm{T}}\mathbf{T}_{\mathrm{active}}\mathbf{q}. \tag{4}$$

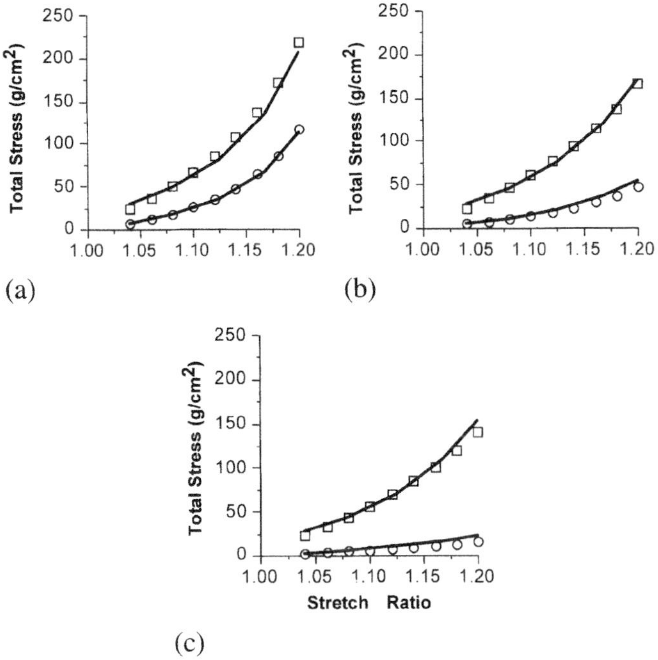

Figure 3. Total stress in fiber (squares) and cross-fiber (circles) directions for equibiaxial (a), off-biaxial (b), and uniaxial (c) stretch protocols compared to model results (solid lines).

The components of the active tensor $\mathbf{T}_{\text{active}}$ were a function of peak intracellular calcium $[\text{Ca}]_i$ and sarcomere length. The parameters of the active model were based on the experimental measurements of sarcomere length and tension [6, 10]. Two variants of the active model were built: one with a uniaxial active stress component T_{ff}^{active}, assuming no transverse stresses ($T_{ss}^{\text{active}} = T_{nn}^{\text{active}} \equiv 0$); and one biaxial model, where transverse active stresses T_{ss}^{active} and T_{nn}^{active} were added as a function of axial stress T_{ff}^{active} and local transverse (λ_s, λ_n) and axial λ_f strains [19] to be consistent with biaxial experimental tests in barium-contracted rabbit myocardium [17]. To reproduce these biaxial observations we made the following assumptions: (1) part of the significant transverse active stress measured in isolated tissue is associated with a 12° angular dispersion (σ) of local myofiber axes about the regional mean [14]; (2) part of the transverse active stress is generated by an unknown mechanism acting at the myofibril level probably associated with cross-bridge geometry [26] and this transverse myofibril stress development is a constant fraction (k) of the axial tension (Appendix B). The experimentally measured active biaxial stresses were also affected by the measured 18° "spay" in mean myofiber angles through the thickness of the left ventricular experimental tissue slices. Appendix B summarizes the equations that incorporate the first two features in an active constitutive law for transversely isotropic active stress. Figure 3 illustrates total stresses for simulated biaxial testing obtained with a three-dimensional finite element model that included the splay of mean fiber angles through the thickness of the test specimen and passive material parameters consistent with resting biaxial stress–strain tests in the rabbit [17]. The comparison in Figure 3 shows excellent agreement with experimental fiber and transverse active stresses [17] for uniaxial, equibiaxial and off-biaxial stretch protocols using constant values of k (0.3) and σ (12°).

Regional three-dimensional strains computed in the left ventricular model were compared with experimental data, measured by Takayama et al. [30]. In these experiments, transmural arrays of radiopaque markers were implanted in six anesthetized dogs at sites approximately one quarter (basal site) and three quarters (apical) of the distance from base to apex along the left ventricular long axis, midway between the left anterior papillary muscle and the anterior ventricular sulcus. Myocardial cleavage plane and muscle fiber angles were measured at each site, and three-dimensional diastolic strains were computed at subendocardium, midwall and subepicardium with respect to fiber, sheet, normal (f, s, n) axes.

Examining these experimental data, we excluded two dogs, because measured strain components for these two studies were significantly different from the other four, which were more representative of their mutual mean.

3. Results

Starting with the transversely isotropic passive material parameters of Guccione et al. [5], we decreased resting fiber stiffness until mean fiber strains agreed with experimentally measured diastolic strains in the basal and apical regions. The three shear coefficients were all increased together to match the measured shear strains on average. The compressibility coefficient was chosen to match mean myocardial volume changes measured in the experiment. The adjusted coefficients for the transversely isotropic nearly incompressible model were: $C = 0.88$ kPa; $b_{ff} = 6.0$; $b_{ss} = 5.0$; $b_{nn} = 5.0$; $b_{fs} = 6.0$; $b_{fn} = 6.0$; $b_{ns} = 6.0$; $C_{\text{compr}} = 3.0$.

As shown in Figures 4–9, end-diastolic strain distributions computed with the passive transversely isotropic model showed quite good agreement for the E_{ff}, E_{nn}, E_{fs} components. However, strain in the sheet direction E_{ss} and the interlaminar shears E_{sn} and E_{fn} were significently underestimated.

Decreasing passive sheet-normal stiffness relative to in-sheet transverse stiffness (b_{nn}/b_{ss}) and keeping the sum $b_{ss}+b_{nn}$ constant, decreased the sheet strain E_{ss} and improved the agreement between model and experiment. Similarly, decreasing interlaminar shear coefficients b_{fn} and b_{ns} and increasing b_{fs} (keeping sum of interlaminar shear coefficients equal) improved the agreement between computed and measuremed interlaminar shear strains E_{fs}, E_{fn}, E_{sn}. The best constant combination was: $C = 0.88$ kPa; $b_{ff} = 6.0$; $b_{ss} = 7.0$; $b_{nn} = 3.0$; $b_{fs} = 12.0$; $b_{fn} = 3.0$; $b_{ns} = 3.0$; $C_{\text{compr}} = 3.0$.

All six three-dimensional passive strain components (calculated with the orthotropic model) at the basal region were within one standard deviation of measure-

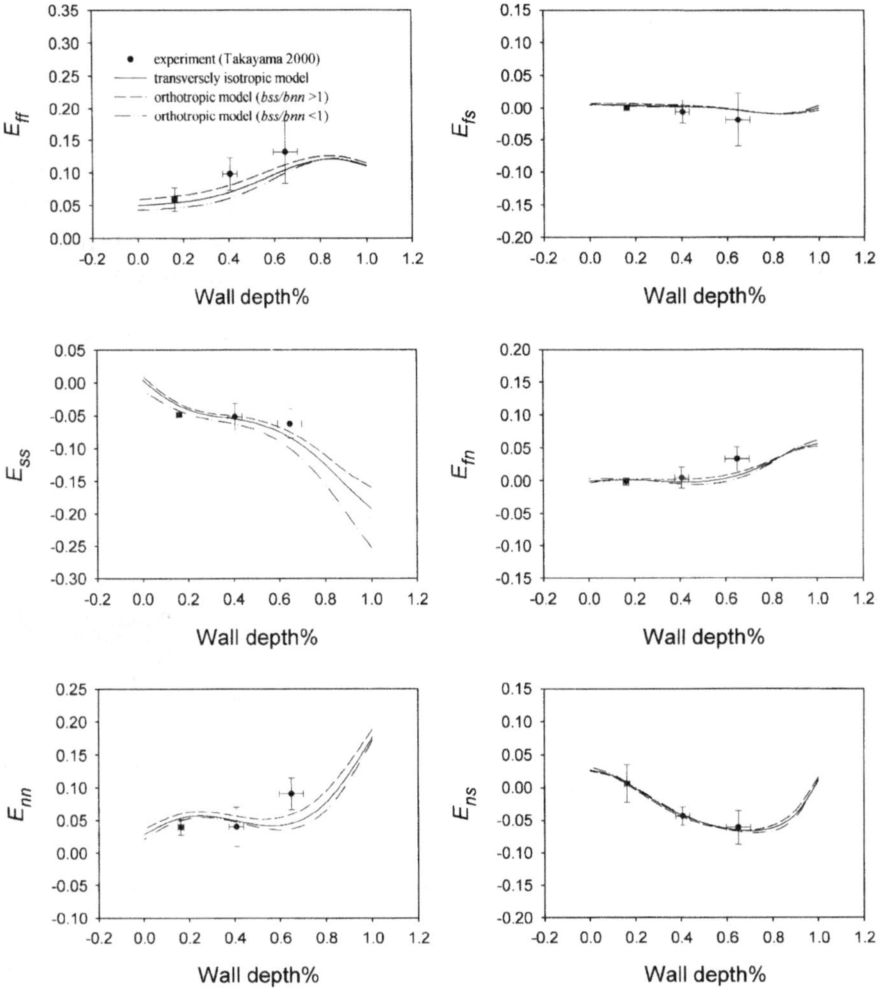

Figure 4. Comparison of experimental mean end-diastolic strains at the basal site [30] with those predicted by the model. Influence of different coefficient ratios b_{ss}/b_{nn} in the constitutive equation on the calculated strains. Preload pressure (reference configuration) was 3 mmHg; end-diastolic pressure was 13 mmHg.

ments reported by Takayama et al. [30] for three different levels of end-diastolic pressure. All constitutive parameters were the same for all three levels of end-diastolic pressure. In the model, the longitudinal coordinate of the basal region was $M = 85°$. In fiber-sheet coordinates, the dependence of the strain components on wall thickness are shown in Figures 4–6 for an end-diastolic pressure of 13 mmHg for different b_{ss}/b_{nn} and $b_{fs}/b_{fn} = b_{fs}/b_{sn}$ ratios. The fiber and normal-sheet strains were positive, and the sheet component was negative. The fiber component did not change significantly from epicardium to endocardium. The shear strains E_{fn} and E_{fs} also did not change significantly, E_{fs} decreased and

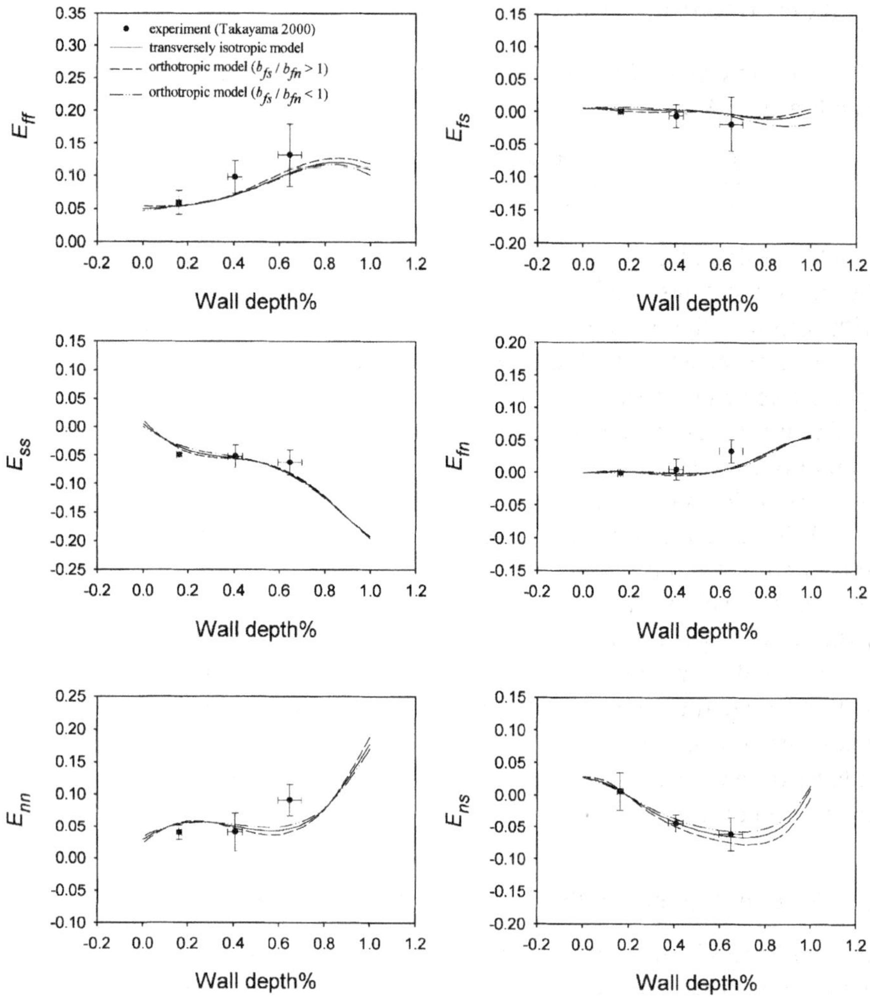

Figure 5. Comparison of experimental mean end-diastolic strains at the basal site [30] with those predicted by the model. The effects of varying the constitutive coefficient ratio $b_{fs}/b_{fn} = b_{fs}/b_{ns}$ on the calculated strains. Preload pressure (reference configuration) was 3 mmHg; end diastolic pressure was 13 mmHg.

E_{fn} increased from epicardium to endocardium. The shear strain E_{sn} was negative and depended significantly on the sheet angle β: in general, E_{sn} was negative when β was positive; E_{sn} was positive when β was negative.

The strain components in fiber sheet coordinates at the apical region ($M = 25°$) were calculated for three different levels of end-diastolic pressure. These strains are shown in Figures 7–9 for an end-diastolic pressure 13 mmHg and different ratios b_{ss}/b_{nn} and $b_{fs}/b_{fn} = b_{fs}/b_{sn}$. All six strain components (orthotropic model) were within or close to one standard deviation of experimental measurements [30] for all three variations of end-diastolic pressure.

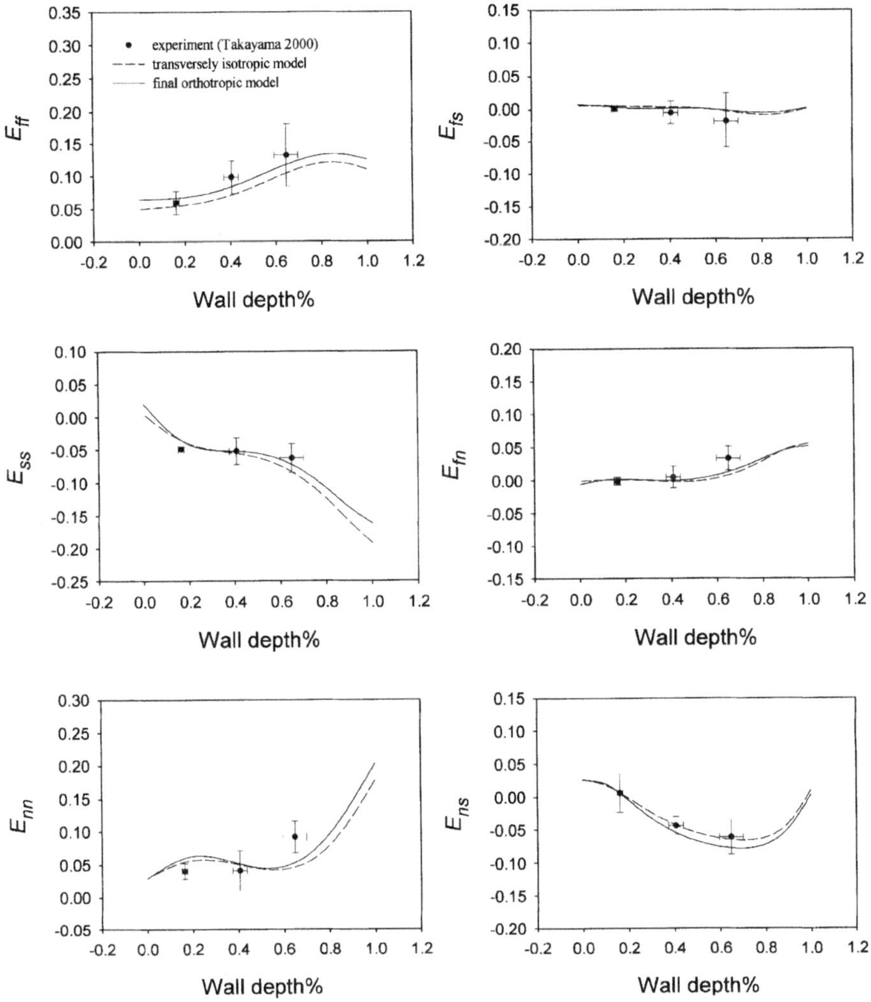

Figure 6. Comparison of experimental mean end-diastolic strains at the basal site [30] with those predicted by the transversely isotropic and final orthotropic models. Preload pressure (reference configuration) was 3 mmHg; end diastole pressure was 13 mmHg.

It can be seen from Figures 4–9 that the ratio b_{ss}/b_{nn} significantly influenced only the E_{ss} strain component, and $b_{fs}/b_{fn} = b_{fs}/b_{ns}$ significantly influenced only the E_{ns} component. Notice that the agreement with experiment was better in the basal region than in the apical region. Strain components in the reference coordinate system were also calculated. The absolute value all of six strain components in fiber-sheet coordinates and reference coordinates were increased, when end-diastolic pressure was increased (in agreement with experimental investigations).

The components of the stress tensor were also calculated in fiber-sheet and reference coordinate systems. Orthotropic passive properties influenced most strongly the fiber stress component and fiber-normal shear stress component. All main stress

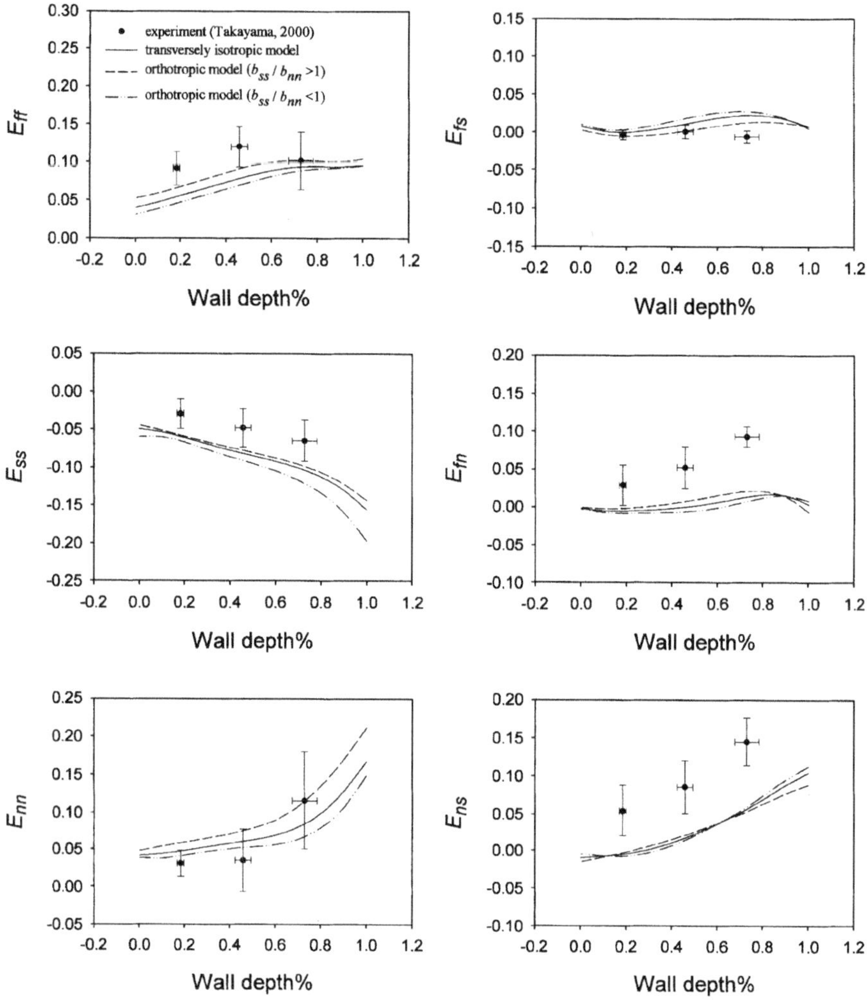

Figure 7. Comparison of experimental mean end-diastolic strains at the apical site [30] with those predicted by the model. Influence of different coefficient ratios b_{ss}/b_{nn} in the constitutive equation on calculated strains. Preload pressure (reference configuration) was 3 mmHg; end-diastolic pressure was 13 mmHg.

components were decreased from endocardium to epicardium. The fiber-normal shear stress component changed significantly from endocardium to epicardium.

End-systolic strains in fiber-sheet coordinates were calculated with both uniaxial active stress and biaxial active stress (see Appendix B). In Figure 10, strain components are shown in fiber-sheet coordinates with uniaxial active stress and either transversely isotropic or orthotropic passive properties. There was little difference between systolic strains in the model with transversely isotropic vs. orthotropic passive properties.

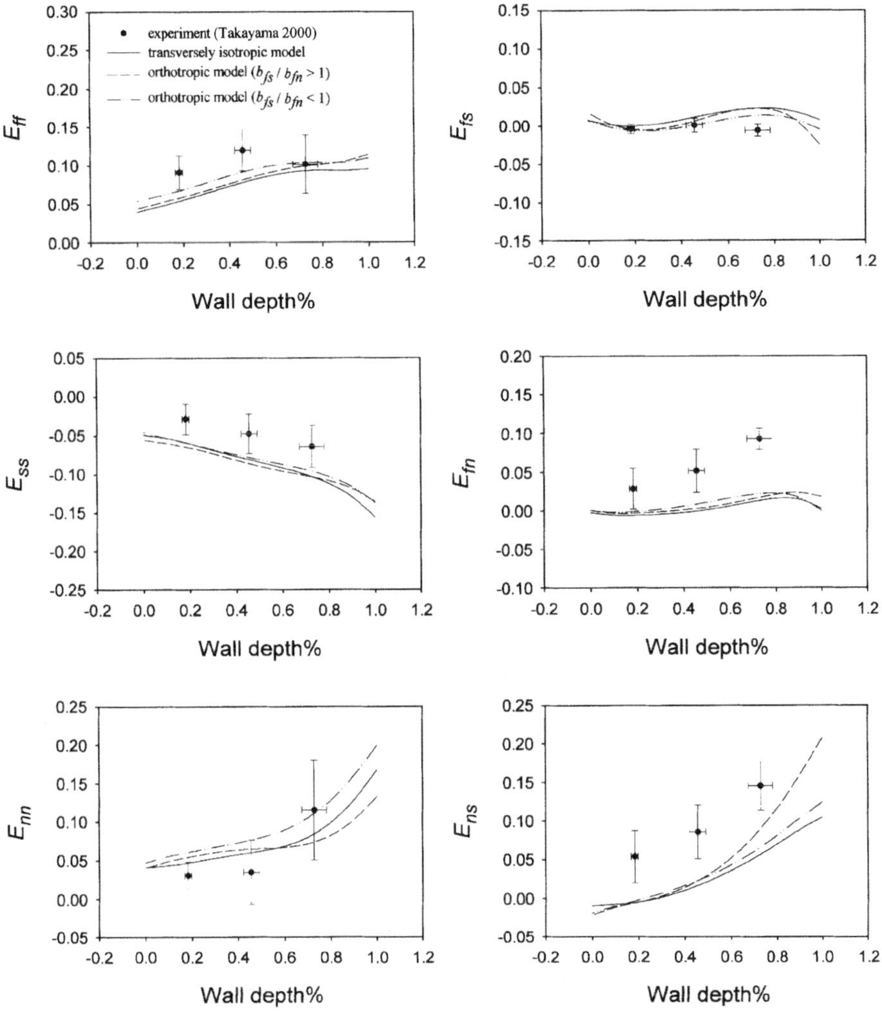

Figure 8. Comparison of experimental mean end-diastolic strains at the apical site [30] with those predicted by the model. Effect of varying the constitutive coefficient ratio $b_{fs}/b_{fn} = b_{fs}/b_{ns}$ on the calculated strains. Preload pressure (reference configuration) was 3 mmHg; end diastole pressure was 13 mmHg.

In Figure 11, the components of the strain tensor are represented in fiber-sheet coordinates with orthotropic passive properties and either uniaxial or biaxial active properties. Agreement with the experimental data of Takayama et al. [30] was significantly better for the model with biaxial active stress than uniaxial active stress. However, there was still a discrepancy in strains (especially E_{fs} and E_{ns}), which may possibly be further reduced by use of an orthotropic active stress.

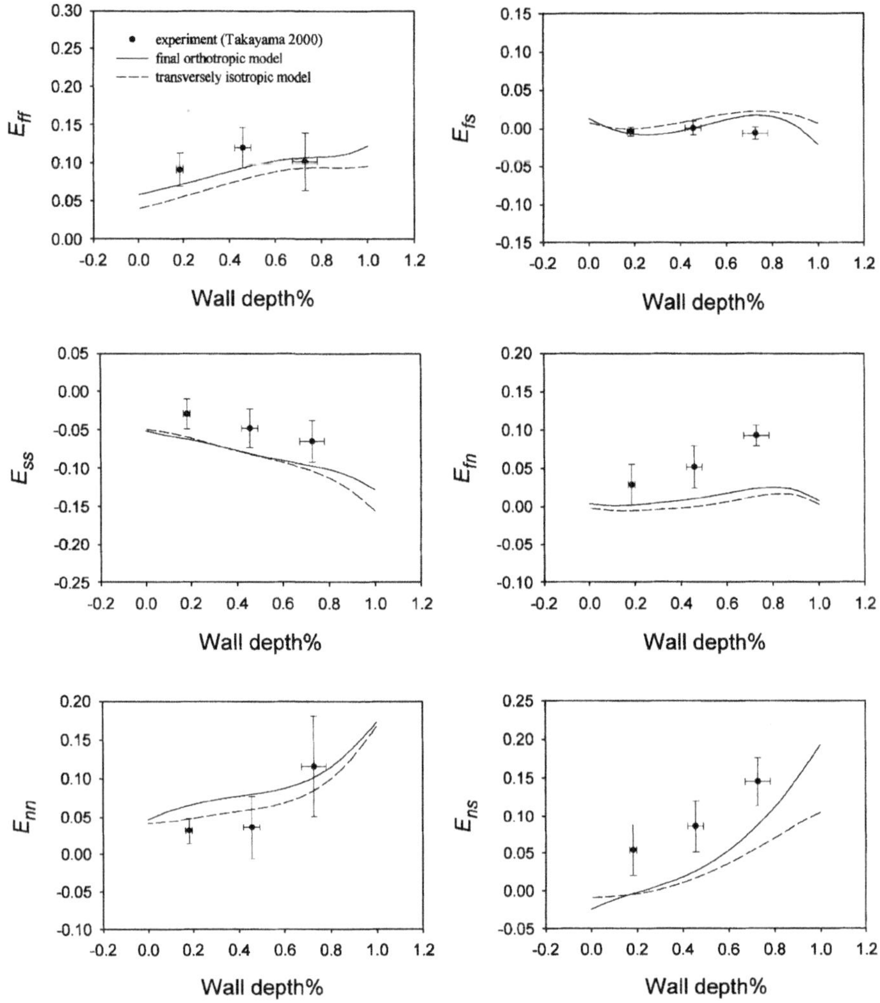

Figure 9. Comparison of experimental mean end-diastolic strains [30] with those predicted by the transversely isotropic and final orthotropic models at the apical site. Preload pressure was 3 mmHg; end-diastolic pressure was 13 mmHg.

4. Discussion

A three-dimensional finite element model of the canine left ventricle was used to compute stress and strain distributions at end-diastole and end-systole. Compared with transverse isotropy, an orthotropic model of resting myocardial properties gave improved agreement with experimentally measured sheet strains and wall thinning when myocardial stiffness transverse to the muscle fibers was increased in the sheet direction relative to the sheet-normal axis. Compared with transverse isotropy, an orthotropic model gave improved agreement with experimentally measured sheet-normal strains when interlaminar shear stiffness was increased in the

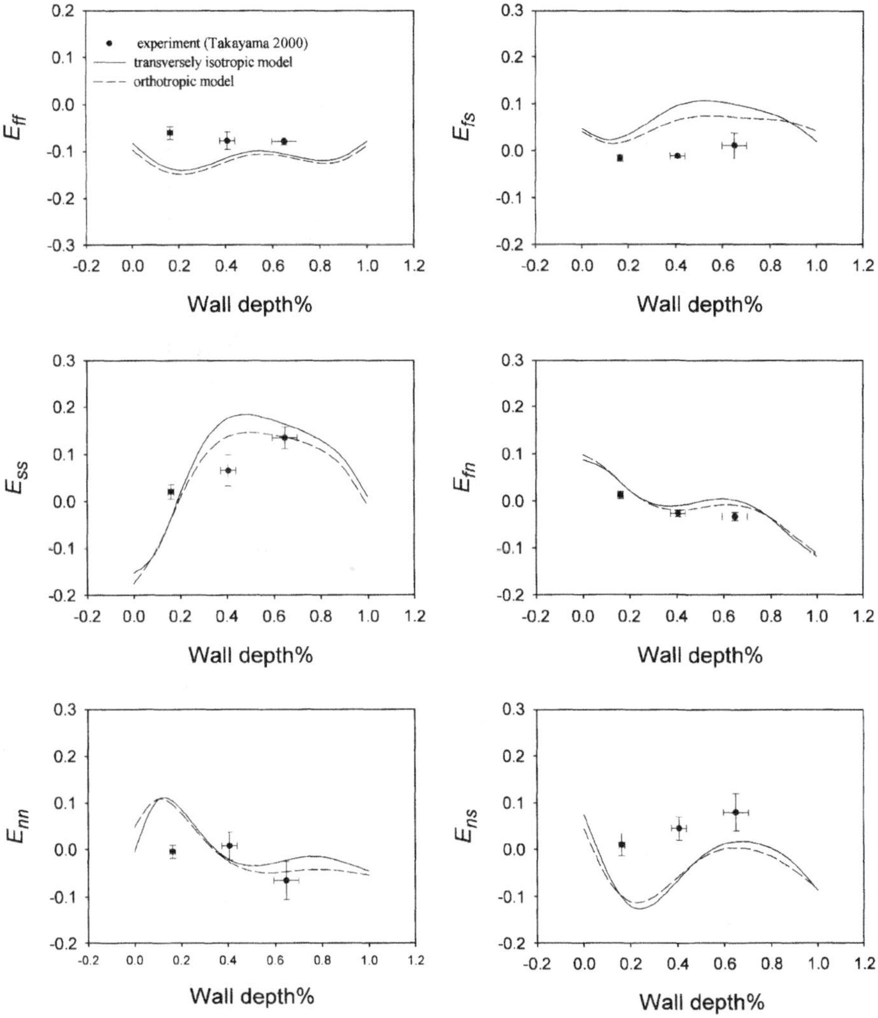

Figure 10. Comparison of experimental mean end-systolic strains at the basal site [30] with those predicted by uniaxial active model with transversely isotropic and final orthotropic passive properties. End-diastolic pressure (reference configuration) was 13 mmHg; end-systolic pressure was 117 mmHg.

fiber-sheet plane relative to the stiffness in planes normal to the sheets. Thus, these results support the hypothesis that normal stiffness transverse to the muscle fibers is greater tangent to the sheets and lower normal to them; shear coefficients are greater within sheet planes and lower transverse to them.

The coefficient b_{ff}, which corresponds to stiffness in the fiber direction, was less than b_{ss}, which corresponds to stiffness in the sheet direction. This may be inconsistent with previous reports of biaxial tissue responses which generally report highest stiffness in the myofiber direction. However, it is likely in those measure-

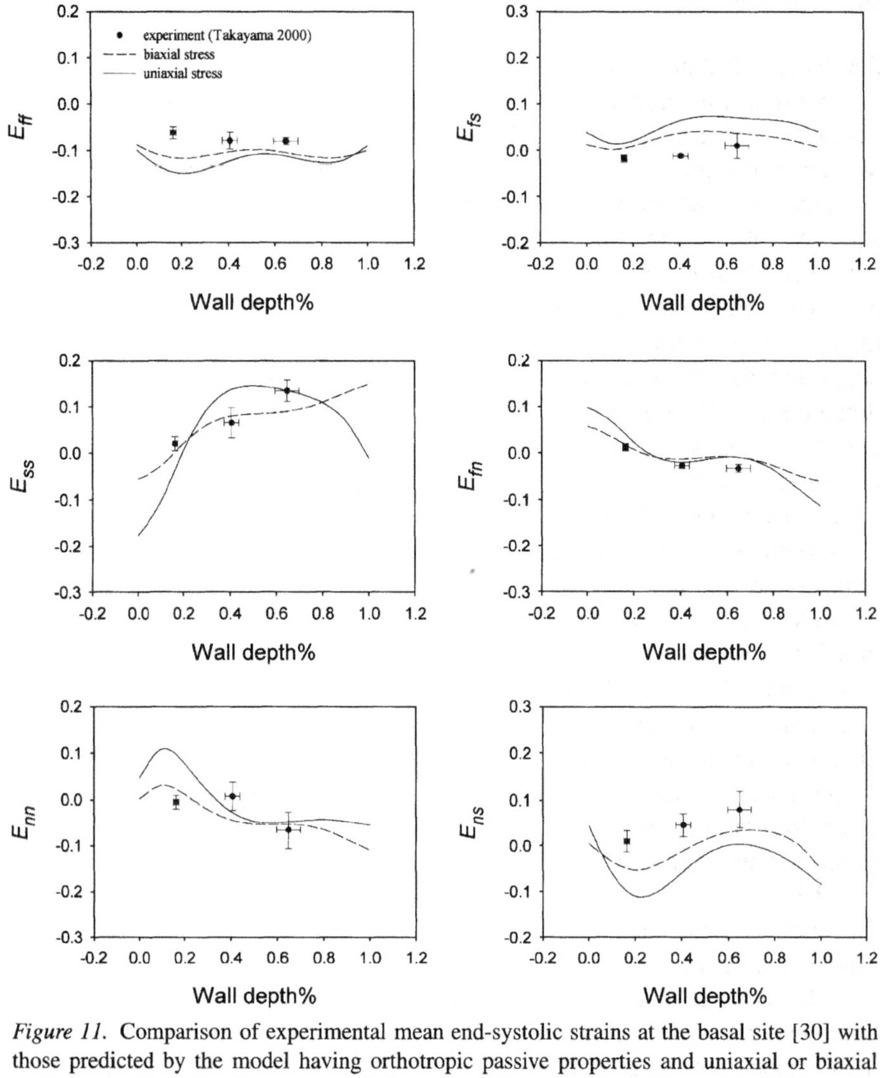

Figure 11. Comparison of experimental mean end-systolic strains at the basal site [30] with those predicted by the model having orthotropic passive properties and uniaxial or biaxial active stress. End-diastolic pressure (reference configuration) was 13 mmHg; end-systolic pressure was 117 mmHg.

ments, that the crossfiber direction was closer to the sheet-normal direction, than the sheet direction. This is in keeping with the finding that average sheet angles in the left ventricle myocardium tend to be less than 45° (Costa et al., 1999) [4]. Hence, in typical test slices of tissue cut parallel to the plane of the wall, the crossfiber axis has a larger component normal than tangent to the sheets. Previous material parameter estimates have been based on measurements in isolated arrested myocardial tissue, where as the experimental strain data [30] were acquired at successive end-diastolic phases of the cardiac cycle in the intact beating heart. This may explain why fiber strains E_{ff} were higher in the study of Takayama et al. [30], then those measured in the isolated perfused heart by May-Newman et al. [18].

As seen in Figures 4–9, the agreement with experimental measurements was generally better at the basal region than the apical region (particularly in the shear strain components). This may be because the ellipsoidal model better approximates left ventricular geometry in the basal and equatorial sites, than at the apex, where the wall thins more rapidly. For example, a change of focal length in the model only significantly affects results in the apical region (particularly the E_{fn} strain component).

In general, agreement with experimental measurements was better in the subepicardium. LeGrice et al. [15, 16] measured the dependence of branch number per square mm for left ventricular myocardium between adjacent layers at different sites of the wall. In the epicardial region, the number of branches was 6 mm^{-2} and increased to endocardium (12 mm^{-2}), suggesting that sheet-normal stiffness may be nonhomogeneous across the wall.

Recently, Cazorla et al. [1] showed, that the sarcomere length-resting tension relationship and sarcomere length-active tension relationship are different for endocardial and epicardial myocytes (rat and ferret), and this difference for active tension was statistically significant. We modeled this inhomogeneity in passive and active material properties, increasing the coefficient of the passive constitutive equation and increasing active tension smoothly from epicardium to endocardium. The most significant effects of these changes were on the endocardial strains E_{ss} and E_{nn}. Our results showed decreasing end-diastolic strains, E_{ss} and E_{nn}, at the endocardial surface and increasing end-systolic strains, E_{ss} and E_{nn}, at the endocardial surface. Using these nonhomogeneous interpretation we reached better agreement with experimental strains, but the improvement was small compared with variations in experimentally measured strains.

By varying coefficients of an orthotropic finite element model of left ventricular mechanics, we were able to investigate the structural basis of regional three-dimensional strains measured in the dog heart relative to local myofiber and sheet axes. The results are consistent with the structurally motivated hypothesis that myofibers are more tightly coupled within the sheets than between adjacent sheets, and that these properties affect regional diastolic function *in vivo*.

Agreement with the experimental end-systolic strains of Takayama et al. [30] was substantially better for the biaxial active model than the conventional uniaxial model of muscle contraction. This observation suggests that the substantial transverse active stresses developed in tonically activated isolated cardiac muscle preparations may be reflective of a similar behavior of intact myocardium *in vivo*. Our constitutive law for biaxial active stress (Appendix B), was derived in part from consideration of the known dispersion of myofiber orientations about the mean. However, the analysis also had to postulate a significant transverse force development at the myofibril level that was a constant fraction (30%) of the active fiber tension. Although some investigators have proposed models to explain this phenomenon based on the crossbridge and myofibril lattice geometry [26], the precise mechanism of this significant transverse force development remains unknown. The influence of orthotropic passive properties on end-systolic strains was small both for the uniaxial active model and the biaxial active model. However, it is possible that making the active muscle stress development orthotropic may further improve the agreement with experimentally observed systolic strains. This possibility remains to be tested in the future.

In summary, an orthotropic material model for resting myocardium improved the agreement between experiment and a three-dimensional model of diastolic left ventricle regional mechanics, when stiffnesses transverse to myocardial laminae were reduced relative to the corresponding coefficients within the plane of the sheet. However, in an active model, these passive material changes had little effect on systolic left ventricular strains. Incorporating a significant component of active stress transverse to the muscle fibers greatly improved the agreement between measured and modeled transverse end-systolic shear strains.

Acknowledgements

We thank Dr Walter O'Dell, Dr. Jim Covell and Dr. Kevin Costa for their valuable advice and contributions to this work. This research was supported by National Science Foundation Grant BES-9634974, the National Biomedical Computation Resource NIH grant P41RR08605 and NIH grant HL 32583-15.

Appendix A

The body/fiber X_i-coordinate system is defined in relation to the ξ_k-coordinate with X_1 lying in the (ξ_1, ξ_2)-coordinate plane. X_2 and X_3 are normal to X_1, with X_2 lying in the (ξ_1, ξ_2) plane. The relationship between these coordinate systems is:

$$\frac{\partial \xi^k}{\partial X^I_{\text{fiber}}} = \begin{bmatrix} \frac{G^{(\xi)}_{22} K_1 - G^{(\xi)}_{12} K_2}{L^2} & \frac{-K_2}{L} & \sqrt{G^{(\xi)33}}\, \frac{G^{(\xi)}_{12} G^{(\xi)}_{23} - G^{(\xi)}_{22} G^{(\xi)}_{13}}{L^2} \\ \frac{G^{(\xi)}_{11} K_2 - G^{(\xi)}_{12} K_1}{L^2} & \frac{K_1}{L} & \sqrt{G^{(\xi)33}}\, \frac{G^{(\xi)}_{12} G^{(\xi)}_{13} - G^{(\xi)}_{11} G^{(\xi)}_{23}}{L^2} \\ 0 & 0 & \sqrt{G^{(\xi)33}} \end{bmatrix}$$

$$\equiv \begin{bmatrix} a_{11} & a_{12} & a_{13} \\ a_{21} & a_{22} & a_{23} \\ a_{31} & a_{32} & a_{33} \end{bmatrix}, \qquad \text{(A.1)}$$

where

$$G^{(\theta)}_{AB} = (\partial Y^R / \partial \theta^A) \cdot (\partial Y^R / \partial \theta^B);$$

$$K_1 = \sqrt{G^{(\xi)}_{11}} \cdot \cos \eta_1; \qquad K_2 = \sqrt{G^{(\xi)}_{22}} \cdot \cos \eta_2;$$

$$L^2 = G^{(\xi)}_{11} G^{(\xi)}_{22} - \left(G^{(\xi)}_{12}\right)^2; \qquad \eta_2 = \cos^{-1}\left(G^{(\xi)}_{12} \left(G^{(\xi)}_{11} G^{(\xi)}_{22}\right)^{-1/2}\right) - \eta_1,$$

where η_1 is defined as the "fiber angle", which in general is a function of (ξ_1, ξ_2, ξ_3).

The fiber ξ_k coordinate system is defined in relation to the sheet $\xi_{k(\text{fiber-sheet})}$ coordinate as:

$$\frac{\partial X^k_{\text{fiber}}}{\partial X^I_{\text{fiber-sheet}}} = \begin{bmatrix} 1 & 0 & 0 \\ 0 & \sin\beta & -\cos\beta \\ 0 & \cos\beta & \sin\beta \end{bmatrix}, \tag{A.2}$$

where β is defined as the "sheet angle".

Finally the relation between ξ_k coordinate system and fiber-sheet coordinates is:

$$\frac{\partial \xi^k}{\partial X^I_{\text{fiber-sheet}}} = \frac{\partial \xi^k}{\partial X^n_{\text{fiber}}} \cdot \frac{\partial X^m_{\text{fiber}}}{\partial X^I_{\text{fiber-sheet}}}$$

$$= \begin{bmatrix} a_{11} & a_{12}\sin\beta + a_{13}\cos\beta & -a_{12}\cos\beta + a_{13}\sin\beta \\ a_{21} & a_{22}\sin\beta + a_{23}\cos\beta & -a_{22}\cos\beta + a_{23}\sin\beta \\ a_{31} & a_{32}\sin\beta + a_{33}\cos\beta & -a_{32}\cos\beta + a_{33}\sin\beta \end{bmatrix}.$$

Appendix B

Fiber, sheet and sheet-normal stresses acting along the mean myofiber, sheet and sheet-normal axes in a ensemble of cells are obtained by integrating myofibril tractions over the distribution of myofiber angles $\mathbf{f}(\theta)$:

$$\mathbf{T}_{\text{active}} = \int_\theta \int_\mu \left[\mathbf{Q}\mathbf{T}^{\text{local}}\mathbf{Q}^{\text{T}}\right] \cdot \mathbf{f}(\theta) \cdot \varphi(\mu) \mathrm{d}\mu \, \mathrm{d}\theta. \tag{B.1}$$

This equation corresponds to our first assumption about the angular dispersion of local fibers about the regional mean. The active stress tensor $\mathbf{T}^{\text{local}}$ in local fiber coordinates for equation (B1) consists only of a fiber component T^{local}_{ff}, which is a function of peak intracellular calcium $[\text{Ca}]_i$ and sarcomere length. The angles θ and μ describe the relationship between the local myofiber axis and the mean fiber axis; $\mathbf{f}(\theta)$ is the fiber orientation probability density distribution which can be described using a von Mises distribution; $\varphi(\mu)$ also is a density distribution which can be described as $\varphi(\mu) = 1/(2\pi)$ $(\mu \in [1; 2\pi])$; $\mathbf{Q}$ defines the relation between the mean fiber-sheet coordinate system and the local cell coordinate system:

$$\mathbf{X}^{\text{fiber-sheet}} = \mathbf{Q}\mathbf{X}^{\text{local}}. \tag{B.2}$$

If $f(\theta)$ is $f^0(\theta)$ in the undeformed reference state, then:

$$\mathbf{f}(\theta) = f^0(\theta) \cdot \begin{bmatrix} 1 & 0 & 0 \\ 0 & \lambda_s/\lambda_f & 0 \\ 0 & 0 & \lambda_n/\lambda_f \end{bmatrix} \equiv f^0(\theta) \cdot \mathbf{\Lambda}, \tag{B.3}$$

where λ_i are fiber, sheet and normal-sheet extensions. In common cases the function $f^0(\theta)$ will depend also on the angle μ.

It will be assumed that the local myofiber orientation may be sufficiently described using a von Mises distribution:

$$f^0(\theta) = \frac{\exp[k_0 \cos(2\theta)]}{2\pi I_0(k_0)}, \tag{B.4}$$

where k_0 is a concentration parameter, and I_0 is the modified Bessel function of order zero, given by:

$$I_0(k) = \frac{1}{2\pi} \int_0^{2\pi} \exp[k_0 \cos\phi]\, d\phi. \tag{B.5}$$

The concentration parameter k_0, can be related to the angular dispersion by:

$$\sigma = \frac{1}{kA_1(k_0)}; \qquad A_1(k_0) = \frac{I_1(k_0)}{I_0(k_0)}; \tag{B.6}$$

where σ is the angular dispersion and I_1 is the modified Bessel function of order one. Series expansions for the modified Bessel functions are given by:

$$I_0(k_0) = \sum_{r=0}^{\infty} (r!)^{-2} \cdot (0.5k_0)^{2r},$$

$$I_p(k_0) = \sum_{r=0}^{\infty} \left[(r+p)! \cdot r!\right]^{-1} \cdot (0.5k_0)^{2r+p}; \quad p = 1, 2, \ldots.$$

The concentration parameter k_0 may assume any value between 0 and ∞ with smaller values associated with more dispersed distributions and larger values associated with narrower distributions.

We may rewrite equation (B1) as follows:

$$\mathbf{T}_{\text{active}} = \frac{1}{2\pi} \left[\int_{\theta=-\pi/2}^{\pi/2} \int_{\mu=0}^{2\pi} \left[\mathbf{Q}\mathbf{T}^{\text{local}}\mathbf{Q}^{\mathrm{T}}\right] \cdot f^0(\theta)\, d\mu\, d\theta \right] \cdot \mathbf{\Lambda}. \tag{B.7}$$

Using our second assumption about transverse mechanisms at the myofibril level, associated with cross-bridge geometry, we obtain T_{ss}^{local} and T_{nn}^{local} as a constant fraction (k) of the active fiber stress T_{ff}^{local}. In this case, the active stress tensor in local fiber coordinates can be written as:

$$\mathbf{T}^{\text{local}} = \begin{bmatrix} T_{ff}^{\text{local}} & 0 & 0 \\ 0 & T_{ss}^{\text{local}} & 0 \\ 0 & 0 & T_{nn}^{\text{local}} \end{bmatrix}; \tag{B.8}$$

$$T_{ss}^{\text{local}} = kT_{ff}^{\text{local}} \frac{\lambda_s}{\lambda_f}; \quad T_{nn}^{\text{local}} = kT_{ff}^{\text{local}} \frac{\lambda_n}{\lambda_f}.$$

The factors λ_s/λ_f and λ_n/λ_f are included to reflect the assumption that k is determined by the angle between the crossbridge and the thin filament, which changes as a function of transverse lattice spacing [18, 24].

After integration of equation (B7) we can obtain the active stress tensor in mean fiber-sheet coordinates:

$$\mathbf{T}_{\text{active}} = \begin{bmatrix} T_{ff}^{\text{active}} & 0 & 0 \\ 0 & T_{ss}^{\text{active}} & 0 \\ 0 & 0 & T_{nn}^{\text{active}} \end{bmatrix}.$$

References

1. O. Cazorla, J. Le Guennec and E. White, Length-tension relationships of sub-epicardial and sub-endocardial single ventricular myocytes from rat and ferret hearts. *J. Mol. Cell Cardiol.* **32** (2000) 735–744.
2. K.D. Costa, P.J. Hunter, J.M. Rogers, J.M. Guccione, L.K. Waldman and A.D. McCulloch, A three-dimensional finite element method for large elastic deformations of ventricular myocardium: Part I – cylindrical and spherical polar coordinates. *ASME J. Biomech. Engng.* **118** (1996) 452–463.
3. K.D. Costa, P.J. Hunter, J.S. Wayne, L.K. Waldman, J.M. Guccione and A.D. McCulloch, A three-dimensional finite element method for large elastic deformations of ventricular myocardium: Part II – Prolate spheroidal coordinates. *ASME J. Biomech. Engng.* **118** (1996) 464–472.
4. K.D. Costa, Y. Takayama, A.D. McCulloch and J.M. Covell, Laminar fiber architecture and three-dimensional systolic mechanics in canine ventricular myocardium. *Am. J. Physiol.* **276** (*Heart Circ. Physiol.* **45**) (1999) 595–607.
5. J.M. Guccione and A.D. McCulloch, Finite element modeling of ventricular mechanics. In: L. Glass, P.J. Hunter and A.D. McCulloch (eds), *Theory of Heart: Biomechanics, Biophysics, and Nonlinear Dynamics of Cardiac Function*, Springer, New York (1991), pp. 121–144.
6. J.M. Guccione and A.D. McCulloch, Mechanics of active contraction in cardiac muscle: Part I – constitutive relations for active fiber stress that describe deactivation. *ASME J. Biomech. Engng.* **115** (1993) 72–81.
7. J.M. Guccione, K.D. Costa and A.D. McCulloch, Finite element stress analysis of the left ventricular mechanics in the beating dog heart. *J. Biomechanics* **28** (1995) 1167–1177.
8. R.M. Huisman, G. Elzinga, N. Westerhof and P. Sipkema, Measurement of left ventricular wall stress. *Cardiovasc. Res.* **14** (1980) 142–153.
9. J.D. Humhrey and F.C.P. Yin, On constitutive relations and finite deformations of passive cardiac tissue: II. Stress analysis in the left ventricle. *Circ. Res.* **65** (1989) 805–817.
10. P.J. Hunter, A.D. McCulloch and H.E.D.J. ter Keurs, Modeling the mechanical properties of cardiac muscle. *Progress in Biophysics & Molecular Biology* **69** (1998) 289–331.
11. B.Ya. Kantor, M.I. Yabluchansky and T.P. Usyk, Cubic potential of the transversely isotropic material taking into account the shear deformation. *School of Fundamental Medicine J.* **1** (1995) 14–18.
12. B.Ya. Kantor, M.I. Yabluchansky and T.P. Usyk, The mathematical model of the left ventricle by transversely isotropic body of revolution. *School of Fundamental Medicine J.* **3** (1997) 19–23.
13. B.Ya. Kantor, M.I. Yabluchansky and T.P. Usyk, The left ventricle mechanics (transversely isotropic body of revolution). In: *Proceedings of the 11th Conference of the European Society of Biomechanics, 8–11 July 1998, Toulouse, France*, J. Biomech. 31, Suppl. 1 (1998), p. 78.

14. W.L. Karlon, Influence of myocardial fiber organization on ventricular function, Ph.D. Thesis, UC San Diego, La Jolla, CA (1998).
15. I.J. LeGrice, B.H. Smill, L.Z. Chai, S.G. Edgar, J.B. Gavin and P.J. Hunter, Laminar structure of the heart: Ventricular myocyte arrangement and connective tissue architecture in the dog. *Am. J. Physiol.* **269** (1995) 571–582.
16. I.J. LeGrice, Y. Takayama and J.W. Covell, Transverse shear along myocardial cleavage planes provides a mechanism for normal systolic wall thickening. *Circ. Res.* **77** (1995) 182–193.
17. D.H.S. Lin and F.C.P. Yin, A multiaxial constitutive law for mammalian left ventricular myocardium in steady-state barium contracture of tetanus. *ASME J. Biomech. Engng.* **120** (1998) 504–517.
18. K.D. May-Newman, J.H. Omens, R.S. Pavelec and A.D. McCulloch, Three-dimensional transmural mechanical interaction between the coronary vasculature and passive myocardium in the dog. *Circ. Res.* **74** (1994) 1166–1178.
19. R. Mazhari and A.D. McCulloch, Three-dimensional mechanics of myocardial contraction: mechanisms of transverse systolic stress. In: *Proceeding of ASME 4th Summer Bioengineering Conference* (1999), pp. 43–44.
20. P.M.F. Nielsen, I.J. LeGrice, B.H. Smaill and P.J. Hunter, Mathematical model of geometry and fibrous structure of the heart. *Am. J. Physiol.* **260** (*Heart Circ. Physiol.* **29**) (1991) 1365–1378.
21. V.P. Novak, F.C.P. Yin and J.D. Humphrey, Regional mechanical properties of passive myocardium. *J. Biomech.* **27** (1994) 403–412.
22. W. O'Dell, C. Moore, W. Hunter, E. Zerhouni and E. McVeigh, Displacement field fitting approach to calculate 3D deformations from parallel-tagged MR images. *Radiology* **195** (1995) 829–835.
23. W. O'Dell, J. Schoeniger, S. Blackband and E. McVeigh, A modified quadrapole gradient set for use in high resolution MRI tagging. *Magnetic Resonance in Medicine* **32** (1994) 246–250.
24. J.H. Omens, K.D. May and A.D. McCulloch, Transmural distribution of three-dimensional strain in the isolated arrested canine left ventricle. *Am. J. Physiol.* **261** (1991) 918–928.
25. S.H. Peng and W.V. Chang, A compressible approach in finite element analysis of rubber-elastic materials. *Computers & Structures* **62** (1997) 573–593.
26. M. Schoenberg, Geometric factors influencing muscle force development, II. Radial forces. *Biophys. J.* **30** (1980) 51–68.
27. H.M. Spotnitz, W.D. Spotnitz, T.S. Cotrell, D. Spiro and E.H. Sonnenblick, Celluar basis for volume-related wall thickness changes in the rat left ventricle. *J. Mol. Cell Cardiol.* **6** (1974) 317–331.
28. D.D. Streeter, Jr., Gross morphology and fiber geometry of the heart. In: *Handbook of Physiology, Section 2: The Cardiovascular System*, Chapter 4. M BR. Bethesda, MD, American Physiological Society, I:61.
29. D.D. Streeter, Jr., H.M. Spotnitz, D.P. Patel, J. Ross, Jr. and E.H. Sonnenblick, Fiber orientation in the canine left ventricle during diastole and systole. *Circ. Res.* **24** (1969) 339–347.
30. Y. Takayama, K.D. Costa and J.W. Covell, Contribution of laminar myofiber architecture to load dependent changes in the mechanics of the left ventricular myocardium (submitted).
31. F.J. Vetter and A.D. McCulloch, Three-dimensional stress and strain in passive rabbit left ventricle: A model study. *Ann. Biomed Eng.* **28** (2000) 781–792.
32. F.C.P. Yin, R.K. Strumpf, P.H. Chew and S.L. Zeger, Quantification of the mechanical properties of noncontracting canine myocardium under simultaneous biaxial loading. *J. Biomech.* **20** (1987) 577–589.
33. G.I. Zahalak, Non-axial muscle stress and stiffness. *J. Theoret. Biology* **182** (1996) 59–84.

Journal of Elasticity **61:** 165–197, 2000.

Modeling Heart Development

LARRY A. TABER[a] and RENATO PERUCCHIO[b]
[a]*Department of Biomedical Engineering, Washington University, St. Louis, MO 63130, U.S.A.*
E-mail: lat@biomed.wustl.edu

[b]*Department of Mechanical Engineering, University of Rochester, Rochester, NY 14627, U.S.A.*
E-mail: rlp@me.rochester.edu

Received 20 April 2000; in revised form 24 October 2000

Abstract. Mechanics plays a major role in heart development. This paper reviews some of the mechanical aspects involved in theoretical modeling of the embryonic heart as it transforms from a single tube into a four-chambered pump. In particular, large deformations and significant alterations in structure lead to highly nonlinear boundary value problems. First, the biological background for the problem is discussed. Next, a modified elasticity theory is presented that includes active contraction and growth, and the theory is incorporated into a finite element analysis. Finally, models for the heart are presented to illustrate the developmental processes of growth, remodeling, and morphogenesis. Combining such models with appropriate experiments should shed light on the complex mechanisms involved in cardiac development.

Key words: development, morphogenesis, cardiac mechanics, finite elements, models.

1. Introduction

During development, the heart transforms from a single tube into a four-chambered pump. This transformation involves a dynamic interaction between genetic and environmental factors that regulate the primary developmental processes of growth (volume change), remodeling (property change), and morphogenesis (shape change) [1, 2]. One of the most important environmental factors is mechanics, which plays both direct and indirect roles throughout development.

On the indirect side, mechanical stress and strain influence the actions of certain genes [3]. This paper, however, focuses on direct effects, as manifested in each of the primary processes:

Growth. During development, the human heart increases in size by two orders of magnitude. In addition to this global growth, regional growth produces residual stress [4–7], which can affect cardiac function. Moreover, a substantial body of evidence suggests that mechanical stress or strain plays a major role in regulating growth of the heart throughout life [1].

Remodeling. Initially, the presumptive heart muscle (myocardium) is a thin layer of epithelial cells that forms the outer layer of the cardiac tube. As these cells differentiate into muscle, the proteins actin and myosin accumulate in the cytoplasm, forming sarcomeres and myofibrils that align gradually to create the highly ordered architecture of the mature heart. In addition, collagen fibers link together to build their own structure in the extracellular matrix [8]. This remodeling alters the stiffness and anisotropy of the myocardium during development.

Morphogenesis. Creating a heart requires extremely complex deformations that result from both passive and active forces. In models for morphogenesis, the large tissue deformations and continuous changes in geometry lead to highly nonlinear and computationally intensive boundary value problems.

The purpose of this paper is to delineate some of the theoretical and computational issues involved in biomechanical studies of heart development. First, we give an overview of some of the critical stages of development. Then, a unified mechanical theory is presented that includes most of the major mechanical features of the three primary processes. (Although viscous forces may be significant during development, we treat the embryonic tissues as pseudoelastic, i.e., the loading and unloading responses are defined by separate elastic constitutive relations [9].) Next, methods for implementing this theory in a finite element (FE) formulation are discussed. Finally, illustrative examples are given for each developmental process. Our intent is to stimulate interest among mechanicists in these and related problems.

2. Overview of Heart Development

The heart is the first functioning organ in the embryo. Remarkably, although its morphology changes dramatically during development, the heart functions without interruption to serve the metabolic needs of the rapidly growing embryo. Thus, the changes that occur during development must be closely coordinated. Any error in position or timing may lead to serious congenital malformations. For background, this section gives a relatively brief overview of some of the main features of cardiac development; more detailed descriptions can be found in [10–12].

This paper focuses primarily on the embryonic chick heart, a popular model in studies of development. Except for the time scale, development of the chick heart parallels that of the human heart. Hamburger and Hamilton [13] divided the 21-day incubation period of the chick into 46 stages based on external characteristics of the developing embryo.

2.1. FORMATION AND FUNCTION OF THE CARDIAC TUBE

During the first day of incubation in the chick (day 19 post-fertilization in humans), heart development begins with membranes forming a pair of tubes on opposite sides of the embryo (Figure 1(a)). These structures then merge and fuse to form a single cardiac tube (about 200 μm in diameter) composed of three layers: a relatively thin outer layer of myocardium, a middle layer of cardiac jelly, and an

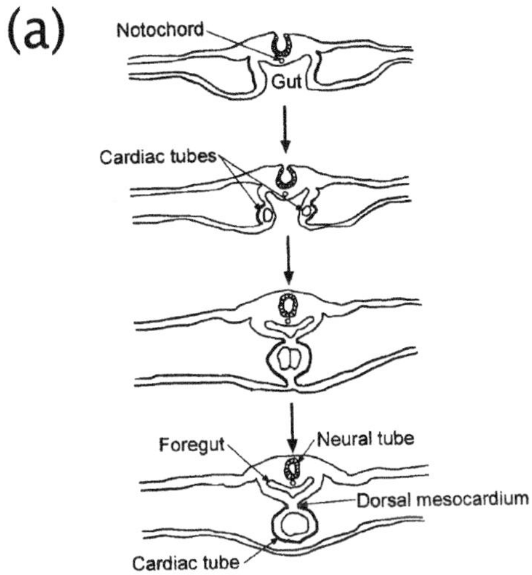

Figure 1. (a) Formation of the cardiac tube. Transverse sections of the chick embryo show bilateral epithelial tubes forming and merging along the centerline of the embryo. Reprinted from [2] with permission from Elsevier Science. (b) Schematic of the structure of the early heart tube, which is composed of three layers: myocardium (MC), cardiac jelly (CJ), and endocardium (EN). The dorsal mesocardium (DM) connects the tube to the embryo. From [55].

inner layer of endocardium (Figure 1(b)). The myocardium is composed of cells containing the only contractile elements in the primitive heart; the cardiac jelly (CJ) is extracellular matrix (ECM); and the one-cell-thick endocardium lines the lumen of the tube.

In the chick, contractions begin at stage 9 (1 day of incubation; 21 days post-fertilization in humans), just after the cardiac tube forms. Effective blood flow, however, does not begin until stage 12 (2 days). Initially, a peristaltic wave of contraction drives the blood through the heart. By day three, however, the flow becomes pulsatile, with regional wall thickenings of CJ called endocardial cushions in the inflow and outflow regions functioning as primitive valves (Figure 2). As the contractile wave passes, the cushions close the lumen locally to prevent backflow.

Despite of the lack of formed valves, the physiologic characteristics of the tubular heart are remarkably similar to those of the mature heart [14–16]. For instance, the ventricular pressure-time curves in the embryo have the distinctive forms of those measured in the mature left ventricle. The peak systolic pressure in the stage 16 (2.3 days) embryonic chick heart is only about 1 to 2 mmHg, however, as compared to about 120 mmHg in the mature heart.

2.2. CARDIAC LOOPING

Soon after the onset of contraction, the morphogenetic process of looping begins, as the heart bends and twists into a curved tube (Figure 3) [17]. Also during this period, a series of bulges develops along the tube (Figure 2): the sinus venosus,

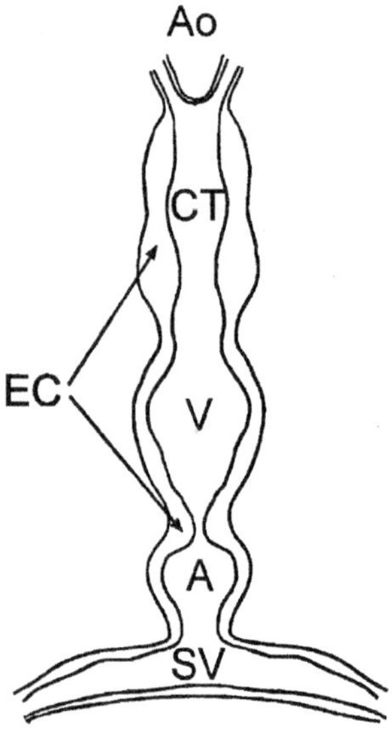

Figure 2. Schematic of the internal structure of the (straightened) cardiac tube after looping. A series of bulges form along the tube, with endocardial cushions acting as primitive valves. Later, the bulges become the four heart chambers. SV: sinus venosus; A: primitive atrium; V: ventricle; CT: conotruncus; Ao: aortic arches; EC: endocardial cushions. Reprinted from [2] with permission from Elsevier Science.

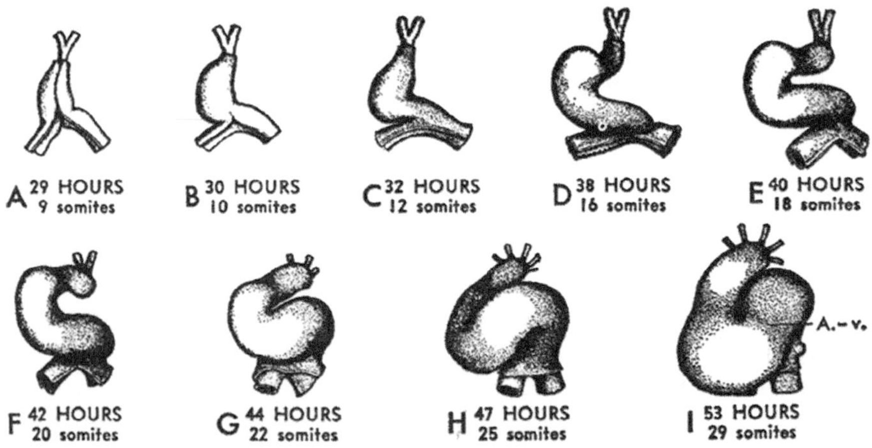

Figure 3. Schematic of cardiac looping in the chick embryo. From [66].

the primitive atrium, the ventricle, and the conotruncus. Looping effectively brings these bulges into their correct relative anatomic positions, setting the basic pattern for the final form of the heart (Figure 3). In the chick, this process is nearly completed by stage 16 (2.3 days).

It is important to note that the tubular heart initially is attached to the embryo along its entire length through a longitudinal structure called the dorsal meso-

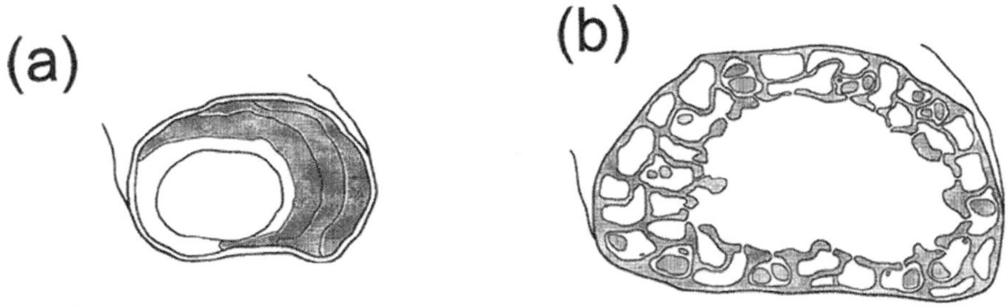

Figure 4. Trabeculation of the embryonic heart. (a) Circumferential endocardial ridges form at stage 18 (3 days) in the chick. (b) By stage 21 (3.5 days), the wall of the heart resembles a sponge with a thin outer layer of compact myocardium. Reprinted from [2] with permission from Elsevier Science.

cardium (DM) (Figure 1). This structure is part of the epithelium that becomes the outer myocardial layer, but it does not differentiate into muscle. During looping, the DM ruptures, leaving the heart tube supported only at its ends.

2.3. MYOCARDIAL TRABECULATION AND COMPACTION

As the heart wall thickens, the metabolic needs of the myocardium no longer can be met entirely by diffusion from the blood in the lumen. Thus, beginning at stage 17 (3 days), the CJ is replaced rapidly by a network of muscular sheets and struts (trabeculae), which begin as a series of circumferential endocardial ridges (Figure 4(a)) [18, 19]. By stage 21 (3.5 days), the ventricular wall is composed of sponge-like muscle surrounded by a thin outer layer of compact myocardium (Figure 4(b)). During these stages, direct blood flow through the trabecular spaces supplies nutrients to the myocardium. Trabeculation continues until stage 25 (4.5 days), when the myocardium begins to compact and the coronary arteries bud.

2.4. SEPTATION AND VALVE FORMATION

Between stages 21 (3.5 days) and 36 (10 days), septation completes the transformation of the cardiac tube into a four-chambered heart (Figure 5). Beginning as coalescing trabeculae, the interventricular septum gradually grows inward near the bottom (apex) of the heart, dividing the tube into the left and right ventricles. In addition, another septum grows to divide the primitive atrium into the left and right atria, and the endocardial cushions are molded into valves. At this time, the heart resembles its mature form. Except for some fine tuning, growth dominates most of the rest of heart development.

2.5. FIBER GEOMETRY

Relatively little is known about the fiber architecture in the developing heart. Scanning electron micrographs suggest that randomly oriented sarcomeres align gradu-

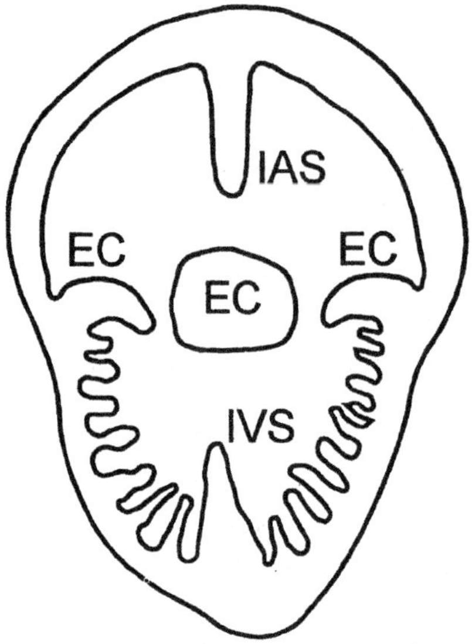

Figure 5. Septation of the cardiac tube. The interventricular septum (IVS), interatrial septum (IAS), and the endocardial cushions (EC) grow inward, dividing the tube into four chambers. Reprinted from [2] with permission from Elsevier Science.

ally during development, and several investigators have described circumferentially arranged actin filaments in the inner layer of the two-cell-thick myocardium as early as stage 10 [20–23]. It is not clear, however, how the fibers become aligned to form the highly ordered pattern of the mature heart [24].

3. Theory

Growth and contractile forces play major roles in cardiac development. This section presents a modified elasticity theory that includes both effects. In essence, the theory combines and expands the theory for volumetric growth of Rodriguez et al. [25] and the theories for tissue morphogenesis of Odell et al. [26] and Oster et al. [27]. (Readers not familiar with elasticity may want to refer to the introductory chapter of [28].)

3.1. KINEMATICS

Active contraction and growth are modeled through changes in the zero-stress configuration for a tissue element. For cellular tissues, it is convenient to consider an "element" as a cell that is approximately stress free when unloaded in isolation. (In acellular tissue, an element is a small material region.) Actually, the internal structures in an isolated cell may be still under self-equilibrating stresses, but we will ignore these stresses here. Furthermore, we assume that contraction and growth change the dimensions of an isolated cell without introducing stress.

Active Contraction. Two types of contractile activity occur in the developing heart. The first type is the rhythmic contraction of the heart that occurs during each cardiac cycle, while the other is a prolonged contraction that drives morphogenesis. The second type of contraction, which occurs in both muscle and nonmuscle cells, is similar to the smooth muscle contraction that occurs in arteries.

Presumably, both types of contraction involve actin-myosin interactions, and we treat them similarly. The physical apparatus for each is probably different, however. For example, primitive myocardial cells (cardiocytes) contain some sarcomeric filaments devoted to beating and other cytoskeletal filaments that may produce shape change. It is notable that most of the major morphogenetic changes in the heart occur before these cells become filled with sarcomeres. Moreover, modelers can take advantage of the fact that sarcomeric contraction occurs on a much shorter time scale (order of a second) than does cytoskeletal contraction and growth (order of hours), i.e., the cyclic stress can be uncoupled computationally from the developmental processes.

Suppose that a rectangular cell contains contractile filaments aligned along the x_1-direction. Then, an **active deformation gradient tensor** can be written in the form

$$\mathbf{F}_a = \lambda_a \mathbf{e}_1 \otimes \mathbf{e}_1 + \lambda_a^{-1/2}(\mathbf{e}_2 \otimes \mathbf{e}_2 + \mathbf{e}_3 \otimes \mathbf{e}_3), \tag{1}$$

where the $\mathbf{e}_i$ are Cartesian unit vectors and λ_a is the activation stretch ratio, i.e., the ratio of the active length to the passive length of the cell in the x_1-direction. During a muscle twitch, for example, λ_a decreases from unity (passive state) to a minimum value (peak activation) and then returns to one. In addition, $\det \mathbf{F}_a = 1$ satisfies the condition of cellular incompressibility. Early in development, when the contractile filaments may be oriented more or less randomly in the x_1x_2-plane, we can take

$$\mathbf{F}_a = \lambda_a(\mathbf{e}_1 \otimes \mathbf{e}_1 + \mathbf{e}_2 \otimes \mathbf{e}_2) + \lambda_a^{-2}\mathbf{e}_3 \otimes \mathbf{e}_3. \tag{2}$$

Growth. Tissue growth is similar kinematically to a negative contraction with one notable exception: growth is not an isochoric process. Volumetric growth can occur by cell division (hyperplasia), increase in cell size (hypertrophy), or accumulation of ECM. Negative growth (atrophy), which also plays an important role during development, is produced by oppositely directed processes. (Negative cell division is cell death.) In terms of mechanics, all of these mechanisms have similar effects, and we make no distinction here between the types of growth. We note, however, that the heart grows primarily by hyperplasia before birth and by hypertrophy after birth [29, 30].

If angles are preserved, then the **growth deformation gradient tensor** for an isolated stress-free element can be written

$$\mathbf{F}_g = \lambda_{g1}\mathbf{e}_1 \otimes \mathbf{e}_1 + \lambda_{g2}\mathbf{e}_2 \otimes \mathbf{e}_2 + \lambda_{g3}\mathbf{e}_3 \otimes \mathbf{e}_3, \tag{3}$$

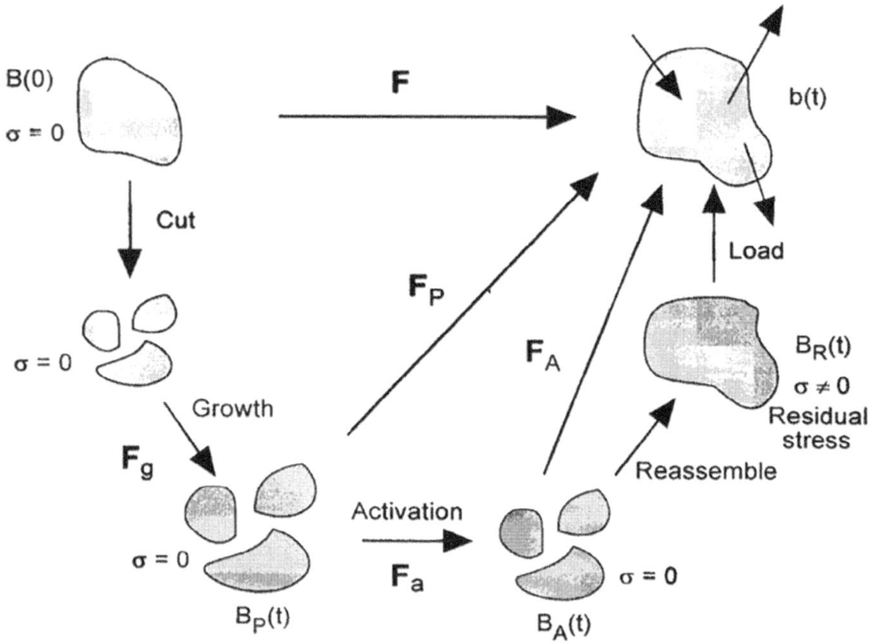

Figure 6. Configurations of a body from the reference state $B(0)$ to the loaded state $b(t)$, including growth and activation.

where the λ_{gi} represent the ratios of lengths after to the lengths before growth ($\lambda_{gi} > 1$ for growth and $\lambda_{gi} < 1$ for atrophy), and $\det \mathbf{F}_g \neq 1$ in general. As discussed later, the λ_{gi} are to be determined from a growth law.

Combined Activation, Growth, and Loading of a Tissue. Thus far, we have considered active contraction and growth on a cellular level. The interaction between these processes on a tissue level is perhaps best visualized through a series of configurations [25]. Consider an elastic body in configuration $B(0)$ that is unloaded and stress-free at $t = 0$ (Figure 6). Imagine now that B is cut into infinitesimal elements, which then grow by $\mathbf{F}_g$ into the passive zero-stress state $B_P(t)$ and activate by $\mathbf{F}_a$ to form the active zero-stress state $B_A(t)$. Next, the elements are reassembled into the configuration $B_R(t)$. Unless the elements of B_R are geometrically compatible, this assembly requires deformation that produces residual stress. Finally, loads are applied to $B_R(t)$, giving $b(t)$.

At this point, we note the following. First, $\mathbf{F}_P$ and $\mathbf{F}_A$ are stress-generating **elastic deformation gradient tensors** relative to the passive and active zero-stress states, respectively (Figure 6). Second, the intermediate configuration B_R is a special case of b for no loads, with $\mathbf{F}_P$ and $\mathbf{F}_A$ providing passive and active residual stresses. Third, the total deformation of b relative to B is $\mathbf{F} = \mathbf{F}_P\mathbf{F}_g = \mathbf{F}_A\mathbf{F}_a\mathbf{F}_g$. Finally, in the absence of activation and growth, $\mathbf{F}_a = \mathbf{F}_g = \mathbf{I}$ with $\mathbf{I}$ being the identity tensor.

3.2. MECHANICAL CONSTITUTIVE RELATIONS

In formulating boundary value problems, since the zero-stress state is time-dependent, it is convenient to refer the field equations to the reference configuration B. Written in terms of $\mathbf{F}$, these equations assume their standard forms. In the presence of activation or growth, however, it is convenient to express the mechanical constitutive equations in terms of $\mathbf{F}_P$ and $\mathbf{F}_A$. Once $\mathbf{F}_g$ and $\mathbf{F}_a$ are known, these tensors can be computed from $\mathbf{F}_P = \mathbf{F}\mathbf{F}_g^{-1}$ and $\mathbf{F}_A = \mathbf{F}\mathbf{F}_g^{-1}\mathbf{F}_a^{-1}$.

General Form. Typically, the strain-energy density (SED) function is assumed to be composed of separate passive and active parts: $W_P(\mathbf{F}_P)$ and $W_A(\mathbf{F}_A)$. Then, the Cauchy stress tensor can be written

$$\boldsymbol{\sigma} = J_P^{-1}\mathbf{F}_P\frac{\partial W_P}{\partial \mathbf{E}_P}\mathbf{F}_P^{\mathrm{T}} + J_A^{-1}\mathbf{F}_A\frac{\partial W_A}{\partial \mathbf{E}_A}\mathbf{F}_A^{\mathrm{T}} - p\mathbf{I}, \tag{4}$$

where p is a Lagrange multiplier and T denotes transpose, while $\mathbf{E}_P = (\mathbf{F}_P^{\mathrm{T}}\mathbf{F}_P - \mathbf{I})/2$ and $\mathbf{E}_A = (\mathbf{F}_A^{\mathrm{T}}\mathbf{F}_A - \mathbf{I})/2$ are Lagrangian strain tensors. For an incompressible material, the *elastic* deformations must satisfy the incompressibility conditions

$$J_P = \det \mathbf{F}_P = 1, \qquad J_A = \det \mathbf{F}_A = 1, \tag{5}$$

and $p = 0$ for a compressible material.

Specific Form. Relatively little is known about the material properties of the embryonic heart. Using available pressure–volume curves, epicardial strain measurements, and other data, we have proposed possible forms for W_P and W_A [31–34]. Models for the developing heart require properties for three basic types of material: compact myocardium (active, incompressible, transversely isotropic), porous myocardium (active, compressible, isotropic), and cardiac jelly (passive, incompressible, isotropic). Consolidating these possibilities yields the proposed functions

$$\begin{aligned} W_P &= \frac{a}{b}\left(\mathrm{e}^{bQ_P} - 1\right) + \frac{a_f}{b_f}\left[\mathrm{e}^{b_f(\lambda_{fP}-1)^2} - 1\right], \\ W_A &= \frac{c(t)}{d(t)}Q_A^{d(t)} + \frac{c_f(t)}{d_f(t)}\left(\lambda_{fA}^{\;d_f(t)} + \lambda_{fA}^{\;-d_f(t)} - 2\right), \end{aligned} \tag{6}$$

where λ_{f_P} and λ_{f_A} are the passive and active stretch ratios, respectively, in the fiber direction. In each of these expressions, the first term represents the isotropic part of the strain energy, and the second term gives the contribution along the fiber direction. Furthermore, a, b, c, d and a_f, b_f, c_f, d_f are material coefficients, and

$$Q = I_1 - 3 + \frac{1-2\nu}{\nu}\left[I_3^{-\nu/(1-2\nu)} - 1\right] \tag{7}$$

in which the I_i are traditional strain invariants, ν is a constant, and $Q_P = Q(\mathbf{F}_P)$ and $Q_A = Q(\mathbf{F}_A)$. This expression for Q corresponds to the SED function for a generalized Blatz–Ko material [35].

Equations (6) can be specialized for each component tissue. For instance, $I_3 = 1$ for an incompressible material, and the second term in Q drops out. In addition, $a_f = c_f = 0$ for isotropic tissues, while $c = 0$ for transversely isotropic muscle. The expressions for W_P and W_A in (6) are reasonable first approximations for the developing heart. As new data become available, however, they likely will require modification.

Activation Parameters. From a phenomenological point of view, activation is characterized primarily by two time-dependent phenomena. One is the local change in zero-stress configuration (as discussed above), and the other is stiffening of the tissue.* Here, the latter effect is included on the local level by letting c, d, c_f, and d_f be specified functions of time t during activation. During sarcomeric contraction, for example, we let $c(t)$ and $c_f(t)$ increase from $c = c_f = 0$ at end diastole to maximum values at end systole and then return to their original values at the onset of passive filling ($d(t)$ and $d_f(t)$ are taken as constants). At the same time, λ_a (see equation (1)) decreases from unity to a minimum value and then returns to one. During cytoskeletal contraction, however, all of these functions increase/decrease asymptotically to their peak activation values.

3.3. GROWTH LAW

In the present models, although $\mathbf{F}_a(t)$ is specified, $\mathbf{F}_g(t)$ is assumed to depend on the local stress field. To determine a possible form for a growth law, we must examine the available biological data.

Heart muscle exhibits two basic types of growth response. To accommodate an increased end-diastolic volume (volume overload), the cavity of the mature left ventricle grows larger while its wall thickness increases relatively little. In response to an elevated end-systolic pressure (pressure overload), on the other hand, the ventricular wall thickens while the cavity volume remains essentially unchanged. The increased wall thickness returns peak stresses to near normal levels [30]. Apparently, the embryonic heart responds similarly to load perturbations [36]. Since myocardial fibers are oriented essentially parallel to the surface of the heart, these observations suggest that heart muscle fibers grow longer in response to elevated passive stress and thicker in response to elevated active stress.

Based on these observations, we postulate that the developing (and mature) heart, adapts to altered loading conditions by growing in such a way as to return wall stresses to optimal set points. Thus, the growth law may have the form

$$\begin{aligned} \dot{\lambda}_{g1} &= A(\sigma_{1P} - \sigma_{0P}), \\ \dot{\lambda}_{g2} &= \dot{\lambda}_{g3} = B(\sigma_{1A} - \sigma_{0A}), \end{aligned} \tag{8}$$

where the λ_{gi} are the growth stretch ratios defined in equation (3), and A and B are constants. In addition, σ_{1P} and σ_{1A} are the respective passive and active Cauchy

* The effects of contraction velocity, calcium concentration, and other factors are neglected.

stresses along the fiber direction x_1, and subscript 0 indicates growth-equilibrium stresses. (Actually, it may be more appropriate to replace the left hand sides of equations (8) by the Eulerian growth rates $\dot{\lambda}_{gi}/\lambda_{gi}$ [25].)

Here, we note that if the growth-equilibrium stresses always are equal to their mature values, then the heart would atrophy early in development, when the blood pressure is extremely small. Since this is not realistic, and if the growth law (8) is valid, then the growth-equilibrium stresses must increase from zero during development and approach their homeostatic values at maturity [37].

4. Finite Element Methods

The primary issues in the FE formulation are finite strains, material nonlinearity, anisotropy, and the modeling of the residual stresses due to activation and growth. To establish the kinematic relations, we consider the motion of a body in a fixed Cartesian coordinate system and let x_i^0 and x_i^t denote the position of point P at times 0 and t, respectively. The components of the displacement vector $\mathbf{u}^t$ are given by $u_i^t = x_i^t - x_i^0$, and the components of the deformation gradient tensor ${}_0^t\mathbf{F}$ are ${}_0^tF_{m,n} = \partial x_m^t/\partial x_n^0$. Hereafter, following the notation used in [38], the left superscript denotes the time at which a quantity is measured while the left subscript refers to the configuration with respect to which the measurement is made. For example, ${}_0^tE_{ij}$ denotes the components of the strains at time t measured with respect to the configuration at time 0.

We adopt a mixed displacement-pressure formulation to model the incompressibility of the cardiac tissue [39]. An independent variable – the so-called hydrostatic pressure – is introduced in the form of a Lagrange multiplier enforcing the incompressibility condition in the augmented SED function

$$W = W_e + {}_0^tP\left({}_0^tI_3 - 1\right), \tag{9}$$

where W_e is the SED function in terms of Green–Lagrangian (GL) strains describing the elastic constitutive relation of the tissue, ${}_0^tP$ is the current hydrostatic pressure (i.e., occurring at time t) measured in the initial configuration (i.e., at time 0); and ${}_0^tI_3$ is the third strain invariant, which must be equal to 1 to enforce the isovolumetric (incompressibility) condition. Then, with inertia neglected for the relatively slow (quasi-static) developmental processes, the total potential energy is given by

$$\Pi\left({}_0^tE_{ij}, {}_0^tP\right) = \int_{{}^0v} \left[W_e + {}_0^tP\left({}_0^tI_3 - 1\right)\right] \mathrm{d}^0v + V, \tag{10}$$

where ${}_0^tE_{ij}$ are the GL strains, ${}^0v$ is the volume of the configuration at time 0, and V is the work performed by the external loads.

Following a variational approach, we derive the finite element matrices by imposing the stationary condition of the total potential energy

$$\delta\Pi = \frac{\partial\Pi}{\partial\,{}_0^tE_{ij}}\delta\,{}_0^tE_{ij} + \frac{\partial\Pi}{\partial\,{}_0^tP}\delta\,{}_0^tP = 0. \tag{11}$$

Since the GL strains and the hydrostatic pressure are independent variables, each of the two variations must vanish separately, yielding

$$\frac{\partial \Pi}{\partial\, {}_0^t E_{ij}} = 0, \tag{12}$$

$$\frac{\partial \Pi}{\partial\, {}_0^t P} = 0. \tag{13}$$

Equations (12) and (13) are the equation of motion (equilibrium) and incompressibility, respectively.

4.1. EQUATION OF MOTION

In order to apply an incremental solution method, we express the equation of motion at time $t + \Delta t$ in the virtual work form

$$\int_{{}^0v} {}^{t+\Delta t}_{\ \ 0}S_{ij}\,\delta\, {}^{t+\Delta t}_{\ \ 0}E_{ij}\,\mathrm{d}^0v = {}^{t+\Delta t}R \tag{14}$$

as given by equations (10) and (12). Here, t denotes the last computed equilibrium configuration, $t + \Delta t$ is the new unknown equilibrium configuration, ${}^{t+\Delta t}_{\ \ 0}S_{ij}$ are 2nd Piola–Kirchhoff (PK) stresses, and ${}^{t+\Delta t}R$ is the current external virtual work. The stress tensor ${}^{t+\Delta t}_{\ \ 0}S_{ij}$ is decomposed in the incremental form as

$${}^{t+\Delta t}_{\ \ 0}S_{ij} = {}_0^t S_{ij} + {}_0S_{ij}, \tag{15}$$

where ${}_0^t S_{ij}$ is the (known) 2nd PK stress at the last equilibrium configuration and ${}_0S_{ij}$ is the (yet to be computed) stress increment. By definition,

$${}_0^t S_{ij} = \frac{\partial W}{\partial\, {}_0^t E_{ij}} = {}_0^t S_{ij}^e + {}_0^t P\, {}_0^t S_{ij}^h, \tag{16}$$

where

$${}_0^t S_{ij}^e = \frac{\partial W_e}{\partial\, {}_0^t E_{ij}}, \qquad {}_0^t S_{ij}^h = \frac{\partial\, {}_0^t I_3}{\partial\, {}_0^t E_{ij}}. \tag{17}$$

The stress increment is defined in terms of the unknown increments ${}_0E_{rs}$ and ${}_0P$ by linearizing the Taylor expansion of ${}^{t+\Delta t}_{\ \ 0}S_{ij}$ to obtain

$$\begin{aligned} {}_0S_{ij} &= \frac{\partial\, {}_0^t S_{ij}}{\partial\, {}_0^t E_{rs}}\, {}_0E_{rs} + \frac{\partial\, {}_0^t S_{ij}}{\partial\, {}_0^t P}\, {}_0P \\ &= \left(\frac{\partial^2 W_e}{\partial\, {}_0^t E_{ij}\, \partial\, {}_0^t E_{rs}} + {}_0^t P \frac{\partial^2\, {}_0^t I_3}{\partial\, {}_0^t E_{ij}\, \partial\, {}_0^t E_{rs}} \right) {}_0E_{rs} + \frac{\partial\, {}_0^t I_3}{\partial\, {}_0^t E_{ij}}\, {}_0P, \end{aligned} \tag{18}$$

where ${}_0E_{ij} = {}^{t+\Delta t}_{\ \ 0}E_{ij} - {}_0^t E_{ij}$ and ${}_0P = {}^{t+\Delta t}_{\ \ 0}P - {}_0^t P$. Finally, we define the elastic constitutive tensor ${}_0C_{ijrs}^e = \partial^2 W_e / \partial\, {}_0^t E_{ij}\, \partial\, {}_0^t E_{rs}$ and the hydrostatic pressure counterpart ${}_0C_{ijrs}^h = \partial^2\, {}_0^t I_3 / \partial\, {}_0^t E_{ij}\, \partial\, {}_0^t E_{rs}$ and rewrite equation (18) as

$${}_0S_{ij} = \left({}_0C_{ijrs}^e + {}_0^t P\, {}_0C_{ijrs}^h\right) {}_0E_{rs} + {}_0^t S_{ij}^h\, {}_0P. \tag{19}$$

4.2. UPDATED LAGRANGIAN FORMULATION

We make at this point an important observation about the modeling of the embryonic cardiac tissue. In theory, any previous equilibrium configuration can be used as the reference configuration in equation (14). In practice, two reference configurations are used: either the stress-free initial configuration (at time 0) or the last computed equilibrium configuration (at time t). The first approach provides the Total Lagrangian (TL) formulation, while the second results in the Updated Lagrangian (UL) formulation. In general, the two approaches differ solely in terms of computational costs, the UL being more efficient because it leads to a simpler strain-displacement relationship. In the particular case of the embryonic heart, there is another reason for adopting an UL approach. The initial passive configuration $B(0)$ may not be stress-free and the active stress-free configuration B_A is purely a theoretical construction (Figure 6), and therefore neither can be used to form the basis of a TL approach.

To formulate a UL approach, we refer stresses and strains at time $t + \Delta t$ to the configuration at time t through the following transformations [38]

$$
\begin{aligned}
{}^{t+\Delta t}_{\;0}S_{ij} &= \frac{{}^{0}\rho}{{}^{t}\rho}\,{}^{0}_{t}F_{i,m}\,{}^{0}_{t}F_{j,n}\,{}^{t+\Delta t}_{\;t}S_{ij}, \\
\delta\,{}^{t+\Delta t}_{\;0}E_{ij} &= {}^{t}_{0}F_{i,m}\,{}^{t}_{0}F_{j,n}\,\delta\,{}^{t+\Delta t}_{\;t}E_{mn},
\end{aligned}
\tag{20}
$$

where ${}^{0}_{t}F_{i,m}$ denotes an element of the deformation gradient tensor ${}^{0}_{t}\mathbf{F} = {}^{t}_{0}\mathbf{F}^{-1}$. Substituting equations (20) into equation (14), we obtain the equation of motion in terms of stresses, strains, and volume all referred to time t

$$
\int_{{}^{t}v} {}^{t+\Delta t}_{\;t}S_{ij}\,\delta\,{}^{t+\Delta t}_{\;t}E_{ij}\,\mathrm{d}\,{}^{t}v = {}^{t+\Delta t}R. \tag{21}
$$

4.3. INCREMENTAL FORM

Stresses and strains in equation (21) are rewritten in incremental form as

$$
\begin{aligned}
{}^{t+\Delta t}_{\;t}S_{ij} &= {}^{t}_{t}S_{ij} + {}_{t}S_{ij} = \sigma_{ij} + {}_{t}\sigma_{ij}, \\
{}^{t+\Delta t}_{\;t}E_{ij} &\equiv {}_{t}E_{ij} = {}_{t}e_{ij} + {}_{t}\eta_{ij},
\end{aligned}
\tag{22}
$$

where ${}^{t}_{t}S_{ij} \equiv \sigma_{ij}$ is the (known) Cauchy stress at the last equilibrium configuration, ${}_{t}\sigma_{ij}$ and ${}_{t}E_{ij}$ are the (yet to be determined) Cauchy stress increment and strain increment, and ${}_{t}e_{ij}$ and ${}_{t}\eta_{ij}$ are the linear and nonlinear components of ${}_{t}E_{ij}$, respectively [38]. Note that, by definition, ${}^{t}_{t}E_{ij} = 0$. The Cauchy stresses and stress increments are related to the corresponding 2nd PK stresses by [38]

$$
\sigma_{ij} = \frac{{}^{t}\rho}{{}^{0}\rho}\,{}^{t}_{0}F_{i,m}\,{}^{t}_{0}F_{j,n}\,{}^{t}_{0}S_{mn}, \qquad {}_{t}\sigma_{ij} = \frac{{}^{t}\rho}{{}^{0}\rho}\,{}^{t}_{0}F_{i,m}\,{}^{t}_{0}F_{j,n}\,{}_{0}S_{mn}. \tag{23}
$$

The other relevant transformations are [38]

$$
\begin{aligned}
\sigma_{ij}^{e} &= \frac{{}^{t}\rho}{{}^{0}\rho}\,{}^{t}_{0}F_{i,m}\,{}^{t}_{0}F_{j,n}\,{}^{t}_{0}S^{e}_{mn}, \qquad \sigma_{ij}^{h} = {}^{t}_{0}F_{i,m}\,{}^{t}_{0}F_{j,n}\,{}^{t}_{0}S^{h}_{mn},\\
{}_{t}C^{e}_{ijrs} &= \frac{{}^{t}\rho}{{}^{0}\rho}\,{}^{t}_{0}F_{i,m}\,{}^{t}_{0}F_{j,n}\,{}^{t}_{0}F_{r,k}\,{}^{t}_{0}F_{s,l}\,{}_{0}C^{e}_{mnkl},\\
{}_{t}C^{h}_{ijrs} &= {}^{t}_{0}F_{i,m}\,{}^{t}_{0}F_{j,n}\,{}^{t}_{0}F_{r,k}\,{}^{t}_{0}F_{s,l}\,{}_{0}C^{h}_{mnkl},\\
{}^{t}_{t}P &= \frac{{}^{t}\rho}{{}^{0}\rho}\,{}^{t}_{0}P; \qquad {}_{t}P = \frac{{}^{t}\rho}{{}^{0}\rho}\,{}_{0}P, \qquad {}_{t}E_{ij} = {}^{0}_{t}F_{i,m}\,{}^{0}_{t}F_{j,n}\,{}_{0}E_{mn}.
\end{aligned}
\tag{24}
$$

Substituting equations (16), (19), and (24) into (23) gives the expressions for the Cauchy stresses and stress increments

$$
\begin{aligned}
\sigma_{ij} &= \sigma_{ij}^{e} + {}^{t}_{t}P\sigma_{ij}^{h},\\
{}_{t}\sigma_{ij} &= \left({}_{t}C^{e}_{ijrs} + {}^{t}_{t}P\,{}_{t}C^{h}_{ijrs}\right){}_{t}E_{rs} + \sigma_{ij}^{h}\,{}_{t}P.
\end{aligned}
\tag{25}
$$

Next, after substituting equations (22) and (25) into equation (21) and linearizing, we obtain the UL equation of motion

$$
\begin{aligned}
&\int_{{}^{t}v}\left({}_{t}C^{e}_{ijrs} + {}^{t}_{t}P\,{}_{t}C^{h}_{ijrs}\right){}_{t}e_{rs}\delta\,{}_{t}e_{ij}\,\mathrm{d}^{t}v + \int_{{}^{t}v}(\sigma_{ij}^{e} + {}^{t}_{t}P\sigma_{ij}^{h})\delta\,{}_{t}\eta_{ij}\,\mathrm{d}^{t}v\\
&\quad + \int_{{}^{t}v}\sigma_{ij}^{h}\,{}_{t}P\delta\,{}_{t}e_{ij}\,\mathrm{d}^{t}v = {}^{t+\Delta t}R - \int_{{}^{t}v}(\sigma_{ij}^{e} + {}^{t}_{t}P\sigma_{ij}^{h})\delta\,{}_{t}e_{ij}\,\mathrm{d}^{t}v.
\end{aligned}
\tag{26}
$$

4.4. FINITE ELEMENT EQUILIBRIUM EQUATION

Following a standard FE approach, the incremental displacement ${}_{t}U$ and the incremental hydrostatic pressure ${}_{t}P$ within each element can be approximated by the product of a set of known base functions and a vector of unknown generalized coordinates

$$
{}_{t}U = \mathbf{N}\,{}_{t}\mathbf{u}, \qquad {}_{t}P = \mathbf{H}\,{}_{t}\mathbf{p}, \tag{27}
$$

where ${}_{t}\mathbf{u}$ are the incremental displacements at the element nodal points, $\mathbf{N}$ are the corresponding isoparametric shape functions, $\mathbf{H}$ are the base functions for the incremental hydrostatic pressure, and ${}_{t}\mathbf{p}$ the unknown pressure coefficients. The strain increments ${}_{t}e_{ij}$ and ${}_{t}\eta_{ij}$ in equation (26) are expressed in terms of ${}_{t}\mathbf{u}$ through the strain-displacement matrices ${}^{t}_{t}\mathbf{B}_{L}$ and ${}^{t}_{t}\mathbf{B}_{NL}$, respectively [38]. For each element, equation (26) is expressed in matrix form as

$$
{}^{t}\mathbf{K}_{u}\,{}_{t}\mathbf{u} + {}^{t}\mathbf{K}_{p}\,{}_{t}\mathbf{p} = {}^{t+\Delta t}\mathbf{R} - {}^{t}\mathbf{Q}_{u}, \tag{28}
$$

where

$$
\begin{aligned}
{}^{t}\mathbf{K}_{u} &= \int_{{}^{t}V}{}^{t}_{t}\mathbf{B}_{L}^{\mathrm{T}}\left[{}_{t}\mathbf{C}^{e} + {}^{t}_{t}P\,{}_{t}\mathbf{C}^{h}\right]{}^{t}_{t}\mathbf{B}_{L}\,\mathrm{d}^{t}V + \int_{{}^{t}V}{}^{t}_{t}\mathbf{B}_{NL}^{\mathrm{T}}\left[\boldsymbol{\sigma}^{e} + {}^{t}_{t}P\boldsymbol{\sigma}^{h}\right]{}^{t}_{t}\mathbf{B}_{NL}\,\mathrm{d}^{t}V,\\
{}^{t}\mathbf{K}_{p} &= \int_{{}^{t}V}{}^{t}_{t}\mathbf{B}_{L}^{\mathrm{T}}\boldsymbol{\sigma}^{h}\mathbf{H}\,\mathrm{d}^{t}V, \qquad {}^{t}\mathbf{Q}_{u} = \int_{{}^{t}V}{}^{t}_{t}\mathbf{B}_{L}^{\mathrm{T}}\left[\boldsymbol{\sigma}^{e} + {}^{t}_{t}P\boldsymbol{\sigma}^{h}\right]\mathrm{d}^{t}V
\end{aligned}
\tag{29}
$$

in which ${}^{t}V$ denotes the element volume and ${}^{t+\Delta t}\mathbf{R}$ the externally applied nodal forces at time $t + \Delta t$.

4.5. EQUATION OF INCOMPRESSIBILITY

At time $t + \Delta t$, equations (10) and (13) yield the equation of incompressibility

$$\int_{{}^{0}v} \left({}^{t+\Delta t}_{\ \ 0}I_3 - 1\right)\delta\, {}^{t+\Delta t}_{\ \ 0}P\, \mathrm{d}^0 v = 0. \tag{30}$$

As for the stresses and strains, we use a truncated Taylor expansion to express ${}^{t+\Delta t}_{\ \ 0}I_3$ as

$${}^{t+\Delta t}_{\ \ 0}I_3 = {}^{t}_{0}I_3 + \frac{\partial\, {}^{t}_{0}I_3}{\partial\, {}^{t}_{0}E_{ij}}\, {}_{0}E_{ij} = {}^{t}_{0}I_3 + {}^{t}_{0}S^{h}_{ij}\, {}_{0}E_{ij}. \tag{31}$$

Substituting equation (31) into (30), and applying the transformations defined in equation (24), we obtain a new linearized equation of incompressibility at configuration t

$$\int_{{}^{t}v} \sigma^{h}_{ij}\, {}_{t}e_{ij}\, \delta\, {}_{t}P\, \mathrm{d}^{t}v = \int_{{}^{t}v} \left(1 - {}^{t}_{0}I_3\right)\delta\, {}_{t}P\, \mathrm{d}^{t}v. \tag{32}$$

At the element level, inserting equation (27) into (32) yields

$${}^{t}\mathbf{K}^{\mathrm{T}}_{p}\, {}_{t}\mathbf{u} = {}^{t}\mathbf{Q}_{p}, \tag{33}$$

where

$${}^{t}\mathbf{K}^{\mathrm{T}}_{p} = \int_{{}^{t}V} \mathbf{H}^{\mathrm{T}} \sigma^{h}\, {}^{t}_{t}\mathbf{B}_{L}\, \mathrm{d}^{t}V, \qquad {}^{t}\mathbf{Q}_{p} = \int_{{}^{t}v} \mathbf{H}^{\mathrm{T}}\left(1 - {}^{t}_{0}I_3\right) \mathrm{d}^{t}v. \tag{34}$$

4.6. FINITE ELEMENT MATRICES

Combining the equilibrium and incompressibility equations, (28) and (33), gives the complete element matrix for the mixed formulation

$$\begin{bmatrix} {}^{t}\mathbf{K}_{u} & {}^{t}\mathbf{K}_{p} \\ {}^{t}\mathbf{K}^{\mathrm{T}}_{p} & \mathbf{0} \end{bmatrix} \begin{Bmatrix} {}_{t}\mathbf{u} \\ {}_{t}\mathbf{p} \end{Bmatrix} = \begin{Bmatrix} {}^{t+\Delta t}\mathbf{R} - {}^{t}\mathbf{Q}_{u} \\ {}^{t}\mathbf{Q}_{p} \end{Bmatrix}. \tag{35}$$

In actual computer implementation, the pressure generalized coordinates ${}_{t}\mathbf{p}$ are local to each element and can be eliminated by static condensation before element assembly. However, the null matrix on the diagonal prevents the direct elimination of the pressure terms. Adopting a penalty approach, we solve this numerical problem by replacing the null matrix with $\epsilon\mathbf{I}$, where $\mathbf{I}$ is the identity matrix, and ϵ is an *ad hoc* small number.

4.7. MODELING ANISOTROPY

The above formulation is suitable for any hyperelastic material law. In the finite element models of this paper, we use a transversely isotropic SED function of the form

$$W_e = \frac{a}{b}\{\exp[b(I_1 - 3)] - 1\} + cE_f^2 \tag{36}$$

for both the myocardium and CJ, where a, b, and c are material constants and E_f is the GL strain in the fiber direction. The two terms in this expression represent W_P and W_A, respectively. (For the CJ, $c = 0$.) However, to simplify the FE implementation, I_1 and E_f always are referred to the same reference configuration (the active zero-stress state). Moreover, since the fiber direction changes from point to point within the heart wall, W_e is defined with respect to a local Cartesian coordinate system, which is given in terms of tangent vectors at each Gauss point within each element. The derivatives of W_e are taken with respect to this local coordinate system, and the results are transformed to the global Cartesian coordinates for the computation of the element stiffness matrix.

The remainder of this paper deals with illustrative examples.

5. Growth

Since another paper in this volume deals with growth in detail, we present here only a simple model for growth of the heart. In this model, the left ventricle is taken as a thin-walled cylindrical membrane of unloaded radius $r_0(t)$ and thickness $h_0(t)$ [40], and the growth law is given by equations (8). During development, the time courses of the active (end systolic) pressure P_a and passive (end diastolic) radius r_p were prescribed (indicated by * in Figure 7). During maturity, the response of the model was computed for pressure overload (50% increase in P_a with constant r_p).

During embryonic development ($\bar{t} < 1$), the computed end diastolic pressure P_p increases monotonically to a stable value around the time of birth ($\bar{t} = 1$, Figure 7). In addition, following rapid increases in $\sigma_{\theta p}$ and $\sigma_{\theta a}$ in the embryo, the wall thickens (r_0/h_0 decreases) and both stresses approach their growth-equilibrium values at maturity ($\bar{\sigma}_{\theta p} = \bar{\sigma}_{\theta a} = 1$). Sudden pressure overloading produces a corresponding jump in $\sigma_{\theta a}$ (Figure 7), which then returns to its growth-equilibrium value as the wall thickens (r_0/h_0 decreases).

Similar growth laws in models for arteries have yielded results that are consistent with observed behavior [37, 41–43]. However, recent studies with a thick-walled model for the heart have revealed that mechanically regulated growth of the heart may be more complex than first thought. In particular, uniform homeostatic wall stresses at both end diastole and end systole, as demanded by equations (8), are likely not possible. Other approaches to this problem have been published [44, 45], but clearly further investigation is warranted.

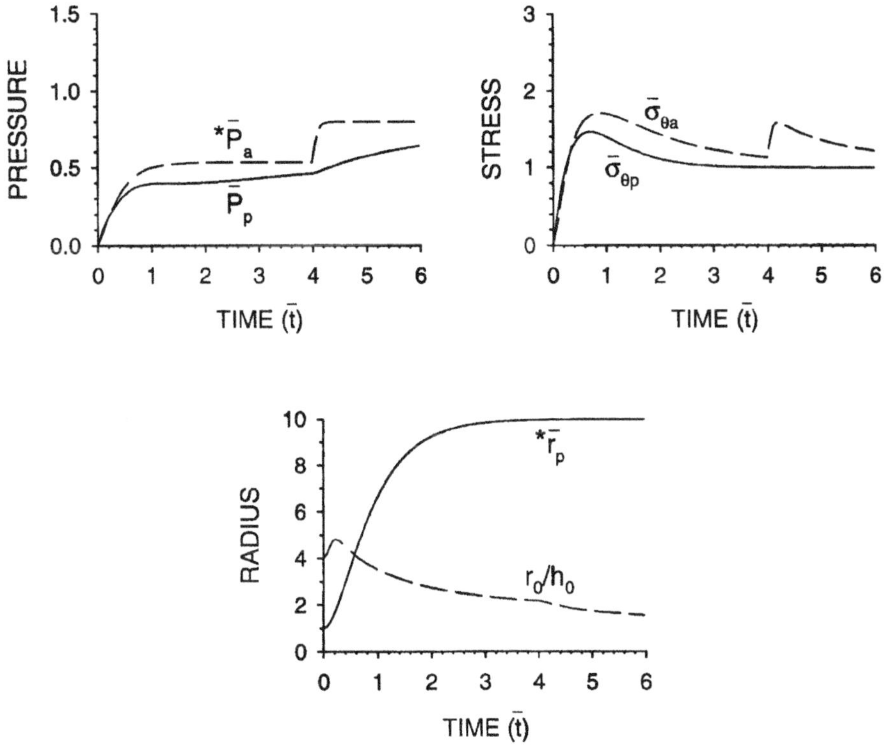

Figure 7. Results from membrane model for the growing left ventricle subjected to a jump in end-systolic pressure at maturity. Nondimensional variables are passive and active blood pressures ($\overline{P}_p$ and $\overline{P}_a$), passive and active wall stresses ($\bar{\sigma}_{\theta p}$ and $\bar{\sigma}_{\theta a}$), end-diastolic radius ($\bar{r}_p$), and unloaded radius-to-thickness ratio (r_0/h_0). During development and following the load perturbation, the stresses approach their mature growth-equilibrium values ($\bar{\sigma}_{\theta p} = \bar{\sigma}_{\theta a} = 1$). Reprinted from [40] with permission (* = prescribed quantities).

6. Remodeling

6.1. EFFECTS OF CARDIAC JELLY ON MECHANICS

Cardiac jelly plays a crucial role in assuring the effective functioning of the heart during the early stages of development, when the cardiac tube is without formed valves (Figsures 1(b) and 2). During these stages, the heart pumps blood through a peristaltic mechanism [10], with the CJ serving a valve-like function. The early cardiac tube contains a thick, non-uniform layer of CJ in the outflow tract, while a much thinner and more uniform layer lines the wall of the ventricle. Using a simplified theoretical model of the heart cross section, Barry [46] has shown that a thick jelly layer is necessary in order to produce lumen closure during contraction, as required for effective pumping action. At later stages, as the flow through the tubular heart becomes pulsatile, apposing mounds of cardiac jelly form at the atrio-ventricular (inflow) and cono-ventricular (outflow) tracts and act as primitive

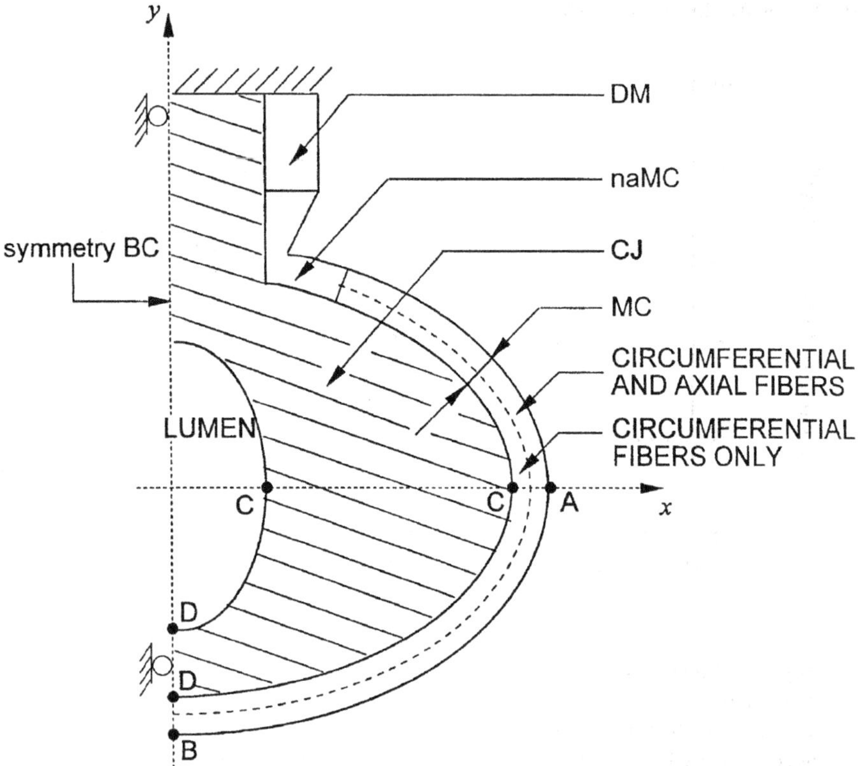

Figure 8. Schematic of cross-section of heart model used to derive the finite element meshes. MC: myocardium, naMC: non-activating myocardium; DM: dorsal mesocardium; CJ: cardiac jelly. The model is symmetric about the y axis.

valves [47]. When the myocardium activates, the jelly mounds come in contact and close the lumen, thereby preventing back flow.

We use our nonlinear FE formulation to investigate how CJ affects the activation induced wall stresses and cross-sectional deformations in the stage 10 chick heart. For each cross-sectional geometry considered, we construct a cylindrical model with uniform cross section, in which the inner lumen and the myocardial layer are elliptical in shape and the dimensions are derived from histological serial sections [48]. Each model consists of four distinct materials: myocardium (MC), cardiac jelly (CJ), dorsal mesocardium (DM) and non-activating myocardium (naMC) (Figure 8). Consistent with observations [23], the MC is modeled as a two-layer sleeve, with the inner layer containing circumferential fibers only and the outer layer containing fibers aligned in both the circumferential and longitudinal directions. The CJ is modeled as an isotropic layer of varying thickness, and the effects of the endocardium are neglected in this model. Each material is pseudoelastic and the material behavior follows the exponential SED function in equation (36).

To examine the variations in transmural stresses due to changes in CJ shape and thickness, we analyze three models all with the same circular outer wall contour and MC thickness (Figure 8). The first model has an elliptical lumen and thus a non-uniform CJ distribution around the wall. In the second model, the lumen is circular and the CJ layer has identical thickness everywhere. The third model has no CJ. Applying activation but no lumen pressure gives the circumferential Cauchy stresses shown in Figure 9. Clearly, the presence of cardiac jelly and the shape of the CJ layer substantially alter the transmural stress distribution. These differences are due primarily to a combination of two effects: the constraining effect of the passive CJ on MC contraction and bending of the wall in regions where the CJ is relatively thin (see Figure 10).

We note at this point that the non-uniform CJ configuration just analyzed corresponds to the cross-sectional morphology found at the level of the outflow tract, where the lumen must be closed or, at least, greatly reduced in size during activation in order to prevent backflow. We compared the above elliptical lumen model with an equivalent circular lumen model containing the same CJ volume but with a uniform CJ distribution. For the same activation, i.e., for the same amount of energy introduced into the system, the area reduction produced by the elliptical lumen is 30% higher than the one due the circular lumen. Clearly, an elliptical lumen produces a more effective pump. Thus, the morphology of the CJ distribution in the outflow tract can be explained in terms of improved mechanical efficiency.

6.2. EFFECTS OF FIBER ORIENTATION IN A CURVED TUBE

We model the embryonic heart as a cut toroidal tube. The cross section is circular and consists of an outer layer of myocardium wrapped around an inner layer of cardiac jelly. A helical fiber architecture is assumed in the myocardium, with γ being the fiber pitch angle relative to the longitudinal direction. With one end of the tube clamped and the other end free, the MC is activated with no applied loads.

Due to muscle activation, the heart model bends and twists, thereby transforming the toroidal tube into a helical tube. Only bending occurs for $\gamma = 0$, but significant twisting also occurs for $\gamma = 45°$ (Figure 11). A similar deformation due to swelling of the CJ may play a role in the looping process [21].

6.3. MODEL FOR TRABECULAE

The geometrical irregularity of the trabeculated wall poses major challenges for the construction of viable FE computational models. Following current approaches in medical image processing, we reconstruct three-dimensional voxel models from stacks of cross-sectional digital images of the stage 21 chick heart generated by confocal microscopy [49]. Because of their inherent discrete nature, voxel models can be easily transformed into valid, albeit very large, finite element meshes.

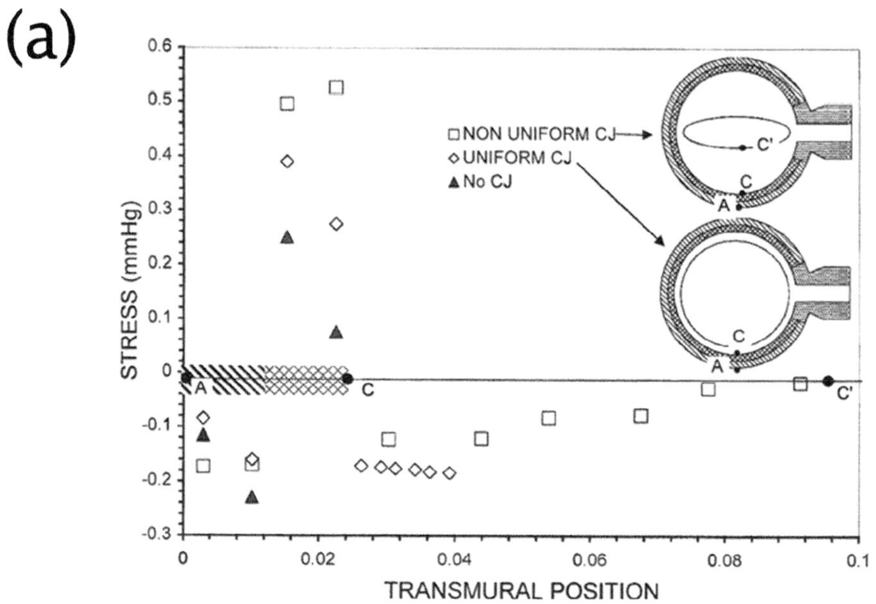

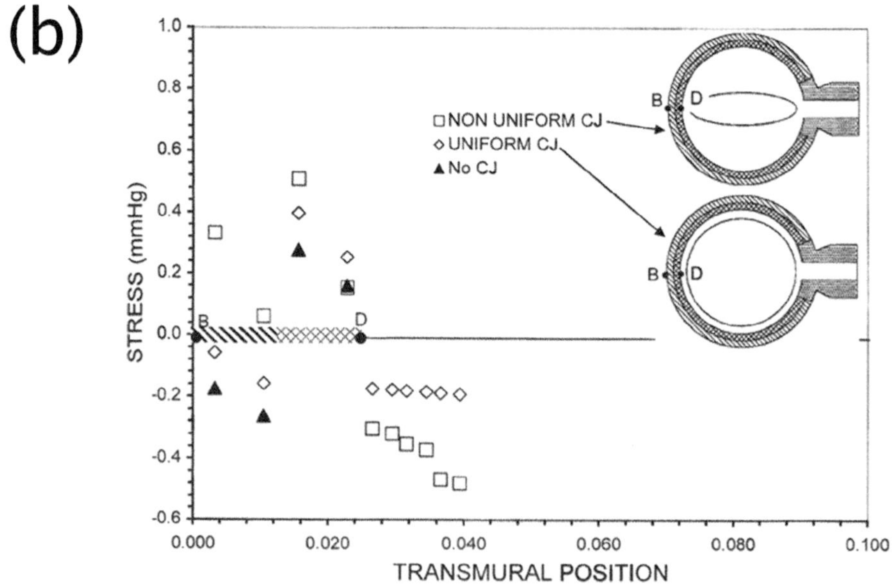

Figure 9. Transmural stresses induced by myocardial activation at the sagittal plane (a) and at dorso-ventral plane (b) with and without cardiac jelly. Cardiac jelly substantially increases peak stresses in the myocardial wall.

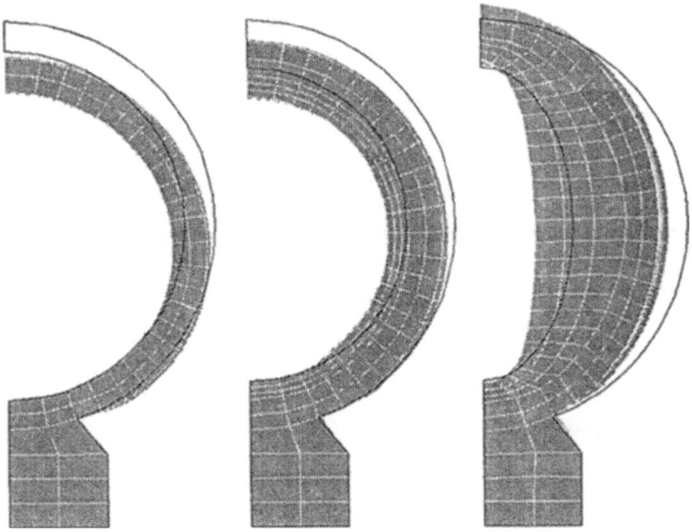

Figure 10. Deformations induced by myocardial activation with no jelly (left), uniformly distributed jelly (center), and elliptically distributed jelly (right). Note the change in the deformation pattern induced by the elliptical distribution.

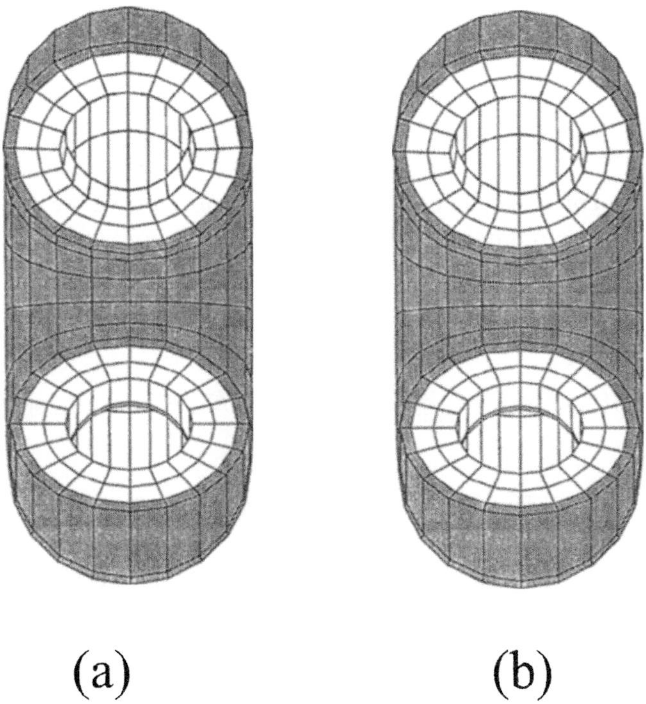

Figure 11. Deformations induced by myocardial activation in the toroidal model with helical fiber architecture. (a) With helical angle $\gamma = 0°$, the heart only bends. (b) For $\gamma = 45°$, fibers induce both bending and twisting resulting in out-of-plane deformed shape.

Due to nonlinearity and incompressibility, however, analysis of very dense meshes representing the entire trabeculated heart can only be performed via direct solution techniques at prohibitive computational costs. To resolve the problem of accurately modeling the mechanical effect of trabeculation while maintaining computational efficiency, we adopt a multi-scale modeling and analysis approach [50, 51].

The trabeculated heart is modeled at two distinct levels: locally, at the level of a small hexahedral volume of trabeculated myocardium – hereafter called Representative Volume Element (RVE) – and globally, with the entire organ geometrically defined by smoothed boundary surfaces. Separate FE analyses are performed at these two levels. First, a local analysis formulates equivalent (pseudo) constitutive laws for trabeculated myocardium for each RVE, and then a global analysis on the smoothed heart incorporates the equivalent material properties previously computed at the local level.

In the FE model representing the RVE, the material properties of the myocardial core in the trabeculae are defined by equation (36). In order to establish fiber directions, we make the assumption that, in each trabecular structure, myofibrils are approximately oriented parallel to the geometrical midline (the geometrical "skeleton") of each trabecula [52]. Finite element testing indicates that the trabeculated myocardium behaves qualitatively like the homogeneous myocardium, but its stiffness is about one order of magnitude smaller and the material is orthotropic in both passive and active states. Based on this finding, we introduce a new SED function for passive trabeculated myocardium in the form of equation $(6)_1$ with $a_f = 0$ and

$$Q_P = b_1 E_{xx}^2 + b_2 E_{yy}^2 + b_3 E_{zz}^2 + b_4 E_{xx} E_{yy} + b_5 E_{yy} E_{zz} + b_6 E_{zz} E_{xx}. \quad (37)$$

Because of the complex, three-dimensional organization of the myofibrils in the trabeculated myocardium, a simple uncoupled expression, as in equation (36) for the active component, is no longer sufficient to represent the mechanical contribution of muscle activation. Thus, we introduce a scaling factor to model the stiffening of the fibers as they become more stretched and assume

$$W_A = (I_1 - 3)\left(c_1 E_{xx}^2 + c_2 E_{yy}^2 + c_3 E_{zz}^2 + c_4 E_{xx} E_{yy} + c_5 E_{yy} E_{zz} + c_6 E_{zz} E_{xx}\right) \quad (38)$$

in which all strains are referred to the active zero-stress configuration.

The numerical testing procedure is organized as follows. We first perform nonlinear FE analyses to simulate two triaxial stretch tests on a specific RVE, producing two independent sets of data. In these tests, all six faces of the hexahedral RVE are attached to straight rigid plates. According to the specific loading protocol, the plates are either fixed or move under a prescribed displacement. The averaged strains are determined by the displacements of these rigid plates and the original model dimensions, while the averaged 2nd PK stresses are approximated by $S = F/(A\lambda)$, where F is the total reaction force on the plate, A is the initial cross-sectional area, and λ is the stretch ratio in the direction orthogonal to the plate. The collected stress-strain data from the first test are used to determine the best-fitted parameters in the SED functions via a modified multidimensional Levenberg–Marquardt nonlinear regression procedure [52, 53]. Then, results from the second set are used to evaluate the predictability of the SED functions previously determined.

A typical RVE undergoing a numerical stretch test is shown in Figure 12, and Figure 13 shows the stress-strain relations computed from the first set of stretch tests superimposed on the theoretical stress-strain curves. Note that, although each stress component is a function of three normal strain components, the figure shows only the relation between a stress component and its corresponding strain due to the difficulty of visualizing a function of three variables. Clearly, the stress-strain relations predicted by the SED functions agree well with the original data used for regression. Furthermore, the results of the second stretch indicate that there is a reasonable agreement between predicted and numerically computed stress-strain data [52]. This confirms that RVE-specific SED functions can be used to reliably and accurately model the local behavior of the trabeculated myocardium in the context of a global analysis scheme.

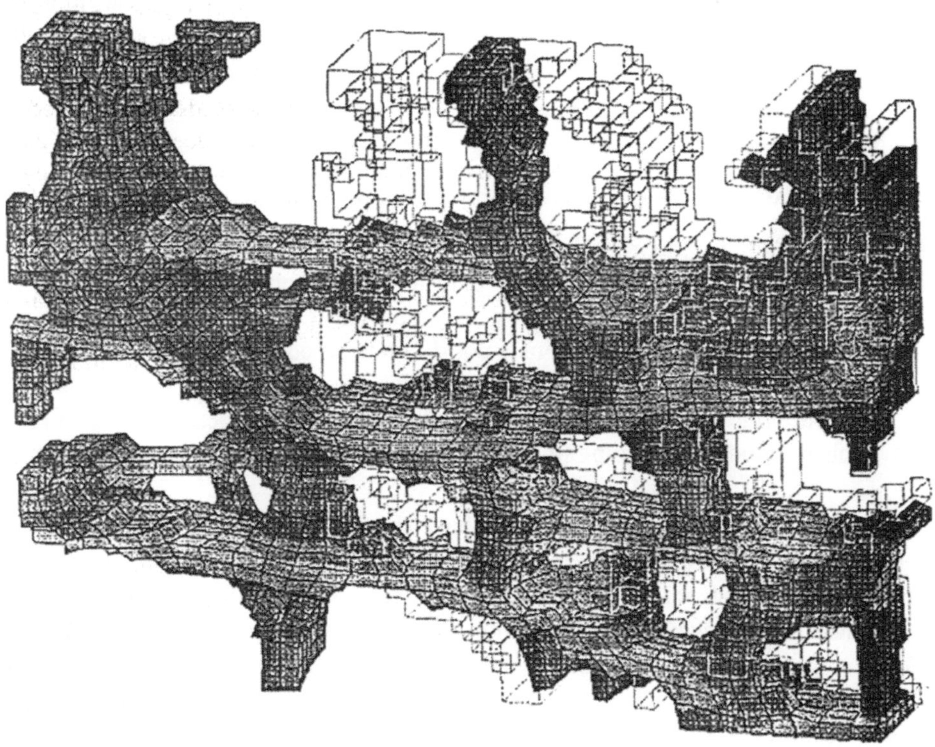

Figure 12. Deformed RVE mesh under a 60% stretch. The undeformed mesh is shown in outline. Reprinted from [52] with permission from Gordon and Breach Publishers, copyright Overseas Publishers Association.

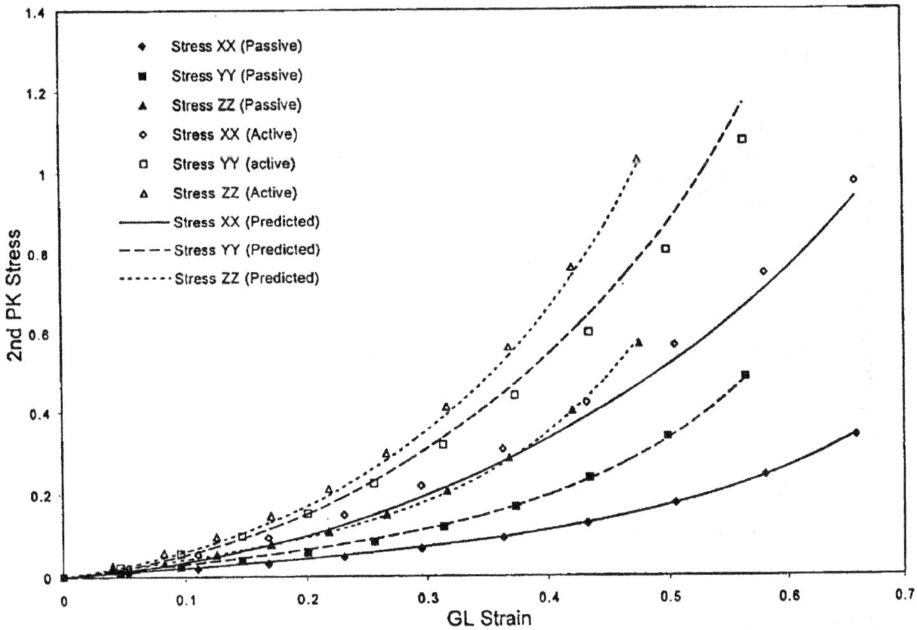

Figure 13. Theoretical predictions (curves) vs. actual finite element results (dots) obtained through numerical simulation following protocol 1. Reprinted from [52] with permission from Gordon and Breach Publishers, copyright Overseas Publishers Association.

7. Morphogenesis

7.1. CARDIAC LOOPING

Looping is a vitally important process during heart development. Abnormal looping causes cardiac malformations in as many as 1% of liveborn and 10% of stillborn human births [54]. Although looping has received a great deal of attention by developmental biologists, it remains a poorly understood process.

Several hypotheses have been proposed for the mechanisms responsible for looping, but none appear to be consistent with all available experimental data [55]. Any realistic model for looping must be consistent with the following observations.

Experimental Considerations. Studies have shown that looping is a process that is intrinsic to the heart and is independent of function. The heart, for example, bends when it is isolated from the embryo and grown in culture [56, 57] and when beating is arrested [58]. However, looping is not always normal during these interventions. For example, the boundary conditions at the ends of the tube affect the final form of the looped heart [59, 60].

Embryological studies suggest that actin microfilaments and the DM play significant roles during the looping process. For example, when actin is disrupted by a global application of the chemical cytochalasin B, looping stops. Moreover, locally applying this chemical to the left side of the ventricle results in (normal) looping

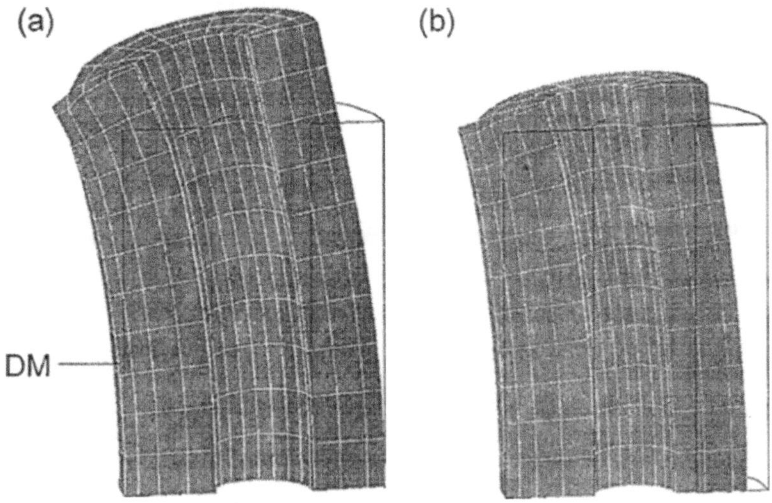

Figure 14. Models for cardiac looping due to (a) swelling of cardiac jelly and (b) contraction of circumferential microfilaments. In both models, the constraint of the relatively stiff dorsal mesocardium (DM) forces the ventricle to bend.

to the right, but application to the right side results in looping to the left [22]. The significance of the DM is suggested by the observation that this structure normally lies along the inner curvature of the looped tube [56].

Models for Looping. To a first approximation, the stage 10 heart is modeled as a thick-walled cylindrical tube composed of an inner layer of CJ and an outer layer of myocardium, with a longitudinal strut along the outside of the tube representing the DM (Figure 14). The modulus of the DM is assumed to be greater than that of the myocardium, which in turn is stiffer than the CJ. In addition, the myocardium in the model contains circumferentially aligned actin microfilaments [23]. Using material properties like those represented by equations (6), we use a finite element analysis to demonstrate two possible mechanisms for looping.

The first mechanism is based on the work of Manasek and colleagues [21], who speculated that the driving force for looping is CJ swelling pressure. In response to this pressure, the expanding tube bends due to the constraint provided by the relatively stiff DM and twists due to the anisotropy of the myocardium. Since the twisting of an anisotropic curved tube already has been discussed, we focus here on the bending component of looping. Swelling of the CJ is simulated by a specified isotropic growth, which induces bending with the DM along the inner curvature (Figure 14(a)). (The myocardium is passive in this model.)

The second mechanism is based on cytoskeletal contraction. Suppose the circumferentially oriented actin microfilaments in the myocardium undergo a prolonged contraction. Then, due to cellular incompressibility, circumferential shortening of the cells would induce longitudinal extension, which would lengthen, but not bend, a uniform tube. The DM, however, restricts the deformation of adjacent cells, and so the tube bends with the DM along the inner curvature (Figure 14(b)).

As discussed in [55], these models are consistent with most, but not all, experimental observations. Moreover, the models show that the amount of bending possible by these mechanisms is limited; thus, other factors likely play a role in the later stages of looping. Clearly, further studies are needed. In particular, measuring regional material properties is crucial. Without a relatively stiff DM region, for example, neither model can work.

7.2. MYOCARDIAL TRABECULATION

Trabeculation of the ventricular wall is an example of the creation of pattern during development. Various chemical and mechanical models have been proposed for biological pattern formation [61]. This section considers a model based on material instability driven by cytoskeletal contraction. The model is related to the morphogenetic models of Odell et al. [26] and Oster et al. [27].

Model for Trabeculation. During trabeculation, the trabeculae grow inward from the myocardium of the heart tube as the CJ disappears. Since the early myocardium is an epithelium (cell sheet), we treat it as a thin membrane composed of an incompressible, elastic material with in-plane contractile properties. Moreover, the effects of curvature are ignored in this simple model, with the membrane assumed to remain flat at all times.

The pattern-forming ability of this model depends on two main assumptions. First, we assume that the actin cytoskeleton behaves like smooth muscle, with the total (active + passive) stress-stretch relation containing a peak and a descending limb (Figure 15(d)). Investigators have speculated that embryonic epithelial cells behave in this manner [62], but to our knowledge, the existence of a descending limb has not been verified experimentally. Second, we assume that embryonic myocardial cells normally operate on the descending limb. There is evidence that this is a valid assumption for vascular smooth muscle [63] and for the sarcomeres in the stage 16–24 embryonic chick heart [31]. However, relevant data for the cytoskeletal contractile proteins is lacking.

The key to the generation of pattern in this and similar models is mechanical instability [61], which is due here to the effectively negative stiffness provided by the descending limb of the stress-stretch curve. Consider, for example, a string of cells (Figure 15), with the lighter cells being slightly stiffer (in the passive state) than the darker cells. When the passive string is stretched (Figure 15(b)), the dark cells stretch a little more than the light cells. (The passive dots in Figure 15(d) indicate the locations of these cells in stress-stretch space.) Note that (1) all cells are stretched to the descending limb of the active stress-stretch curve and (2) the passive engineering (first Piola–Kirchhoff) stress T is the same in all cells, as required by equilibrium.

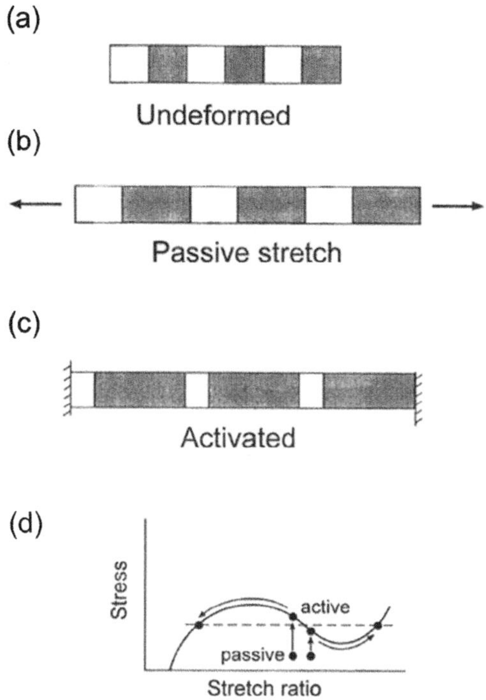

Figure 15. Schematic of pattern formation in a row of cells. (a) Passive, undeformed row with the light cells being stiffer than the dark cells. (b) The passive cells are stretched to the descending limb of the stress-stretch curve. The dark cells stretch more than the light cells. (c) When activated isometrically, the stronger light cells shorten, forcing the weaker dark cells to lengthen. (d) Motions of cells in stress-stretch space. Reprinted from [64] with kind permission of Kluwer Academic Publishers.

Next, with the string held at a fixed length, all cells are activated simultaneously. Since the shorter light cells generate more active stress than the dark cells (Figure 15(d)), the light cells contract to even shorter lengths, while the weaker dark cells are forced to lengthen (Figure 15(c)). Since all cells must have the same T, the light cells jump to the ascending limb of the curve, while the dark cells move to the overstretched portion of the curve (Figure 15(d)). In this way, pattern arises, with short light cells alternating with much longer dark cells (Figure 15(c)).

Model Analysis. The analysis of the model is based on nonlinear membrane theory, modified to include contractile activity. For illustration, we consider a square epithelium with edges along the x_1 and x_2 axes that undergoes isotropic or transversely isotropic activation in the plane of the membrane.

For isotropic activation, the active first Piola–Kirchoff stress tensor is taken in the form

$$\mathbf{T}_a = T_a \mathbf{I}, \tag{39}$$

where

$$T_a = \begin{cases} c_a(J_A - 1)(J_{A0} - J_A) & \text{for } 1 \leqslant J_A \leqslant J_{A0}, \\ 0 & \text{for other } J_A. \end{cases} \tag{40}$$

Here, $J = \lambda_1\lambda_2$ is a two-dimensional stretch invariant, $J_A = \lambda_1\lambda_2/\lambda_a^2$ is computed relative to the active zero-stress state, and $J_{A0} = \lambda_{10}\lambda_{20}/\lambda_a^2$ is the overstretch cut-off point, with the λ_{i0} being specified parameters. In addition, the coefficient c_a increases asymptotically in time toward a peak value (while λ_a decreases). For transversely isotropic activation relative to the fiber direction $\mathbf{e}_f$,

$$\mathbf{T}_a = T_a \mathbf{e}_f \otimes \mathbf{e}_f, \tag{41}$$

where

$$T_a = \begin{cases} c_a(\lambda_{fA} - 1)(\lambda_{A0} - \lambda_{fA}) & \text{for } 1 \leqslant \lambda_{fA} \leqslant \lambda_{A0}, \\ 0 & \text{for other } \lambda_{fA} \end{cases} \tag{42}$$

with $\lambda_{A0} = \lambda_{f0}/\lambda_a$ being the overstretch cut-off. The T–λ curves given by equations (40) and (42) resemble that shown in (Figure 15(d)).

The membrane equations were solved using a finite difference method with explicit time integration [64]. Results are shown for a membrane with a random distribution of passive modulus within 2% of the mean. First, the passive membrane was stretched equibiaxially ($\lambda_1 = \lambda_2$) to the descending limb of the stress-stretch curve, and then, with the edges fixed, all cells were activated simultaneously.

In the isotropic membrane, the distribution of membrane thickness (λ_3) shows a gradual evolution of randomly distributed trabecular-like ridges as activation

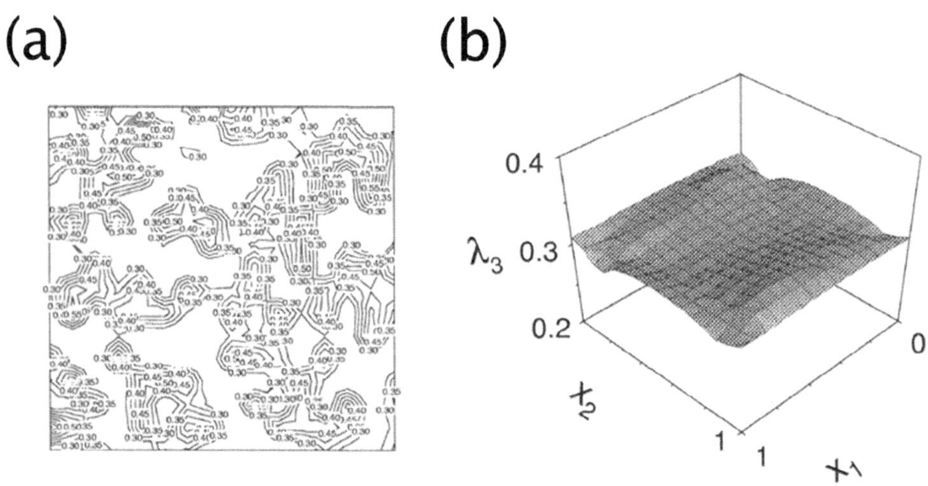

Figure 16. Plots of transverse normal stretch (λ_3) in a membrane model for myocardial trabeculation. (a) Isotropic activation: a random trabecular-like pattern forms. Reprinted from [64] with kind permission from Kluwer Academic Publishers. (b) Anisotropic activation with long-range stresses included in the fiber (x_1)-direction: ridges form along the fiber direction.

proceeds (Figure 16(a)). Although resembling the morphology during the later stages of myocardial trabeculation, such a random pattern is not consistent with the circumferential ridges that first appear in the embryonic heart (Figure 4(a)). When transverse isotropy is included, elongated ridges appear normal to the fiber direction (results not shown), similar to a buckling type of pattern. These results, however, still are not consistent with observations that these ridges form along the fiber direction. The model gives more realistic results when so-called "long-range effects" are included.

Long-range effects may arise if cells exert forces, possibly through the protein matrix of the CJ, directly on non-neighboring cells. To a first approximation, long-range stresses can be represented by a term involving the second spatial derivative of strain [61]. In general, the long-range second Piola–Kirchhoff stress tensor can be written in the form

$$\mathbf{S}_L = -\nabla \cdot \mathbf{C} \cdot \nabla \mathbf{E}, \tag{43}$$

where ∇ is the gradient operator, $\mathbf{E}$ is the Lagrangian strain tensor, and $\mathbf{C}$ is a second-order coefficient tensor.

For illustration, both the local and long-range properties are taken as transversely isotropic relative to the x_1-direction. (Some fibers in the CJ are oriented in the same direction as the cardiocytes [65].) Again, the distribution of passive stiffness is random, the membrane is stretched to the descending limb of the stress-stretch curve, and all cells are activated simultaneously. Here, however, the boundary conditions are taken as periodic. As activation proceeds, two ridges form along the fiber direction (Figure 16(b)), consistent with the first trabeculations that appear in the embryonic heart [18].

Although this mechanism may initiate trabeculation, further development includes trabecular growth [18]. Such growth may be stimulated by stress concentrations near the bases of the trabeculae (results not shown).

8. Conclusions and Future Work

Biomechanics plays an important role in many aspects of cardiac development. This subject, however, remains largely untouched by engineers. One objective of this article is to stimulate interest in this area of research.

Developing realistic models for the transforming embryonic heart is an extremely challenging but worthwhile endeavor. Models can be useful tools in studying the mechanics of normal and abnormal heart development, especially as part of an interdisciplinary research effort involving engineers, physiologists, and molecular biologists. The real value of a theoretical model is in its predictive capacity, but such predictions generally cannot be trusted until the model is validated experimentally.

An important problem not addressed in this paper is remeshing of the finite element grid. The large changes in geometry that occur during morphogenesis may lead to highly distorted elements, causing the mesh to become unviable [38]. Thus, periodic remeshing may be necessary to model correctly cardiac morphogenesis. Of course, an automatic remeshing scheme would be highly desireable.

Future work on the mechanics of heart development should include developing more realistic models for cardiac looping and myocardial trabeculation. Theoretical models also would be useful for studying the formation of the cardiac tube, fluid-solid interactions, and valve formation. Ultimately, models linking mechanical and molecular events will be needed.

Acknowledgements

This work was supported by NIH grant R01 HL46367. The authors thank Professor George Zahalak for helpful suggestions concerning long-range stresses in pattern formation. In addition, the assistance of Srinivasan Varahoor in conducting finite element computations is gratefully acknowledged.

References

1. L.A. Taber, Biomechanics of growth, remodeling, and morphogenesis. *Appl. Mech. Rev.* **48** (1995) 487–545.
2. L.A. Taber, Mechanical aspects of cardiac development. *Progress in Biophysics and Molecular Biology* **69** (1998) 255.
3. J. Sadoshima, L. Jahn, T. Takahashi, T.J. Kulik and S. Izumo, Molecular characterization of the stretch-induced adaptation of cultured cardiac cells. *J. Biological Chem.* **267** (1992) 10551–10560.
4. J.H. Omens and Y.C. Fung, Residual strain in rat left ventricle. *Circulation Research* **66** (1990) 37–45.
5. J.H. Omens, H.A. Rockman and J.W. Covell, Passive ventricular mechanics in tight-skin mice. *Amer. J. Physiology* **266** (1994) H1169–H1176.
6. E.K. Rodriguez, J.H. Omens, L.K. Waldman and A.D. McCulloch, Effect of residual stress on transmural sarcomere length distributions in rat left ventricle. *Amer. J. Physiology* **264** (1993) H1048–H1056.
7. L.A. Taber, N. Hu, T. Pexieder, E.B. Clark and B.B. Keller, Residual strain in the ventricle of the stage 16–24 chick embryo. *Circulation Research* **72** (1993) 455–462.
8. J.B. Caulfield and T.K. Borg, The Collagen network of the heart. *Laboratory Investigation* **40** (1979) 364–372.
9. Y.C. Fung, *Biomechanics: Mechanical Properties of Living Tissues*, 2nd edn. Springer, Berlin (1993).
10. A.L. Romanoff, *The Avian Embryo: Structural and Functional Development*. Macmillan, New York (1960).
11. J.M. Icardo and F.J. Manasek, Cardiogenesis: Development mechanisms and embryology. In: H.A. Fozzard (ed.), *The Heart and Cardiovascular System*, 2nd edn. Raven Press (1992) pp. 1563–1586.
12. W.J. Larsen, *Human Embryology*. Churchill, Livingston (1993).
13. V. Hamburger and H.L. Hamilton, A series of normal stages in the development of the chick embryo. *J. Morphology* **88** (1951) 49–92.

14. E.B. Clark, Growth, Morphogenesis and function: The dynamics of cardiac development. In: J.H. Moller, W. Neal and J. Lock (eds), *Fetal, Neonatal, and Infant Heart Disease*. Appleton-Century-Crofts (1989) pp. 3–23.
15. B.B. Keller, N. Hu and E.B. Clark, Correlation of ventricular area, perimeter, and conotruncal diameter with ventricular mass and function in the chick embryo from stages 12 to 24. *Circulation Research* **66** (1990) 109–114.
16. B.B. Keller, N. Hu, P.J. Serrino and E.B. Clark, Ventricular pressure-area loop characteristics in the stages 16–24 chick embryo. *Circulation Research* **68** (1991) 226–231.
17. J. Manner, Cardiac looping in the chick embryo: A morphological review with special reference to terminological and biomechanical aspects of the looping process. *Anatomical Record* **259** (2000) 248–262.
18. D. Sedmera, T. Pexieder, N. Hu and E.B. Clark, Developmental changes in the myocardial architecture of the chick. *Anatomical Record* **248** (1997) 421–432.
19. D. Sedmera, T. Pexieder, M. Vuillemin, R.P. Thompson and R.H. Anderson, Developmental patterning of the myocardium. *Anatomical Record* **258** (2000) 319–337.
20. A. Nakamura, R.R. Kulikowski, J.W. Lacktis and F.J. Manasek, Heart looping: A regulated response to deforming forces. In: R. van Praagh and A. Takao (eds), *Etiology and Morphogenesis of Congenital Heart Disease*. Futura Publishing (1980) pp. 81–98.
21. F.J. Manasek, R.R. Kulikowski, A. Nakamura, Q. Nguyenphuc and J.W. Lacktis, Early heart development: A new model of cardiac morphogenesis. In: R. Zak (ed.), *Growth of the Heart in Health and Disease*. Raven Press (1984) pp. 105–130.
22. N. Itasaki, H. Nakamura, H. Sumida and M. Yasuda, Actin bundles on the right side in the caudal part of the heart tube play a role in dextro-looping in the embryonic chick heart. *Anatomy and Embryology* **183** (1991) 29–39.
23. I. Shiraishi, T. Takamatsu, T. Minamikawa and S. Fujita, 3-D observation of actin filaments during cardiac myofibrinogenesis in chick embryo using a confocal laser scanning microscope. *Anatomy and Embryology* **185** (1992) 401–408.
24. D.D. Streeter, Gross morphology and fiber geometry of the heart. In: R.M. Berne, N. Sperelakis and S. R. Geiger (eds), *Handbook of Physiology, Section 2: The Cardiovascular System, Vol. I: The Heart.* Amer. Physiolog. Soc. (1979) pp. 61–112.
25. E.K. Rodriguez, A. Hoger and A.D. McCulloch, Stress-dependent finite growth in soft elastic tissues. *J. Biomechanics* **27** (1994) 455–467.
26. G.M. Odell, G. Oster, P. Alberch and B. Burnside, The mechanical basis of morphogenesis. I. Epithelial folding and invagination. *Developmental Biology* **85** (1981) 446–462.
27. G.F. Oster, J.D. Murray and A.K. Harris, Mechanical aspects of mesenchymal morphogenesis. *J. Embryology Exper. Morphology* **78** (1983) 83–125.
28. L.A. Taber and C.E. Miller, Overview: Biomechanics of cardiac development. In: E.B. Clark, R.R. Markwald and A. Takao (eds), *Developmental Mechanisms of Heart Disease*. Futura (1995) pp. 387–419.
29. R.R. Burton, Heart growth and size in homeotherms. In: R.J. Goss (ed.), *Regulation of Organ and Tissue Growth.* Academic Press, New York (1972) pp. 101–125.
30. W. Grossman, Cardiac hypertrophy: Useful adaptation or pathologic process? *Amer. J. Medicine* **69** (1980) 576–584.
31. L.A. Taber, B.B. Keller and E.B. Clark, Cardiac mechanics in the stage-16 chick embryo. *J. Biomechanical Engrg.* **114** (1992) 427–434.
32. I.E. Lin and L.A. Taber, Mechanical effects of looping in the embryonic chick heart. *J. Biomechanics* **27** (1994) 311–321.
33. M. Yang, L.A. Taber and E.B. Clark, A nonlinear poroelastic model for the trabecular embryonic heart. *J. Biomechanical Engrg.* **116** (1994) 213–223.
34. C.E. Miller, M.A. Vanni, L.A. Taber and B.B. Keller, Passive stress-strain measurements in the stage-16 and stage-18 embryonic chick heart. *J. Biomechanical Engrg.* **119** (1997) 445–451.

35. P.D. Blatz and W.L. Ko, Application of finite elasticity to the deformation of rubbery materials. *Trans. Soc. Rheology* **6** (1962) 223–251.
36. E.B. Clark, N. Hu, P. Frommelt, G.K. Vandekieft, J.L. Dummett and R.J. Tomanek, Effect of increased pressure on ventricular growth in stage 21 chick embryos. *Amer. J. Physiology* **257** (1989) H55–H61.
37. L.A. Taber, A model for aortic growth based on fluid shear and fiber stresses. *J. Biomechanical Engrg.* **120** (1998) 348–354.
38. K.J. Bathe, *Finite Element Procedures*. Prentice-Hall, Englewood Cliffs, NJ (1996).
39. R. Srinivasan and R. Perucchio, Finite element analysis of anisotropic nonlinear incompressible elastic solids by a mixed model. *Internat. J. Numer. Methods Engrg.* **37** (1994) 3075–3092.
40. L.A. Taber, Biomechanical growth laws for muscle tissue. *J. Theoret. Biology* **193** (1998) 201–213.
41. L.A. Taber and D.W. Eggers, Theoretical study of stress-modulated growth in the aorta. *J. Theoret. Biology* **180** (1996) 343–357.
42. A. Rachev, N. Stergiopulos and J.J. Meister, A model for geometric and mechanical adaptation of arteries to sustained hypertension. *J. Biomechanical Engrg.* **120** (1998) 9–17.
43. A. Rachev, N. Stergiopulos and J.J. Meister, Theoretical study of dynamics of arterial wall remodeling in response to changes in blood pressure. *J. Biomechanics* **29** (1996) 635–642.
44. T. Arts, F.W. Prinzen, L.H.E.H. Snoeckx, J.M. Rijcken and R.S. Reneman, Adaptation of cardiac structure by mechanical feedback in the environment of the cell: A model study. *Biophys. J.* **66** (1994) 953–961.
45. I.E. Lin and L.A. Taber, A model for stress-induced growth in the developing heart. *J. Biomechanical Engrg.* **117** (1995) 343–349.
46. A. Barry, The functional significance of the cardiac jelly in the tubular heart of the chick embryo. *Anatomical Record* **102** (1948) 289–298.
47. B.M. Patten, T.C. Kramer and A. Barry, Valvular action in the embryonic chick heart by localized apposition of endocardial masses. *Anatomical Record* **102** (1948) 299–311.
48. R. Perucchio, R. Srinivasan, A.C. Gittenberger-de Groot and R.E. Poelmann, Finite element analysis of the early embryonic chick heart during contraction. In: *Proc. of the 1995 Bioengineering Conf. (ASME)* (1995) pp. 119–120.
49. N. Baba, Computer-aided three-dimensional reconstruction from serial section images. In: D.P. Hader (ed.), *Image Analysis in Biology*. CRC Press, Boca Raton, FL (1992) pp. 251–270.
50. P. Suquet, Elements of homogenization theory for inelastic solid mechanics. In: *Homogenization Techniques for Composite Media.* Springer, Berlin (1987) pp. 194–278.
51. S.J. Hollister and N. Kikuchi, A homogenization sampling procedure for calculating trabecular bone effective stiffness and tissue level stress. *J. Biomechanics* **27** (1994) 433–444.
52. W. Xie and R. Perucchio, Multiscale finite element modeling of the trabeculated embryonic heart: Numerical evaluation of the constitutive relations for the trabeculated myocardium. *Comput. Methods Biomechanics Biomedical Engrg.* (2001) (in press).
53. W.H. Press, S.A. Teukolsky, W.T. Vetterling and B.P. Flannery, *Numerical Recipes in C.* Cambridge Univ. Press, Cambridge (1992).
54. R.P. Harvey, Cardiac looping – An uneasy deal with laterality. *Cell & Developmental Biology* **9** (1998) 101–108.
55. L.A. Taber, I.E. Lin and E.B. Clark, Mechanics of cardiac looping. *Developmental Dynamics* **203** (1995) 42–50.
56. J.K. Butler, An experimental analysis of cardiac loop formation in the chick. M.S. Thesis, University of Texas (1952).
57. A. Manning and J.C. McLachlan, Looping of chick embryo hearts in vitro. *J. Anatomy* **168** (1990) 257–263.
58. F.J. Manasek and R.G. Monroe, Early cardiac morphogenesis is independent of function. *Developmental Biology* **27** (1972) 584–588.

59. M.E. Flynn, A.S. Pikalow, R.S. Kimmelman and R.L. Searls, The mechanism of cervical flexure formation in the chick. *Anatomy and Embryology* **184** (1991) 411–420.
60. J. Manner, W. Seidl and G. Steding, Correlation between the embryonic head flexures and cardiac development. *Anatomy and Embryology* **188** (1993) 269–285.
61. J.D. Murray, *Mathematical Biology*, 2nd edn. Springer, Berlin (1993).
62. R. Gordon and G.W. Brodland, The cytoskeletal mechanics of brain morphogenesis. Cell state splitters cause primary neural induction. *Cell Biophysics* **11** (1987) 177–238.
63. Y.C. Fung, *Biodynamics: Circulation*, 2nd edn. Springer, Berlin (1997).
64. L.A. Taber, Pattern formation in a nonlinear membrane model for epithelial morphogenesis. *Acta Biotheoretica* **48** (2000) 47–63.
65. I. Shiraishi, T. Takamatsu and S. Fujita, Three-dimensional observation with a confocal scanning laser microscope of fibronectin immunolabeling during cardiac looping in the chick embryo. *Anatomy and Embryology* **191** (1995) 183–189.
66. B.M. Patten, The formation of the cardiac loop in the chick. *Amer. J. Anatomy* **30** (1922) 373–397.

Journal of Elasticity **61:** 199–246, 2000.

Biaxial Mechanical Evaluation of Planar Biological Materials

MICHAEL S. SACKS
Tissue Mechanics Laboratory, Department of Bioengineering, University of Pittsburgh, PA 15261, U.S.A. E-mail:msacks@pitt.edu

Received 15 May 2000; in revised form 23 February 2001

Abstract. A fundamental goal in constitutive modeling is to predict the mechanical behavior of a material under a generalized loading state. To achieve this goal, rigorous experimentation involving all relevant deformations is necessary to obtain both the form and material constants of a strain-energy density function. For both natural biological tissues and tissue-derived soft biomaterials, there exist many physiological, surgical, and medical device applications where rigorous constitutive models are required. Since biological tissues are generally considered incompressible, planar biaxial testing allows for a two-dimensional stress-state that can be used to characterize fully their mechanical properties. Application of biaxial testing to biological tissues initially developed as an extension of the techniques developed for the investigation of rubber elasticity [43, 57]. However, whereas for rubber-like materials the continuum scale is that of large polymer molecules, it is at the fiber-level ($\sim$1 μm) for soft biological tissues. This is underscored by the fact that the fibers that comprise biological tissues exhibit finite nonlinear stress-strain responses and undergo large strains and rotations, which together induce complex mechanical behaviors not easily accounted for in classic constitutive models. Accounting for these behaviors by careful experimental evaluation and formulation of a constitutive model continues to be a challenging area in biomechanics. The focus of this paper is to describe a history of the application of biaxial testing techniques to soft planar tissues, their relation to relevant modern biomechanical constitutive theories, and important future trends.

Key words: biaxial mechanical testing, constitutive modeling of planar biomaterials, homogeneity, mechanical properties of collagenous tissues.

1. Introduction

A fundamental goal in constitutive modeling is to predict the mechanical behavior of a material under a generalized loading state. To achieve this goal, rigorous experimentation involving all relevant deformations is necessary to obtain material constants and to evaluate the model's predictive capabilities. Depending on the specific theoretical approach, specific experimental protocols are also used to guide the specific formulation of the model.

For incompressible materials in thin planar (membrane) configurations, planar biaxial testing allows for a two-dimensional stress-state that is sufficient to develop a constitutive equation. For example, in the case of isotropic rubber materials that undergo finite deformations, Rivlin et al. [43] developed the following generalized strain energy formulation:

$$W = \sum_{i=0,j=0}^{\infty} C_{ij}(I_1 - 3)^i (I_2 - 3)^j, \qquad C_{00} = 0, \tag{1}$$

where I_1 and I_2 are the first and second coordinate invariants measures of deformation, defined as:

$$\begin{aligned} I_1 &= \operatorname{tr}\mathbf{C} = \operatorname{tr}\mathbf{B}, \\ I_2 &= \frac{1}{2}\left[(\operatorname{tr}\mathbf{C})^2 - \operatorname{tr}\mathbf{C}^2\right] = \frac{1}{2}\left[(\operatorname{tr}\mathbf{B})^2 - \operatorname{tr}\mathbf{B}^2\right], \\ \mathbf{C} &= \mathbf{F}^{\mathrm{T}}\mathbf{F}, \quad \mathbf{B} = \mathbf{F}\mathbf{F}^{\mathrm{T}}, \end{aligned} \tag{2}$$

where $\mathbf{C}$ and $\mathbf{B}$ are the right and left Cauchy–Green deformation tensors, respectively, and $\mathbf{F}$ is the deformation gradient tensor. For pure homogenous strain, the principal stresses t_i are expressed as [42]:

$$t_i = 2\left(\lambda_i^2 \frac{\partial W}{\partial I_1} - \frac{1}{\lambda_i^2}\frac{\partial W}{\partial I_2}\right) + p, \quad i = 1, 2, 3, \tag{3}$$

where p is an arbitrary hydrostatic pressure stress.

For both natural biological tissues and tissue-derived soft biomaterials, there exist many physiological, surgical, and medical device applications where rigorous constitutive models are required. Examples include skin, myocardium, tendons, natural and prosthetic heart valves, and blood vessels. For these biological materials particular challenges in constitutive modeling are encountered due to their complex mechanical behavior. For example, because of their oriented fibrous structures they often exhibit pronounced mechanical anisotropy. In addition, they exhibit highly nonlinear stress-strain relationships, large deformations, viscoelasticity, and strong axial coupling. Taken as a whole, soft biomaterial defy simple material models.

Mechanical investigations have usually been confined to uniaxial studies because of the difficulties in controlling two or even three-dimensional boundary conditions. However, due to anisotropy uniaxial data cannot be used to extrapolate to the fully generalized three-dimensional constitutive equations, even if multi-dimensional strain data from the uniaxial experiment is available. There have also been investigations using inflation of circular membranes, which under assumption of isotropy can provide the necessary experimental data [61, 62]. Virtually all tissues are anisotropic, however, hence this method cannot be applied. Further, when attempting to determine material constants for complex material models, biaxial testing methods are required that include complex testing protocols that allow large variations in stress and strain states for full characterization [5, 31, 38].

Application of biaxial testing of biological tissues originally developed using approaches developed from studies on rubber elasticity [43, 57]. In 1948, Treloar [57] pioneered techniques to apply two independently variable strains in two perpendicular directions with simultaneous measurement of the stresses. However, in Treloar's device the axial stretch ratios λ_1 and λ_2 could not be separately controlled. This experimental limitation did not facilitate precise determination of the form of W and its derivatives, $\partial W/\partial I_1$ $\partial W/\partial I_2$, which occur in (3). In 1951, Rivlin developed a modified biaxial device that allowed for applied biaxial loads to rubber sheets, which allowed more precise experimental control. Using this device, Rivlin developed an integrated theoretical-experimental methodology in which the form of W in (1) could be derived and evaluated directly from multi-axial experimental data. This approach, and those that followed, greatly clarified the complex mechanical behavior of rubber, which can be confounded by such factors as material instabilities, physical aging, and Mullin's effect.

Similar to rubber materials, biaxial experiments on soft biological tissues are generally difficult to perform, and also present additional challenges unique to biological tissues. Just a few of the experimental problems include: small specimen sizes, structural and compositional heterogeneity, difficulty in gripping (without doing damage), dramatic effects of different gripping techniques (St. Venant-like effects), difficulty in precisely identifying material axes, difficulty in assuring constant forces along specimen edges, large specimen-to-specimen variability, and time-dependent changes due to biological degradation. In addition, a question of homogeneity of deformation within the specimen is paramount. These issues can often frustrate the application of even the most straightforward attempts to develop a constitutive model.

Biaxial testing devices have to be much more elaborate than uniaxial ones because of the need to control two boundary conditions. In particular, the edges must be able to expand freely in the lateral direction, and in the central target region the stress and strain states should be uniform so that data analysis can be performed simply. The target region must be small and located away from the outer edges to avoid the effects of specimen grips or tethers. Strain is always measured optically to avoid any mechanical interference.

In addition to the above experimental issues, relating the observed mechanical response to tissue structure is perhaps more paramount than in other more traditional material applications, where the continuum scale is usually at the size of large polymer molecules. In contrast, biological soft tissues are comprised of a dense network of primarily collagen and elastin fibers, which indicates a continuum scale at the fiber-scale (typically $\sim$1 μm). In addition, the fibers can undergo large rotations and exhibit a nonlinear stress-strain behavior that can induce complex behaviors at the macro-specimen scale not easily accounted for in classic material models. Accounting for these behaviors in both experimental evaluation and formulation of appropriate constitutive models continues to be challenging.

The focus of this paper is to review the history of the application of biaxial testing techniques to soft planar tissues and their relation to relevant biomechanical constitutive theories, with a further aim to underscore important future trends. The perspective of this review is that of an applied mechanician interested in quantifying and modeling the multi-axial mechanical response of soft biological tissues and biologically-derived biomaterials (i.e., biological tissues chemically or otherwise modified for use in medical devices). While not an exhaustive review, the works of all investigators utilizing biaxial testing techniques for biological tissues known to the author have been included. The review first covers the basic theory and kinematic analysis of a biaxial test. Next, the initial groundbreaking and more recent studies of biological tissues are presented. Finally, since biaxial testing and constitutive modeling has been the focus of much of the author's recent work, the final section focused on the results of his studies.

2. Biaxial Testing: Kinematics, Stresses, and Energetics

2.1. BASIC TECHNIQUES FOR BIAXIAL TESTING OF SOFT BIOLOGICAL MATERIALS

In general, biaxial testing of biological tissues are performed using thin specimens, which are either a membrane in its native form or a thin section prepared from a thick tissue slab. The specimen is mounted to the biaxial device in trampoline-like fashion using thin threads, which allows the edges to expand freely in the lateral direction (Figure 1(a)). Testing is generally performed with the specimen completely immersed in phosphate buffered normal saline (pH 7.4) at room or body (37°C) temperature. The central target region must be sufficiently small and located away from the outer edges to avoid the tethering effects (Figure 1(a)). Thus, in the central target region the stress and strain field is generally considered homogeneous, so that the components of **F** are independent of position.

2.2. KINEMATICS OF A BIAXIAL TEST

We consider the following homogeneous biaxial deformation

$$x_1 = \lambda_1 X_1 + \kappa_1 X_2, \qquad x_1 = \lambda_2 X_2 + \kappa_2 X_1, \qquad x_3 = \lambda_3 X_3, \tag{4}$$

where $\mathbf{X}$ and $\mathbf{x}$ are the locations of material particles in the reference and deformed states, respectively, and λ_i and κ_i are the components of **F** (i.e., λ_i are the stretch ratios and κ_i measures of in-plane shear). Since soft tissues are composed primarily of water and have negligible permeability [16], they can be considered incompressible so that λ_3 is calculated from $\det \mathbf{F} = 1$.

Experimentally, strain is measured optically to avoid any mechanical interference with the specimen. This is typically done by tracking the position of markers mounted on the upper specimen surface that delimit the central target region using optical tracking software [14, 48] (Figure 1(a)). The optical tracking software will produce the marker positions $\mathbf{x}_n$ as a function of time, where n the marker number (Figure 1(b)). In both our laboratory [48] and in others [18, 19], finite element shape functions are used to approximate the position vector field within the central

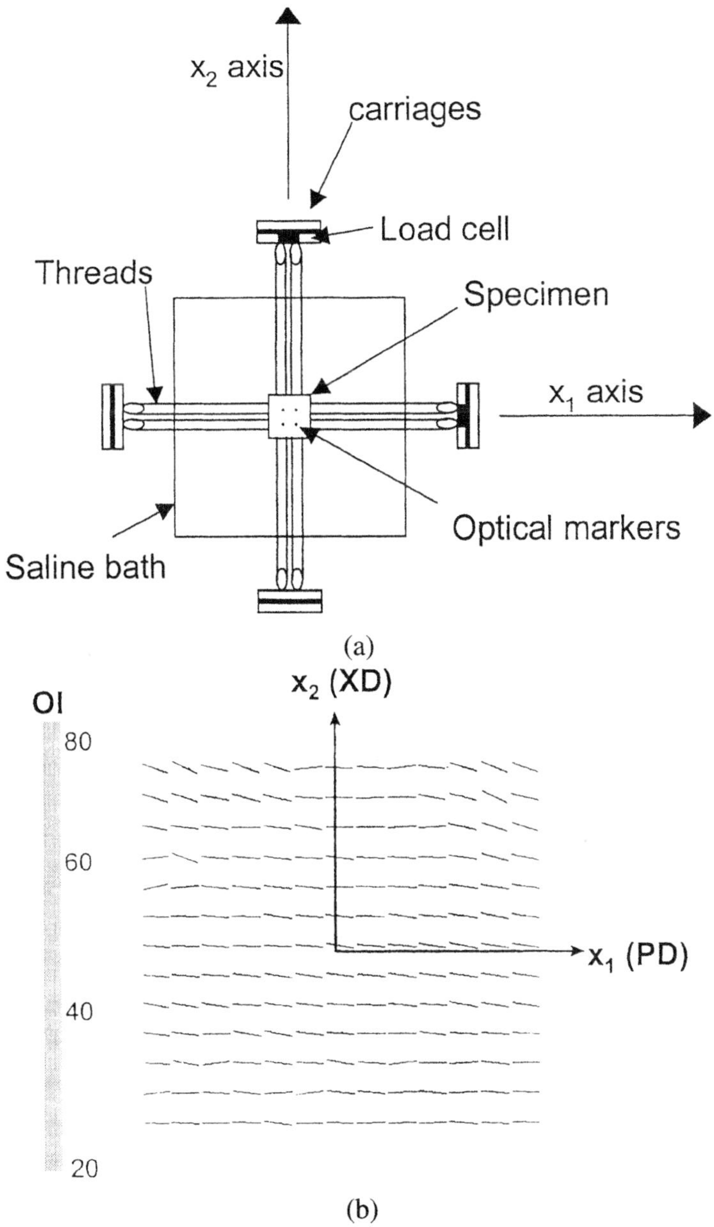

Figure 1. (a) Schematic of a biaxial testing device, and (b) a biaxial test specimen overlaid on a gray scale representation of the degree of collagen fiber alignment using an orientation index (OI), demonstrating high uniformity of both fiber preferred directions and OI, along with the definition of the PD and XD axes.

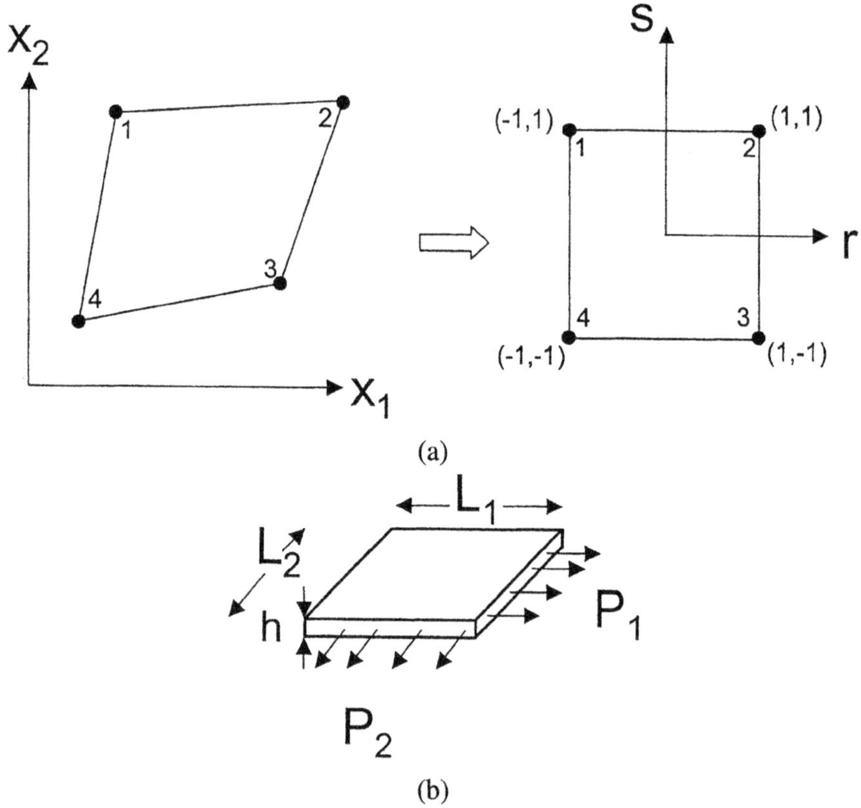

Figure 2. Mapping of the real-time positions of four tissue markers from the original device coordinates (x_1, x_2) to the isoparametric coordinates (r, s), allowing for bi-linear interpolation of the displacement field. (b) Schematic of a biaxial specimen.

target regions. Thus, the displacement field $\mathbf{u}(= \mathbf{x} - \mathbf{X})$ at any time or location within the central target region can be determined using

$$\mathbf{u}(r, s) = \sum_{n=1}^{m} f_n(r, s)\mathbf{u}_n, \tag{5}$$

where f_n is the shape function of node (or equivalently marker) n, m is the total number of nodes (markers), and r, s are the isoparametric coordinates [2] where the displacement is to be determined (Figure 2(a)). Since (5) represent a linear sum of m equations where f_n are functions of only r, s, the spatial derivatives of u_i with respect to r, s are given by

$$\frac{\partial \mathbf{u}}{\partial r} = \sum_{n=1}^{m} \frac{\partial f_n}{\partial r}\mathbf{u}_n, \qquad \frac{\partial \mathbf{u}}{\partial s} = \sum_{n=1}^{m} \frac{\partial f_n}{\partial s}\mathbf{u}_n. \tag{6}$$

The spatial derivatives with respect to $\mathbf{x}$ are determined for u_1 and u_2 by substitution into the following relations and inverting [24]:

$$\begin{bmatrix} \dfrac{\partial}{\partial r} \\ \dfrac{\partial}{\partial s} \end{bmatrix} = \begin{bmatrix} \dfrac{\partial u_1}{\partial r} & \dfrac{\partial u_2}{\partial r} \\ \dfrac{\partial u_1}{\partial s} & \dfrac{\partial u_2}{\partial s} \end{bmatrix} \begin{bmatrix} \dfrac{\partial}{\partial x_1} \\ \dfrac{\partial}{\partial x_2} \end{bmatrix}. \tag{7}$$

The components of $\mathbf{F}$ at each time point can be determined using

$$\mathbf{F} = \mathbf{G} + \mathbf{I} = \begin{bmatrix} \dfrac{\partial u_1}{\partial X_1} & \dfrac{\partial u_1}{\partial X_2} \\ \dfrac{\partial u_2}{\partial X_1} & \dfrac{\partial u_2}{\partial X_2} \end{bmatrix} + \begin{bmatrix} 1 & 0 \\ 0 & 1 \end{bmatrix} = \begin{bmatrix} \lambda_1 & \kappa_1 \\ \kappa_2 & \lambda_2 \end{bmatrix}, \tag{8}$$

where $\mathbf{G} = \partial\mathbf{u}/\partial\mathbf{X}$. Thus, from (8) the components of $\mathbf{F}$ can be directly obtained at each time point during the test. The components of the in-plane Green's strain tensor $\mathbf{E}$ are then computed using

$$\begin{aligned} E_{11} &= \frac{1}{2}\left(\lambda_1^2 + \kappa_2^2 - 1\right), \qquad E_{12} = \frac{1}{2}\left(\lambda_1\kappa_1 + \lambda_2\kappa_2\right), \\ E_{22} &= \frac{1}{2}\left(\lambda_2^2 + \kappa_1^2 - 1\right). \end{aligned} \tag{9}$$

2.3. STRESS AND EQUILIBRIUM

As mentioned above, biaxial testing of biological tissues are performed using thin specimens (no more than ~3 mm, usually <1 mm) and acted on by only in-plane loads. A state of plane stress is thus assumed so that the components t_{i3} ($i = 1, 2, 3$) of the Cauchy stress $\mathbf{t}$ are 0. For the deformation in (4), equilibrium is satisfied in the absence of body forces (i.e., $\operatorname{div}\mathbf{t} = 0$), even thought it can vary through the thickness. Experimentally, one measures the initial specimen dimensions, so that the Lagrangian stresses $\mathbf{T}$ (force/unit original cross-sectional area) are used for convenience. The components of $\mathbf{T}$ are computed from the measured axial forces $\mathbf{P}$ using

$$T_{11} = \frac{P_1}{hL_2}, \qquad T_{22} = \frac{P_2}{hL_1}, \tag{10}$$

where h the specimen thickness and L_i the specimen lengths (Figure 2(b)). Since experimentally applied loads are normal to the edges, $T_{12} = T_{21} = 0$. The second Piola–Kirchhoff stress $\mathbf{S}$ is determined using $\mathbf{S} = \mathbf{T} \cdot \mathbf{F}^{-1}$, and the Cauchy stress $\mathbf{t}$ is determined using $\mathbf{t} = \mathbf{F} \cdot \mathbf{T}/\mathbf{J}$, which in component form are given by (with $T_{12} = T_{21} = 0$):

$$t_{11} = \lambda_1 T_{11}, \quad t_{22} = \lambda_2 T_{22}, \quad t_{12} = \kappa_1 T_{22}, \quad t_{21} = \kappa_2 T_{11}. \tag{11}$$

2.4. ENERGY BALANCE IN A BIAXIAL TEST FOR ANISOTROPIC MATERIALS

For completeness of presentation, we can write the total free energy of a material governed by $W(\mathbf{F})$ for a generally anisotropic, nonlinear elastic material as

$$E = \int_{\Omega} W[\nabla \mathbf{y}(\mathbf{x})]\mathrm{d}x + L, \tag{12}$$

where L is the energy of the biaxial loading device, and $\mathbf{y}(\mathbf{x})$ is the deformation of $x \in \Omega$, Ω being the region defined by the specimen. L will generally have the form

$$L = \int_{\partial\Omega} \mathbf{y}(\mathbf{x}) \cdot \mathbf{Rn}\,\mathrm{d}A, \tag{13}$$

where $\mathbf{R}$ is any constant matrix, which in the biaxial case has the form

$$\mathbf{R} = \sigma_1 \mathbf{e}_1 \otimes \mathbf{e}_1 + \sigma_2 \mathbf{e}_2 \otimes \mathbf{e}_2, \tag{14}$$

where σ_i are the Piola–Kirchhoff stresses, $\mathbf{e}_1$, $\mathbf{e}_2$ orthonormal, and no kinematic constraints. Then, with L having this form, one can use the divergence theorem to write the total energy as

$$E = \int_{\Omega} \{W[\nabla \mathbf{y}(\mathbf{x})] - \mathbf{R} \cdot \nabla \mathbf{y}(\mathbf{x})\}\,\mathrm{d}x. \tag{15}$$

Although not used in biological tissue analyses, one can use this approach to find W by finding $\mathbf{F}$ that minimizes this expression. The interested reader is referred to [1, 10] for further details.

3. First Biaxial Mechanical Studies of Biological Tissues

3.1. FUNG AND CO-WORKERS

The first investigators to develop and utilize planar biaxial testing for soft biological tissues were Lanir and Fung in 1974 [30, 31] who investigated the mechanical properties of rabbit skin. Briefly, a 3 cm to 6 cm square skin specimen was mounted in a trampoline-like fashion with up to 68 individual attachments distributed equally over the four specimen sides (17/side). Similar to Rivlin [43], the tension on each line could be individually adjusted insure a reasonably uniform stress was applied to each specimen side. Actuator motion was controlled utilizing a function generator. To avoid the effects of local stress concentrations of the suture attachments, bidirectional tissue strain was measured in a central region by monitoring the distance between pairs of lines separated by ~5 mm along each axis video dimensional analyzers (VDA) [64]. Briefly, a VDA is an electronic device the works with a video camera signal to convert the distance between two dark-light or light-dark transitions in the image to a linearly proportional voltage, which can be recorded and converted to displacement. Typically, soft tissues are light in color so that the distance between dark lines applied to the tissue surface are tracked in real time. The advantage of the VDA is that it allows for non-contacting, real-time displacement measurements directly using conventional video. For biaxial studies, Lanir and Fung utilized two orthogonally positioned VDA with a common optical path to allow for simultaneous, synchronized displacement measurements along each stretch axis.

Experimental results demonstrated that skin exhibited a nonlinear, orthotropic stress-strain response, whose material axis depended on the specimen's anatomic orientation. While differences between the loading and unloading curves were observed due to hysteresis, the loading and unloading stress-strain responses were essentially independent of strain rate. It is important to note that these results *underscore the major phenomenon found in all subsequent biaxial mechanical investigations of soft planar tissues.*

Based on these experimental observations and data, Tong and Fung [56] used the above biaxial data to develop a constitutive model. Because of the insensitivity to strain rate, separate pseudo strain-energy functions could be developed for the loading and unloading phases of the stress-strain curve [16]. Thus, for the loading and unloading phases the in-plane second Piola–Kirchhoff stress components are derived from a two-dimensional strain energy function W:

$$S_{ij} = \frac{\partial W}{\partial E_{ij}}. \tag{16}$$

For the form of W, Tong and Fung observed from the experimental data that the stress-strain curves had a very shallow slope followed by an abrupt transition to a very high stiffness. Due to this biphasic-like behavior, they started with the generalized form

$$\begin{aligned}\rho_0 W = {} & \frac{1}{2}\left(\alpha_1 E_{11}^2 + \alpha_2 E_{22}^2 + \alpha_3\left(E_{12}^2 + E_{21}^2\right) + 2\alpha_4 E_{11} E_{22}\right) \\ & + \left[\frac{c}{2} \exp\left(a_1 E_{11}^2 + a_2 E_{22}^2 + a_3\left(E_{12}^2 + E_{21}^2\right) + 2a_4 E_{11} E_{22}\right.\right. \\ & \left.\left. + \gamma_1 E_{11}^3 + \gamma_2 E_{22}^3 + \gamma_4 E_{11}^2 E_2 + \gamma_5 E_{11} E_{22}^2\right)^{-1}\right],\end{aligned} \tag{17}$$

where ρ_0 is the initial tissue density, α_i, a_i, and γ_i are material constants, and E_{ij} is the Green strain tensor. This form was able to model both the low (first term on the right hand side) and high stress (second term on the right hand side) regions of the stress-strain curve.

In practice, (17) contains many more terms than is actually necessary to model the stress-strain curve. For all practical purposes the α_i and γ_i terms are not necessary to obtain a satisfactory fit to the majority of the stress-strain curve, especially the higher stress regions. Thus, (17) can be reduced to the "Fung" type (which incidentally was found to fit the data almost as well)

$$\rho_0 W = \frac{c}{2}\left(\mathrm{e}^{Q} - 1\right), \tag{18}$$

where $Q = c_{ijkl} E_{ij} E_{kl}$. In practice in biaxial testing the shear strain E_{12} is nearly zero, so that (18) can be written as

$$\rho_0 W = \frac{c}{2}\left[\exp\left(a_1 E_{11}^2 + a_2 E_{22}^2 + 2a_4 E_{11} E_{22}\right) - 1\right]. \tag{19}$$

This is perhaps the most broadly used constitutive model to date for the biaxial response of soft biological tissues (as well as other loading states), including skin [56], pericardium [8], epicardium [22], visceral pleura [23], and myocardium (see below).

One of the difficulties in applying (18) to biological tissue is the high amount of inter-specimen variability, which in-turn translated into wide variability in material parameter values. The sources of variability have been attributed to experimental noise, numerical instability of the fitting algorithms resulting from the nonlinearity of (18), and strain history dependence of the tissue. These problems can confound the ability to obtain a unique set of material constants either for a given specimen or a class of biomaterials. Further, the residuals in nonlinear regression may not be normally distributed, disallowing conventional statistical analysis.

In their study of canine pericardium, Yin et al. [65] developed a statistical-based approach to assess the sources of and account for the variability in material constants in describing biaxial stress-strain data. Using experimental data for canine pericardium [8], they determined a strain energy function (including exploring the use of non-integer powers of the Green strain). They performed a residual analysis to determine if standard statistical methods could be used to assess the variability, and if not, then they used nonparameteric methods (boot strapping). Using a five parameter exponential strain energy function, pericardial tissue was found to be strain-history dependent and anisotropic, which could not be attributed to either experimental noise or instability in the numerical algorithms.

3.2. VITO AND CO-WORKERS

Another group active in developing multi-axial constitutive relationships was Vito and co-workers. Among the technical improvements of their device were the use of multi-particle tracking to allow computation of the complete in-plane strain tensor and the use of real-time computer control [58]. Perhaps their main contribution was the development of a technique to identify the specimen's material axis [9]. Generally, identification a material axis is based on observations of the gross specimen shape (e.g., long axis of a blood vessel) or gross fiber architecture (e.g., myocardium). However, in many tissues the fibers are too small to be visually observed and up to the time of the study there were no rapid, non-destructive techniques for quantification of fiber architecture.

In Choi and Vito's technique, they identified the material axis by determining which directions, when loaded to the same stress, demonstrated the greatest and least strain values. To demonstrate their approach, Choi and Vito [9] utilized canine pericardium, a thin membrane that surrounds the heart. The pericardium functions to restrict excessive dilation of the heart during filling, is involved with the hemodynamic interaction of the heart's right and left ventricles, and in humans provides mechanical support to the diaphragm. It is composed primarily of collagen, a fibrous protein (the most common in the body) that possesses high tensile strength, and is thus found in tissues that require significant mechanical strength or provide structural support. To determine the orientation of the material axes, round pericardial tissue specimens were prepared, with opposing pairs of small clamps placed throughout the specimen's perimeter in 15° increments. For each opposing pair, the specimen was preloaded, and two marks were made aligned to the stretch axes. This procedure was repeated for each successive attachment pairs. When fully unloaded, the markers produced an ellipsoidal pattern whose semi-axes were aligned to the material axes.

To perform the biaxial tests, a square biaxial test specimen was cut from the original circular specimen, with edges aligned parallel and perpendicular to the material axis as determined above. Multiple test protocols were used to obtain stress-strain data under multiple loading states. Practically, this was accomplished using constant ratios of strain or stress during each protocol, with a sufficient number of protocols and ratios chosen to cover the complete E_{11}–E_{22} or S_{11}–S_{22} plane. Choi and Vito [9] then used the following strain energy function for the canine pericardium biaxial mechanical data:

$$\rho_0 W = b_0\left[\exp\left(b_1 E_{11}^2\right) + \exp\left(b_2 E_{22}^2\right) + \exp(2b_3 E_{11} E_{22}) - 3\right], \tag{20}$$

where b_i are material constants. They demonstrated that when data from a single test protocol was used to determine the values for b_i, different values were obtained for each protocol. Only when the data from all protocols was used simultaneously were the "true" material constants obtained for the specimen. This was shown to be due to the presence of multiple collinearities due to the use of constant $E_{11} : E_{22}$ or $S_{11} : S_{22}$ ratios. This concept was extended to the multi-axial testing of blood vessels [7].

Although the values for the b_i were rigorously obtained and the model was shown to work well under strain controlled tests, it did not work as well under load control tests. The reason underlying this disagreement is at present unknown, and suggests a need for experimental and theoretical investigations of constitutive theories that can better handle mixed boundary conditions. Another problematic finding with canine pericardium is the substantial variability in both degree of anisotropy (varying from quasi-isotropic to moderately anisotropic), which translated into significant inter-specimen variability in material constants. Thus, while reliable material parameters for (20) may be reliably obtained for an individual specimen, generic material parameters for canine pericardium could not be obtained. This problem was addressed by Sacks [48] and is described in detail in Section 3.

3.3. VENTRICULAR MYOCARDIUM

There is considerable clinical and basic interest in calculating regional stresses in the heart, primarily from the prevalence of heart disease in the western world. Regional stress computation requires accurate constitutive models of both the active and passive tissue states. The left ventricular wall of the heart is a thick-walled structure with a complex three-dimensional fibrous structure, with fibers consisting of both myocytes (muscle fibers) and connective tissue fibers, which are primarily collagen. Because of its structural and functional complexity, the application of biaxial mechanical analysis to the muscle of the thick ventricular wall of the heart (the myocardium) has been extremely challenging. However, biaxial tests on thin sections removed from the mid-wall have been used as a first step towards developing the full three-dimensional stress-strain constitutive relationship.

The first such studies were conducted by Demer and Yin [12] and Strumpf et al. [54]. Excised sheets of canine myocardium from the mid-wall of the left ventricle were subjected to cyclic loading and unloading in the predominant fiber and cross-fiber directions to determine passive mechanical properties. During each stretch the ratio of the orthogonal strains was kept constant and the corresponding stresses remained proportional. Myocardium under biaxial loading exhibited both nonlinear elasticity and viscoelasticity with some strain-rate dependence, but very little rate dependence in the area enclosed by the loading and unloading portions of the stress-strain curves (i.e. hysteresis). Fiber and cross-fiber directions demonstrated anisotropic behavior, with both the degree and direction of the anisotropy being dependent upon the region of the heart from which specimens are obtained. Shacklock [53] has also performed biaxial testing with no mechanical preconditioning, and found similar mechanical responses. Novak et al. [39] also explored regional differences in the biaxial mechanical properties, and determined that the mechanical properties of the heart are qualitatively similar from region to region, but quantitatively different. Sacks and Chuong [46] performed similar studies on right ventricular myocardium, which also was qualitatively similar to left ventricular myocardium but exhibited regional differences and more pronounced anisotropy.

Yin et al. [63] fitted the myocardial biaxial stress-strain data both with exponential strain-energy functions with quadratic powers of strains as well as with an alternative function with nonintegral powers of strains. They also used the non-parametric methods developed for pericardium [65] to assess the reliability of the coefficients for each of these functions. The quadratic strain-energy functions resulted in wide intra- and interspecimen variability in the coefficients. Moreover, both their absolute and relative values demonstrated marked load history dependence such that interpretation of the direction of anisotropy was difficult. Fitting the data with the alternative non-integral strain-energy function seemed to alleviate these problems. This alternative strain-energy function may provide more self-consistent results than the more commonly used quadratic strain-energy functions.

3.4. ALTERNATIVE APPROACHES TO DETERMINING STRAIN ENERGY FUNCTIONS

While based on rigorous experimental data and able to capture the 2D in-plane biaxial response well, the constitutive models described above present certain difficulties both in terms of their form and in parameter determination. In particular, while based on fundamental mechanics principals, there is no additional knowledge to guide the particular choice of form of the model, and is somewhat arbitrary. Models are generally evaluated for the degree of over parameterization using comprehensive statistical methods (e.g., [65]) and subsequently modified, generally in the reduction of the number of parameters.

Following methods developed by Rivlin et al. for rubber elasticity [43], Humphrey et al. developed a new functional form for myocardium biaxial mechanical properties. In this approach, Humphrey defined a subclass of transversely isotropic materials whose strain energy function is a function of two strain invariants [20]:

$$W = W(I_1, I_4), \tag{21}$$

where I_1 and I_4 are the first and fourth strain invariants, with $I_4 = \alpha^2$, where α is the stretch ratio along the muscle fiber direction. Similar to the approach by Rivlin, this form allows determination of the dependence of W_1 and W_α on I_1 and α directly from the experimentally obtained stress and deformation data [20]:

$$2W_1 = \frac{\xi_4 t_{11} - \xi_2 t_{22}}{\xi_1\xi_4 - \xi_2\xi_3}, \qquad W_\alpha = \alpha\frac{\xi_1 t_{22} - \xi_3 t_{11}}{\xi_1\xi_4 - \xi_2\xi_3}, \tag{22}$$

where $W_1 = \partial W/\partial I_1$, $W_\alpha = \partial W/\partial \alpha$, t_{ij} denotes the physical components of the Cauchy stress tensor, and

$$\begin{aligned}
\xi_1 &= \lambda_1^2 + \kappa_1^2 - \lambda_3^2,\\
\xi_2 &= \lambda_1^2\cos^2(\theta) + 2\lambda_1\kappa_1\cos(\theta)\sin(\theta) + \kappa_1^2\sin^2(\theta),\\
\xi_3 &= \lambda_2^2 + \kappa_2^2 - \lambda_3^2,\\
\xi_4 &= \kappa_2^2\cos^2(\theta) + 2\lambda_2\kappa_2\cos(\theta)\sin(\theta) + \lambda_2^2\sin^2(\theta),\\
\xi_5 &= \kappa_1\lambda_2 + \kappa_2\lambda_1,\\
\xi_6 &= \lambda_1\kappa_2\cos^2(\theta) + (\lambda_1\lambda_2 + \kappa_1\kappa_2)\cos(\theta)\sin(\theta) + \kappa_1\lambda_2\sin^2(\theta),
\end{aligned} \tag{23}$$

where λ_i and κ_i are components of the deformation gradient tensor and θ is the fiber angle with respect to the x_1 axis. Next, based on experimental plots of W_1 and W_α vs. I_1 and α, they assumed the following functional form for $W(I_1, \alpha)$:

$$W(I_1, \alpha) = \sum_{i=0}^{n}\sum_{j=0}^{m} c_{ij}(I_1 - 3)^i(\alpha - 1)^j, \tag{24}$$

where c_{ij} are material parameters. The interested reader is referred to [20] for details of the derivation.

To apply this approach, Humphrey et al. [20] modified their biaxial testing device to perform constant strain invariant tests. From the experimental data generated, plots of W_1 and W_α vs. either I_1 varied and α held constant or I_1 held constant and α varied were generated using (22) and (23). Based on the resulting response functions, as well theoretical restrictions on the values of c_{ij} (e.g., at zero strain $W(3,1) = 0$, requiring $c_{00} = 0$), the following form for passive myocardium was derived:

$$W(I_1, \alpha) = c_1(\alpha - 1) + c_2(\alpha - 1)^3 + c_3(I_1 - 3) + c_4(I_1 - 3)(\alpha - 1) + c_5(I_1 - 3)^2, \tag{25}$$

where c_i are the material constants. In obtaining the material constants, additional empirical inequalities were applied to set bounds on the values for c_i.

The constant invariant tests used to derive (25) require that the tissue specimen be subjected to simultaneous loading and unloading, which strictly violates pseudoelasticity [16]. Thus, only data from the loading portion of the equi-biaxial strain and constant α tests were used to determine values for c_i. The data from constant I_1 tests were used only to find the functional form, and were hence excluded since the tissue experiences simultaneous loading and unloading during these tests. The resulting model was found to fit the biaxial data quite well, and was also found not to be over-parameterized [21]. The results also emphasized the need for good data, including accurate measurement of the applied forces, original dimensions and experimental deformations. Sacks and Chuong [47] later successfully applied (25) to right ventricular myocardium, where the effects of a fiber splay within the specimen were incorporated. May–Newman also applied a similar approach to the biaxial mechanical properties of the mitral valve [37]. It is also interesting to note that the overall approach of using theoretically guided experiments is not restricted to the form of (24), but has also been successfully applied to the epicardium (a thin connective tissue layer surrounding the heart) [22] using the Fung model (18) and constant E_{11} and E_{22} tests.

Clearly, the strength of the above approach is that it allows derivation of the functional form of W to be rigorously based on direct evaluation of the experimental data. This avoids the limitations of the trial-and-error approach of earlier work, and is an elegant example on how theory and experiment can be integrated. However, while shown to model passive tissue properties well, the approach has several limitations. In principal, the approach can be applied to specimens where there is a nonzero distribution of fiber orientations. However, the author has found the model to be weakly dependent with respect to trans-mural layer orientations, potentially due to the transverse isotropy assumption (Sacks, unpublished data). More detailed, realistic inter-layer models may need to be developed. Further, in order to determine the form of W pseudo-elasticity must be violated (i.e., the tissue must be subjected to simultaneous loading and unloading protocols). While the loading and unloading curves are qualitatively similar, differences do exist and may influence the sensitivity and final choice of the form.

3.5. ACTIVE TISSUE PROPERTIES

The above mechanical tests were performed on passive myocardial tissue properties only. Biaxial properties of actively contracting tissues have been investigated in a very limited fashion, due to the significant experimental challenges of keeping muscular tissue viable and stable while performing all testing protocols. The first known study known to the author was by Strumpf et al. [55] for the diaphragm. The architecture, vascular supply, and ease of tetanization (electronically-stimulated contraction) make the diaphragm an ideal structure in which to assess multidimensional mechanical properties of active and passive muscle. Test specimens were approximately 3 × 3 cm in size with a single predominant fiber direction. Highly nonlinear behavior was seen in the passive state with a limit of extensibility in both directions. The specimens were also markedly anisotropic, with the cross-fiber direction being stiffer than the fiber direction. During tetanization, the extent and distribution of anisotropy were significantly altered, with most specimens either isotropic or stiffer in the fiber direction.

The diaphragm is a tri-layered structure, consisting of an inner thick muscular layer and two thin collagenous membranes. In order to determine the mechanical contributions of the outer membrane layers, Strumpf et al. [55] disrupted the membranes covering each surface. They found increased extensibility and decreased the anisotropy, thereby suggesting that these membranes bear most of the passive load and contribute greatly to the cross-fiber stiffness and anisotropy of the intact diaphragm. Both before and after disruption of the surface membranes, there was still a consistent increase in cross-fiber stress during tetanization, implying active force generation perpendicular to the fiber direction.

In a later study, Lin and Yin [35] subjected thin midwall slices of rabbit myocardium to biaxial stretching. Assuming transverse isotropy in both the passive and active conditions, they used the method of Humphrey et al. [20, 21] to obtain both passive and active constitutive laws. They concluded that the functional forms of the constitutive law for the actively contracted states were the same, but differ from those in the passive state. Hence, *one cannot simply substitute differing values for the coefficients of the passive law to describe the active tissue properties* [3]. Similar to the diaphragm, there were significant stresses developed in the cross-fiber direction (more than 40% of those in the fiber direction) that cannot be attributed to either deformation effects or nonparallel muscle fibers. These studies underscore the potentially complex micro-structural interactions which result in substantially complex behavior of actively contracting tissue behavior.

4. More Recent Developments

4.1. CONTROL OF SPECIMEN STRUCTURE

A reoccurring difficulty in many of the above studies, particularly so for collagenous tissues such as skin and pericardium, is the substantial degree of interspecimen variability. This variability underscores the need to determine the source of the underlying biological variability for accurate and meaningful determination of material constants. This is a particular problem when using biologically-derived tissues in medical devices (e.g., bioprosthetic heart valves), where accurate constitutive models are essential for device design, determining effects of novel chemical treatments, and in the understanding simulation of fatigue damage. One such tissue is chemically treated bovine pericardium, which although mechanically anisotropic [33, 66], there is no evidence that bioprosthetic heart valves are constructed to accommodate or take advantage of this anisotropy in a systematic way. Finite element stress analyses of bioprosthetic heart valves suggest that high flexural stresses during valve opening and high tensile stresses during valve closure are associated with failure locations [6, 17, 27]. Their accuracy, however, is limited by the use of simplistic isotropic material approximations, since in reality chemically treated bovine pericardium is an anisotropic, biocomposite material.

To quantify planar tissue fibrous structure, Sacks et al. [50] have developed a small angle light scattering technique (SALS). SALS allows rapid quantification of the angular distribution of fibers at each point in the tissue, from which the preferred fiber direction and degree of orientation can be determined. A SALS-based tissue sorting procedure was used to guide the selection of bovine pericardial specimens to minimize structural variability. Note that in this study, we use designations preferred fiber (PD) and cross-preferred fiber (XD) for the x_1 and x_2 axes, respectively (Figure 1(b)). An extensive biaxial test protocol was then used and the resulting stress-strain data fitted to (20). Results indicated that (20) was able to reproduce the mechanical response of chemically treated bovine pericardium over a wide range of biaxial test protocols (Figures 3(a), (b)). When applied to data from test protocols not used to derive the material constants, (20) fit the data well (Figures 3(c), (d)). The high structural uniformity resulted in both a consistent mechanical response and low variability in the material constants.

Because of this consistency, the data from all specimens were combined into a single data set. From this dataset "group" material constants were determined, which represent a more generalized estimate of tissue properties. The individual specimen constants showed predictably better agreement with the data, but the group constants were able to represent the data reasonably well (Figure 3(c),(d)). This study demonstrated that much of the past difficulties with tissue variability were a direct result of uncontrolled variabilities in tissue structure. Previous studies on myocardium [21, 47, 54] controlled specimen structure by visual selection and alignment to the overall preferred fiber direction. This was an important step in the analysis of orthotropic biological tissues. The use of SALS, however, not only allowed similar specimen fiber alignment for collagenous tissues, but more importantly the quantitative relation between the degree of fiber alignment and degree of mechanical anisotropy. To the author's knowledge, this is the first time such a comparison was undertaken for fibrous collagenous tissues.

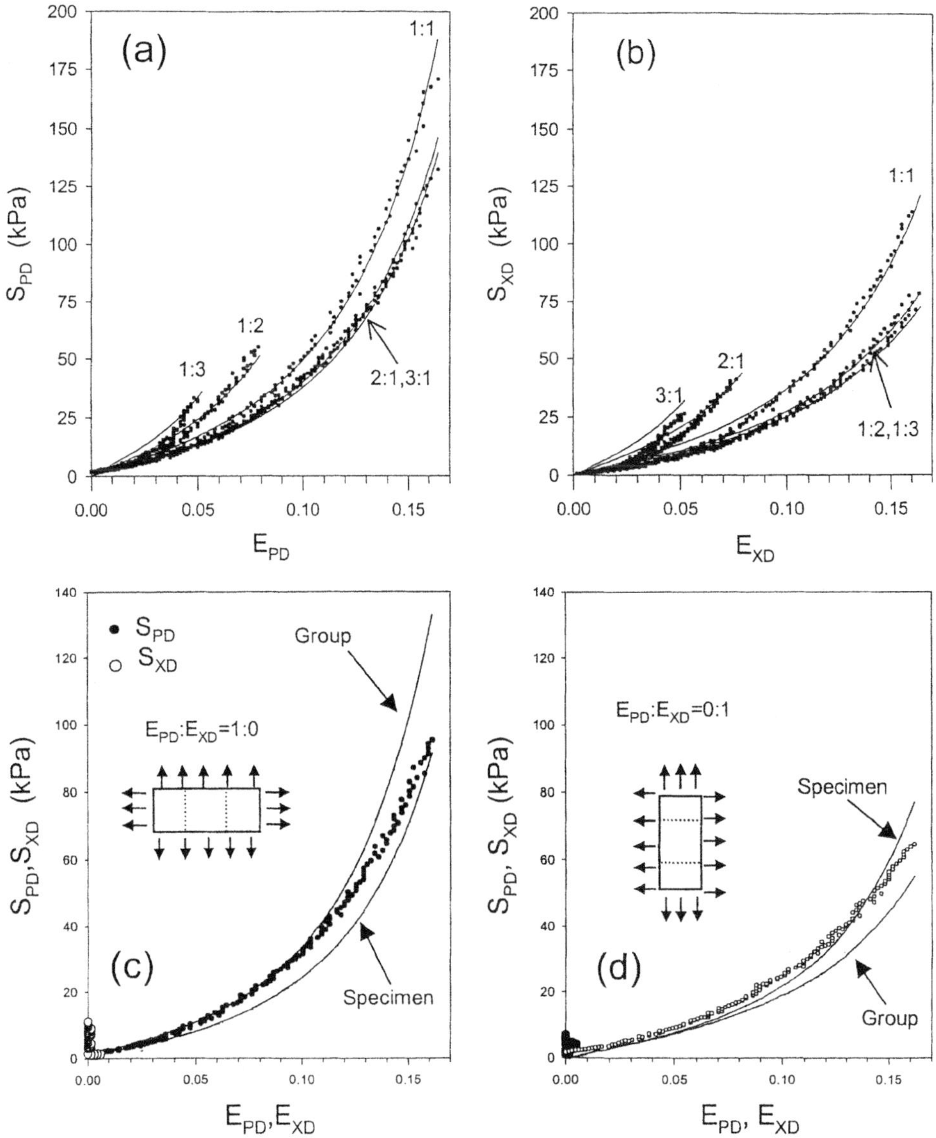

Figure 3. (a), (b) Biaxial mechanical behavior of chemically treated bovine pericardium for the five strain protocols (indicated beside each curve) for the (a) PD and (b) XD axes. Also shown is the fit of equation (20). The predictive behavior of (20) is demonstrated using the data from the $E_{PD} : E_{XD} = 1 : 0$ (c) and (d) $0 : 1$ tests, that were *not used* to determine the material constants. Although the individual specimen constants showed predictably better agreement with the data, both the specimen and group constants showed a good fit to the data.

In addition to minimizing tissue variability, tight control of tissue structure allows elucidation of the more subtle aspects of tissue mechanical properties. In our previous studies on pericardial mechanical properties, we observed that the strain level along the x_1 axis (or preferred fiber direction, Figure 1(a)) axis has a stronger effect on the x_2 axis stress level (or cross-preferred fiber, Figure 1(a)) than the x_2 axis strain level has on the x_1 axis stress level [49]. While this phenomenon has been termed "coupling," it is distinct from the expression

$$\frac{\partial}{\partial E_{11}}\left(\frac{\partial W}{\partial E_{22}}\right) = \frac{\partial}{\partial E_{22}}\left(\frac{\partial W}{\partial E_{11}}\right),$$

which holds at a particular strain state (E_{11}, E_{22}). In contrast, the phenomena referred to here deals with how stress magnitudes along one axis are affected by the strain level along the perpendicular axis. Similar mechanical coupling properties have reported for mitral valve leaflets [36], and for passive myocardium (e.g., [63, Figure 4]).

This coupling phenomenon was examined by normalizing the peak stresses from the non equibiaxial strain tests to the equibiaxial peak stresses for each axis. This approach allows for a simple means to quantify and graphically compare the mechanical coupling properties. The x_1 axis demonstrated a general trend that as the x_2 axis was subjected to lower maximum strain levels, there was only a small reduction in the x_1 maximum stress (Figure 4(a), $E_{11} : E_{22} = 1 : 0, 3 : 1$, and $2 : 1$). In contrast, when the x_1 axis was subjected to lower maximum strain levels, the normalized x_2 maximum stresses decreased to 0.25–0.55 (Figure 3(b), $E_{11} : E_{11} = 1 : 2, 1 : 3$, and $0 : 1$). At the time of the study, the exact cause for these complex coupling properties was unknown. Later work by our group using a structural constitutive approach has demonstrated that this effect is a direct result of the particular architecture of the pericardium's collagen fibers. This architecture can produce complex mechanical behaviors at the tissue level due to large fiber strains and rotations, along with a nonlinear fiber stress-strain relationship [4, 45].

4.2. ERRORS WITH ISOTROPIC MATERIAL ASSUMPTIONS

Although demonstrated to have orthotropic mechanical properties, isotropic models are frequently used to simply subsequent analysis. To evaluate the errors in the use of an isotropic material model in modeling pericardial mechanical properties, the biaxial data was fit to the following isotropic strain energy function [13]:

$$W = c_1\left(\exp^{c_2(I_1-3)} - 1\right), \tag{26}$$

where c_1 and c_2 are material constants, and I_1 is the first strain invariant in terms of stretch ratios. This model predicted a stress-strain response intermediate between the x_1 and x_2 responses for the equibiaxial test [48]. Thus, an isotropic model essentially predicts an average axial stress-strain response for an equibiaxial strain

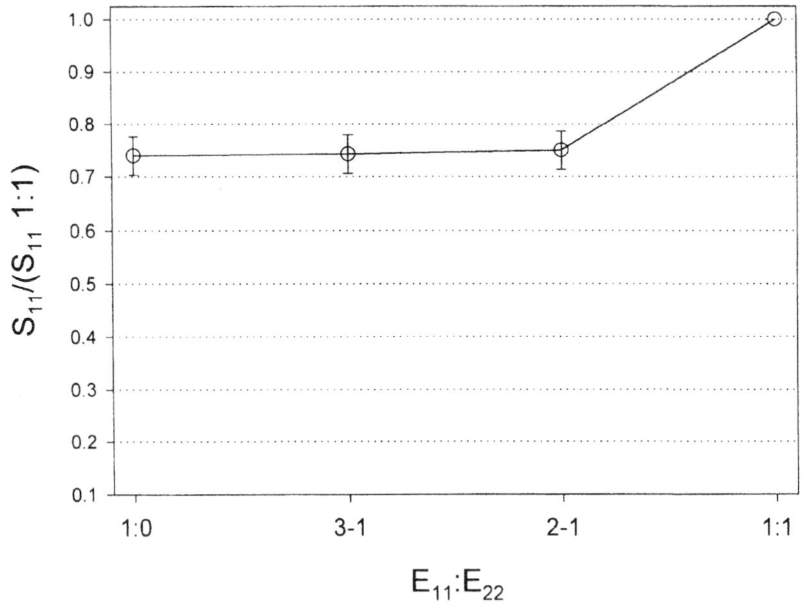

(a)

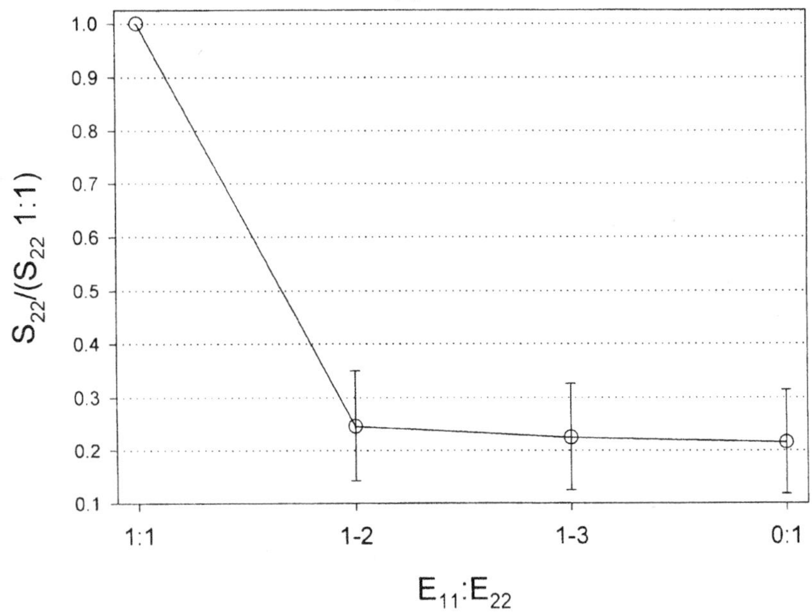

(b)

Figure 4. A graphical demonstration of asymmetric mechanical coupling using the peak stresses from the non-equibiaxial tests normalized to the equibiaxial peak stresses for each axis. (a) The x_1 material axis demonstrated a general trend that as the x_2 axis was subjected to lower maximum strain levels, there was only a small reduction in the x_1 maximum stress. (b) In contrast, when the x_1 axis was subjected to lower maximum strain levls, the normalized x_2 maximum stresses decreased to $\sim$0.25.

state. The agreement with the other test protocols was in general worse. Thus, although providing an approximate estimation of stresses under an equibiaxial strain state, isotropic material models clearly cannot reproduce the complex orthotropic mechanical responses found for pericardium.

4.3. HOMOGENEITY OF THE STRESS AND STRAIN FIELD

A critical aspect of biaxial testing is the need for homogeneity of the stress and strain field within the central region of the specimen, as originally stated in (4). The homogeneity of the stress and strain can be affected by several parameters, the most critical being the use of discrete point loading to allow unconstrained transverse deformation on the specimen boundaries. In small samples, the effects of stress concentrations generated at the suture attachment sites may propagate into the measurement region, negating reliable estimates of the stress. For example, the aortic heart valve cusp is smaller and more structurally heterogeneous than most tissues examined with biaxial mechanical testing techniques [5]. Typical widths for biological tissues previously tested with biaxial techniques are two to three times the size of a cusp and many times larger than a single structural region in a cusp (i.e. belly region) [8, 9, 25, 48]. Using finite element analysis and biaxial tests on synthetic membranes, Nielsen and colleagues [38] attempted to determine the validity of biaxial testing methods for small specimens ($\sim$10 mm square). The authors found that, for homogeneous isotropic materials loaded at four discrete locations per side, the strain and stress were relatively uniform in the central 25% of the sample. However, the authors correctly note that the stress and strain distributions would be less uniform for anisotropic and/or heterogeneous materials. For complex anisotropic tissues, direct experimental evidence is clearly needed to demonstrate that a sufficiently homogeneous area for strain measurement exists and that the stress distribution in the central region is sufficiently homogeneous despite the point loading at the boundary.

In the case of the aortic valve cusp tissue, we tested the hypothesis that the highly aligned cuspal fiber architecture rather than the boundary stress concentrations induced by suture lines dominated the local strain field [5]. This hypothesis is especially important when testing small heterogeneous tissue specimens such as the aortic valve cusp, where boundary tethering effects may propagate toward the interior of the specimen, affecting the homogeneity of the stress field within the region delimited by the markers. In addition, there are also practical experimental concerns including:

(1) determining the degree of heterogeneity of the strain field, hence the maximum size, of the area delimited by the optical markers, and
(2) determining optimal location for strain measurements (i.e., the placement of the optical markers).

To address these questions, we performed experiments where the biaxial specimen fiber orientation was varied and examined its effects on the local strain field (details can be found in [5]). Specimens had their edges aligned "on-axis" with the fiber direction in the center of the cusps, whereas three other specimens had their edges cut at ~45° to the fiber direction ("off-axis"). To track the tissue displacements, nine small graphite markers (75–150 μm diameter) were affixed to the ventricular surface in a square array. Nine markers were used to achieve quadratic interpolation of the displacement field **u**.

The strong effect of the underlying fiber architecture on the two-dimensional strain distribution was clear in all samples. This relationship is clearly demonstrated when the local fiber preferred directions and the corresponding principal strains are superimposed (Figures 5(a), (b)). In both the on-axis (Figure 5(a)) and off-axis cases (Figure 5(b)), the maximum principal strains were approximately perpendicular to the preferred fiber directions, with the second (lesser) principal strains aligned with the fiber directions. These results demonstrate quantitatively that the local tissue is much more compliant perpendicular to the preferred fiber direction than along the fiber direction, and appears to be relatively unaffected by the stress concentrations due to suture tethering forces.

As a further example on how to address heterogeneity of tissue structure, we computed the direction and magnitude of the principal Green's strains at the locations of the SALS structural measurements. Details of the procedure are given in [5]. Briefly, the principal Green's strains were superimposed on the map of local preferred fiber directions to graphically compare the strain field and structural data. Results demonstrated by the structure-strain studies is that the strain field was most uniform in the center of the belly and most variable near the nodulus (Figure 5(c)). Note too the profound difference in the stress-strain data between the nodulus and central belly region (Figure 5(c)). Since the central belly region is structurally most similar to the bulk of the tissue specimen, and the fibers are approximately aligned with the test axes in this region, the lower belly region (B in Figure 5(c)) was determined as the optimal strain measurement site. Further, analysis of the strain field indicated that the values for E_{ij} did not vary by more the 5% of the mean value (typically 2%–3%) within a 3 mm × 3 mm area in the lower belly region.

Although the above results clearly demonstrated the difficulties and presented solutions for addressing questions of heterogeneity of the stress and strain fields within a biaxial specimen, alternative methods of gripping are also possible. Recently, Waldman and Lee [59] demonstrated that substantially different stress-strain curves can be generated in the same tissue by changing from tethering to a clamping method. Clamping was used as an attempt to improve stress-transfer from fiber to matrix. In a tethered specimen, the point loadings have been shown to induce dramatic local reorientation of the fiber structure at the point of attachment [3]. Using SALS, Waldman et al. [60] demonstrated that the change in stress-strain curves was due to a dramatically different strain (and hence stress) field in the vicinity of the clamps, as evidence by substantial fiber reorientation and alignment. Interestingly, within the region of the specimen where the optical makers were located, fiber realignment was only a function of the local tissue

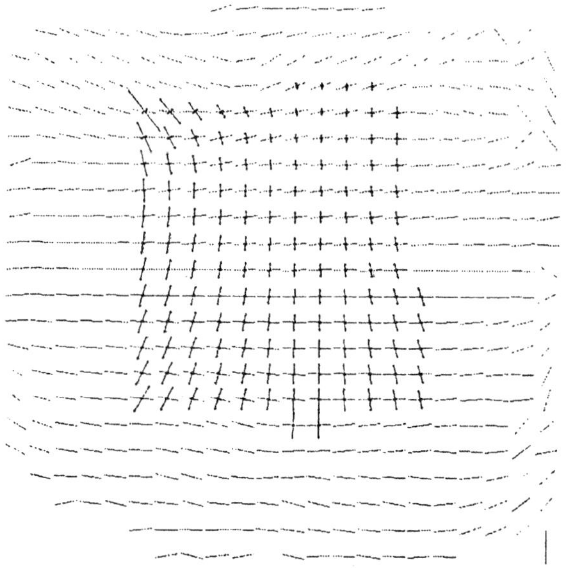

(a)

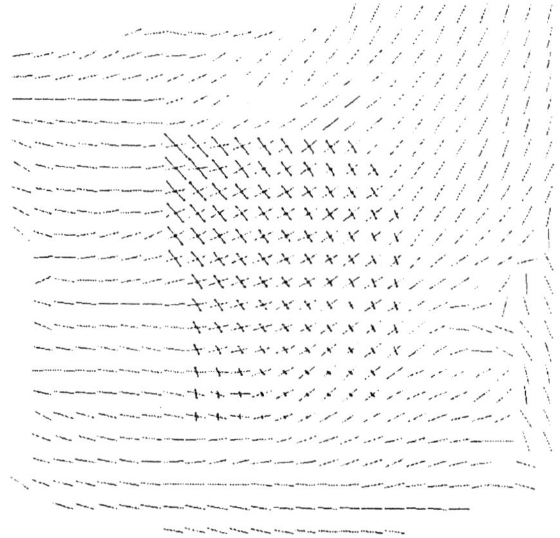

(b)

Figure 5. Peak principal Green's strains under 60 N/m equibiaxial tension for the aortic valve cusp superimposed over the collagen fiber directions for an (a) on-axis and (b) 45° specimen. Lines represent local collagen fiber preferred directions and arrows the principal strains, with lengths proportional to strain magnitude. The principal strains were closely oriented to the local preferred fiber directions throughout most of the region for both orientations. (c) Plot of collagen fiber orientation (lines) superposed over the local degree of orientation (OI). Here, the white "holes" in the plot correspond to the displacement markers. (d) Corresponding tension-strain plots for different portions with N = nodulus, M = midregion, B = belly, highlighting the heterogeneity of the mechanical properties.

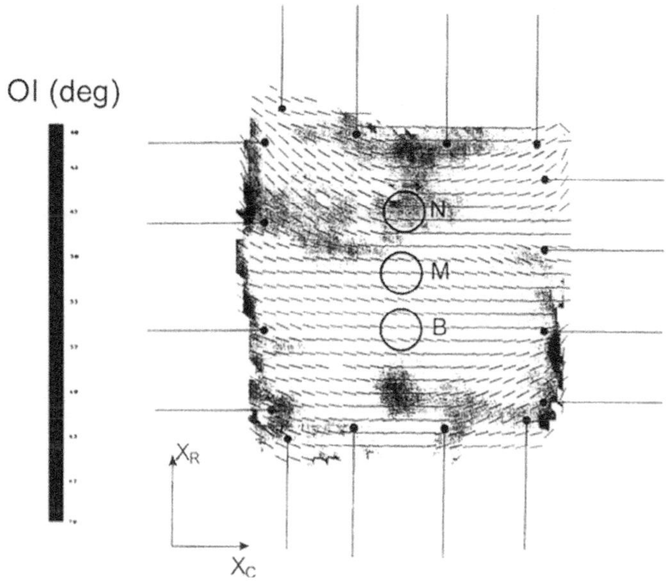

(c)

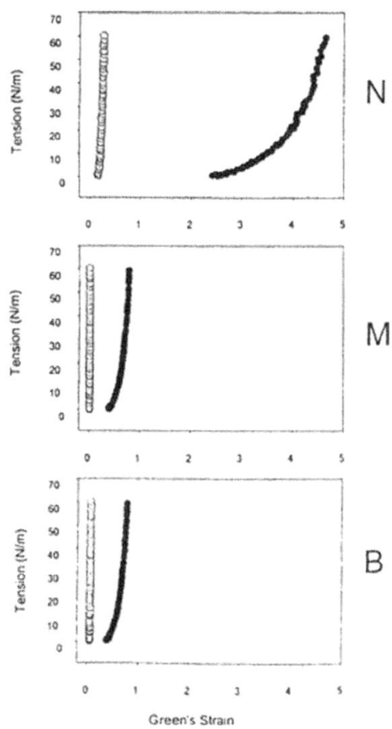

(d)

Figure 5. (Continued.)

strains, and was unaffected by the method of attachment. These results underscore the importance of addressing tissue attachment and important role of boundary conditions in determining biaxial stress-strain data.

4.4. INCORPORATION OF THE EFFECTS OF IN-PLANE SHEAR

A limitation in all of the above cited biaxial studies of soft tissues has been the inability to include the effects of in-plane shear. This is due to the inability of current mechanical testing devices to induce a state of in-plane shear, due to the added cost and complexity. In the typical planar biaxial mechanical test, specimens are mounted with their material axes aligned to the test axes (e.g., [9, 46, 48, 63]) and subjected to a wide range of strain states. While sufficient for characterizing on-axis anisotropic mechanical properties, this method cannot predict the tissue's response in the presence of planar shear strains.

To induce shearing strains in planar biaxial experiments, the biaxial testing system must be able to apply a non-uniform system of forces along each of the test specimen's boundaries. Recently, Khalsa et al. [26] developed a novel biaxial testing system to induce in-plane shear. When applied to skin, the device was able to accurately induce controlled states of normal and shear stresses [15]. However, this device required a more complex biaxial testing system, including twelve load transducers and actuators (other designs could potentially include more). Another consideration in planar biaxial testing with shear is when the loads along one axis are zero, there is the potential for out-of-plane deformations (i.e., wrinkling). A simpler, more economical approach was clearly needed for the derivation of planar biaxial mechanical properties that include shear.

The author recently developed a method to induce a combined state of in-plane biaxial normal and shear strains [44]. The method takes advantage of rotation of the test specimen's material axis to the test axes (Figure 6(a)). The method required only modest modifications to the biaxial testing device [48] to allow the specimen to undergo unrestricted shear deformations. Specifically, the device was modified to allow in-plane shearing through the use of rotating carriages (Figure 6(b)). Two loops of 000 nylon suture of equal length were attached to each side of the specimens with four stainless steel surgical staples. Symmetrically placed pulleys insured equal tension on each pair of suture lines, with the pulleys mounted on a stainless steel rod attached to a bearing to insure each pulley supported equal tension (Figure 6(b)). The complete assembly was able to rotate freely (both pulleys and bearing) with negligible friction for applied loads up to 50 N. This straightforward design eliminated the use of individual actuators and force transducers for each suture line, as well the need for adjustment of individual suture line tensions. Small styrofoam floats were attached to each staple to make the mounted sample neutrally buoyant (Figure 6(c)).

Biaxial data from five glutaraldehyde treated bovine pericardium are presented to demonstrate the feasibility of the approach, including a fit to an orthotropic strain

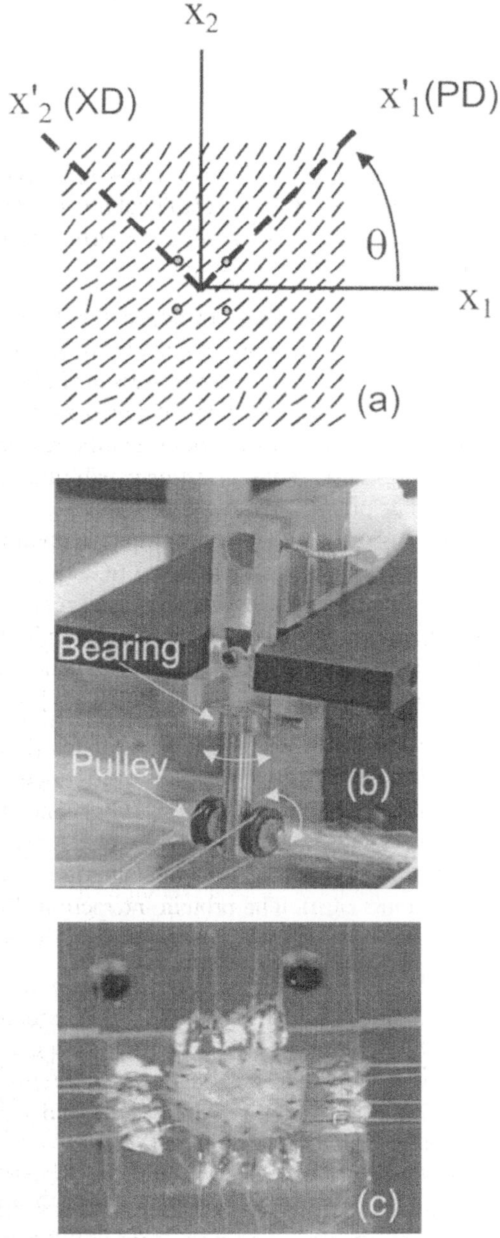

Figure 6. (a) A schematic of a biaxial test specimen where the lines represent the preferred fiber direction, which are rotated 45° from the x_1–x_2 biaxial testing device axes and are represented by the x'_1–x'_2 axes. The small circles indicate where the four markers are located for strain measurements. (b) and (c) show the specialized carriages and suture attachment placement that allow the specimen to shear freely. Note that the small white objects in (c) are the styrofoam floats attached to the staples.

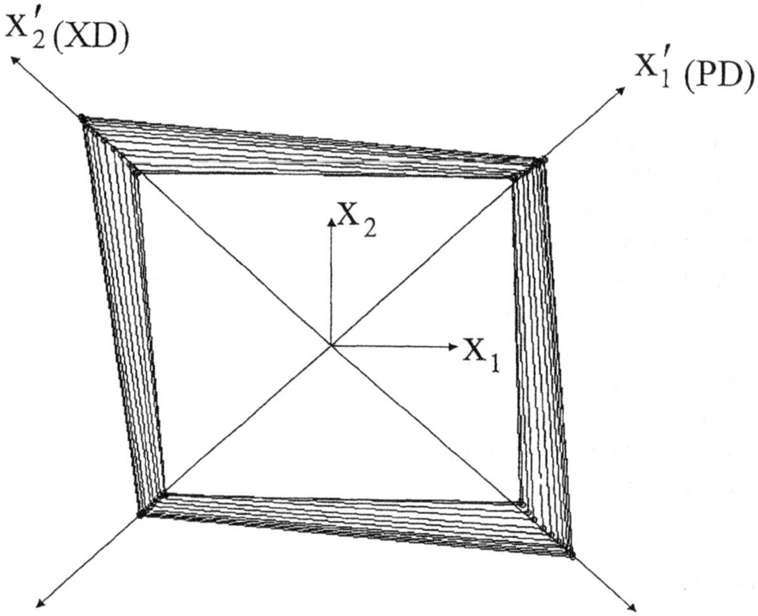

Figure 7. Video marker trajectories for an equibiaxial experiment ($E_{11} = E_{22}$) demonstrating $E_{12} \neq 0$. Consistent with the direction of the specimen's material axes, the smallest and largest strains were along the PD and XD directions (see Figure 6), respectively.

energy function that included shear and shear-normal terms. To demonstrate the method, five glutaraldehyde treated bovine pericardium specimens were prepared with their preferred fiber directions (defining the material axes) oriented at 45° to the device axes to induce a maximum shear state. The test protocol included a wide range of biaxial strain states, and the resulting biaxial data re-expressed in material axes $\mathbf{x}'$ coordinate system (Figure 6(a)). The relative magnitude of the induced shearing strain is apparent when the specimen's markers trajectories are plotted under equibiaxial normal strain (i.e., $E_{11} \sim E_{22}$, $E_{12} \neq 0$, Figure 7). For this specimen, the shear angle α at maximum strain was 10.9°. Consistent with the direction of the material axes, the smallest and largest strains were along the x_1'(PD) and x_2'(XD) directions, respectively (Figure 6).

The resulting biaxial data were then fit to the following generalized Fung model (18) with the in-plane shear terms retained:

$$W = \frac{c}{2}\big[\exp\big(A_1 E_{11}'^2 + A_2 E_{22}'^2 + 2A_3 E_{11}' E_{22}' + A_4 E_{12}'^2 + 2A_5 E_{11}' E_{12}' + 2A_6 E_{22}' E_{12}'\big) - 1\big], \tag{27}$$

where E_{ij}' is the Green's strain tensor in the material axes coordinate system and c and A_i are material constants. While W was able to fit the data very well (Figure 8), the constants A_5 and A_6 were found not to contribute significantly to the fit and were considered unnecessary to model the shear strain response. While not

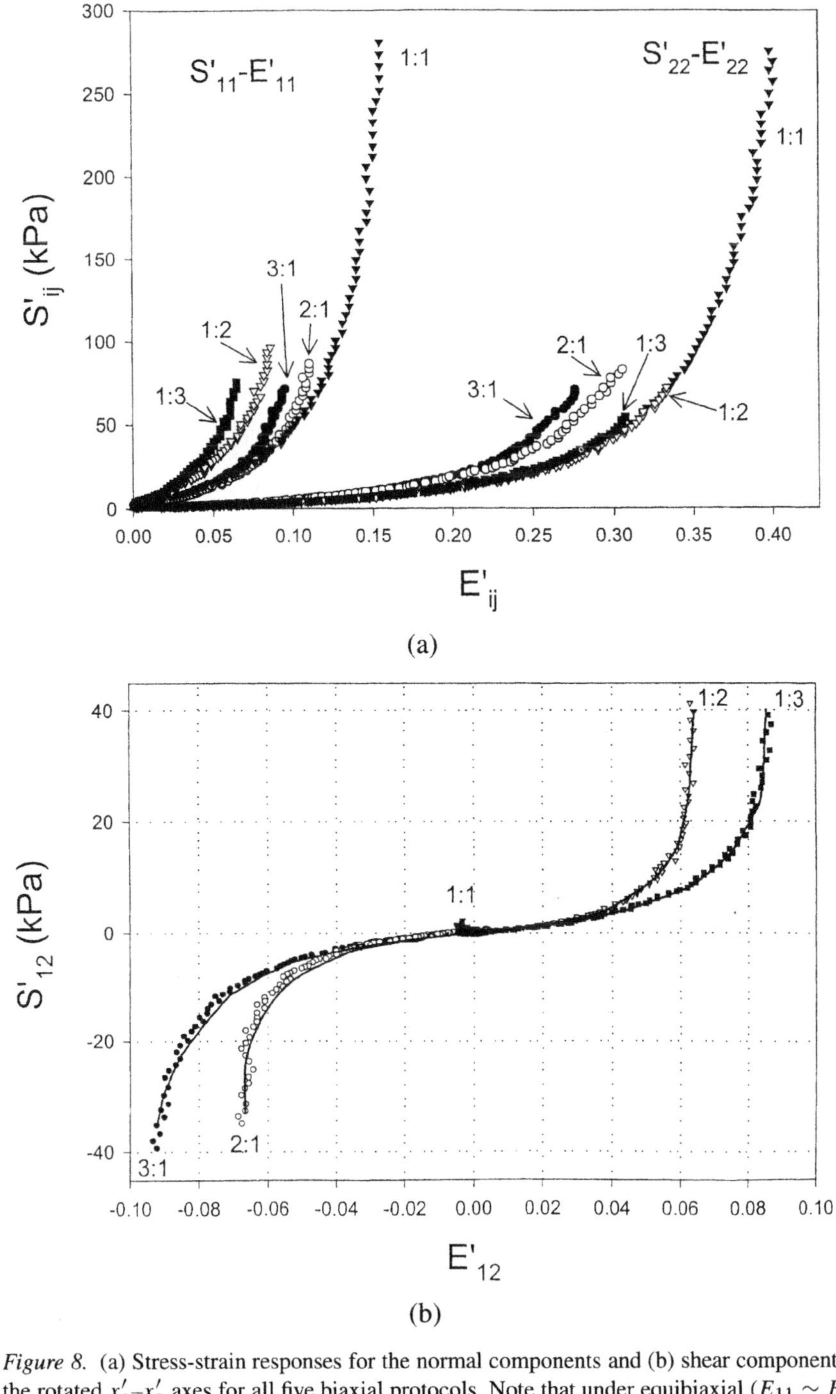

Figure 8. (a) Stress-strain responses for the normal components and (b) shear components in the rotated x'_1–x'_2 axes for all five biaxial protocols. Note that under equibiaxial ($E_{11} \sim E_{22}$) loading, E'_{12} and S'_{12} are ~0 as shown in (b). Also shown for the shear components is the fit of (27) (with $A_5 = A_6 = 0$), showing good agreement with the data. The fit was also good for the normal components E_{11} and E_{22} in (a), but was not shown for clarity.

able to independently control the amount of shear strain or induce a state of pure shear, the method presented readily produces a state of simultaneous in-plane shear and normal strains. Further, the method is very general and can be applied to any anisotropic planar tissue that has identifiable material axes.

4.5. STRUCTURAL CONSTITUTIVE MODELS

While the phenomenological constitutive models described previously are able to fit the mechanical data well, they are limited in that they cannot elucidate the underlying cause of mechanical behavior. Alternatively, structurally based constitutive models not only avoid ambiguities in material characterization but also offer insights into the function, structure, and mechanics of tissue components. Perhaps the most complete approach has been by Lanir et al. [28, 29, 32]. In this approach, the tissues net or total strain energy is assumed to be the sum of the individual fiber strain energies, linked through appropriate tensor transformation from the fiber coordinate to the global tissue coordinates.

As an example of how a structural model can be implemented for the aortic valve (AV). On the most basic functional level, the aortic valve is essentially a check-valve that serves to prevent retrograde blood flow from the aorta back into the left ventricle. This seemingly simple function belies the structural complexity and the elegant solid-fluid mechanical interaction necessary for normal aortic valve function. The aortic valve is capable of withstanding 30–40 million cycles per year, resulting in a total of ~3 billion cycles in single lifetime.

To achieve the combination of low flexural rigidity necessary to allow normal valve opening and high tensile strength required to resist trans-valvular pressures of 80 mmHg, nature has evolved a tri-layered cuspal structure. These layers are the ventricularis, spongiosa, and fibrosa. As its name implies, the ventricularis layer faces the left ventricular chamber and is composed of a dense network of collagen and elastin fibers. The spongiosa layer contains a high concentration of proteoglycans. The fibrosa layer is composed predominantly of a dense network of collagen fibers, and is thought to be the major stress-bearing layer. Interest in understanding the diseases of the aortic valve and the need for improvement in replacement valves has prompted our lab to develop a constitutive model for the native aortic valve.

The fiber architecture [51] and biaxial mechanical data [5] suggest that a structural approach is the most suitable method for the formulation of a constitutive model for the AV cusp. For example, we have demonstrated that strong axial coupling produced nonmonotonic relationships between stress and strain, including negative strains [5]. Our related work on native and chemically treated bovine pericardium suggests that a structural approach is both feasible and attractive for bioprosthetic heart valve biomaterials [45].

Details of the model have been previously presented [4]. Briefly, we assume that the planar biaxial mechanical properties of the cusp can be represented as a planar array of collagen fibers. Anatomically, these fibers most closely represent the dense, highly aligned collagen fibers in the fibrosa layer. Next, the angular fiber distribution and the density of the fibers are assumed constant throughout the tissue. Based on our SALS results for the aortic valve cusp [51], we utilize the fact that the angular distribution of the collagen fibers, $R(\theta)$, can be represented by a Gaussian distribution,

$$R(\theta) = \frac{1}{\sigma\sqrt{2\pi}} \exp\left[\frac{-(\theta - M)^2}{2\sigma^2}\right], \tag{28}$$

where θ is the direction with respect to the x_1 axis (Figure 1(a)), σ is the standard deviation and M is the mean of the distribution. M was determined experimentally for each specimen by using the preferred fiber directions as determined by SALS [51]. The "effective" fiber stress-strain properties were represented using

$$S_f = A\left[\exp(BE_f) - 1\right], \tag{29}$$

where S_f is the second Piola–Kirchhoff fiber stress, E_f is the fiber Green's strain. This formulation for the fiber stress-strain law avoids detailed descriptions of complex crimp distributions.

For valvular tissue, it is more convenient to work with membrane stresses due to considerations such as variable total and layer thickness, and heterogeneous layer structure [5]. Further, since the biaxial mechanical tests are run using membrane stress control using the specimen's unloaded dimensions, a Lagrangian membrane stress measure is used in the constitutive formulation. We also assume that interspecimen variations in fiber volume fraction V_f and thickness h are negligible, so that the product hV_f can be conveniently absorbed into the material constant A. The resulting expressions for the Lagrangian membrane stresses T_{ij} are:

$$\begin{aligned} T_{11} &= \int_{-\pi/2}^{\pi/2} \mathrm{S}_f^*(E_f) R(\theta)\left(\lambda_1 \cos^2\theta + \kappa_1 \sin\theta\cos\theta\right) \mathrm{d}\theta, \\ T_{22} &= \int_{-\pi/2}^{\pi/2} \mathrm{S}_f^*(E_f) R(\theta)\left(\lambda_2 \sin^2\theta + \kappa_2 \sin\theta\cos\theta\right) \mathrm{d}\theta, \end{aligned} \tag{30}$$

where $A^* = hV_f A$ and $\mathrm{S}_f^* = A^*[\exp(BE_f) - 1]$. The parameters A^*, B, and σ were estimated by fitting (15) to the complete biaxial data set [5].

The fit of the model to the data was good despite the complexity of the mechanical response over the broad range of biaxial loading states (Figure 9(a)). The model fit the data from all seven protocols well even though the data from the outer protocols (1 and 7, see inset in Figure 9(a)) were not used in the parameter estimation. Using only three material parameters, the quantitative "goodness of fit" was comparable to phenomenological models of other tissues [9, 21, 37].

It is informative to demonstrate the utility of a structural approach in minimizing the number of parameters necessary to model the in-plane behavior of the cusp. In the case of the Fung type (18), a three parameter version would be

$$W = \frac{c}{2}\left[\exp(\mathrm{A}_1 E_{11}^2 + \mathrm{A}_2 E_{22}^2) - 1\right]. \tag{31}$$

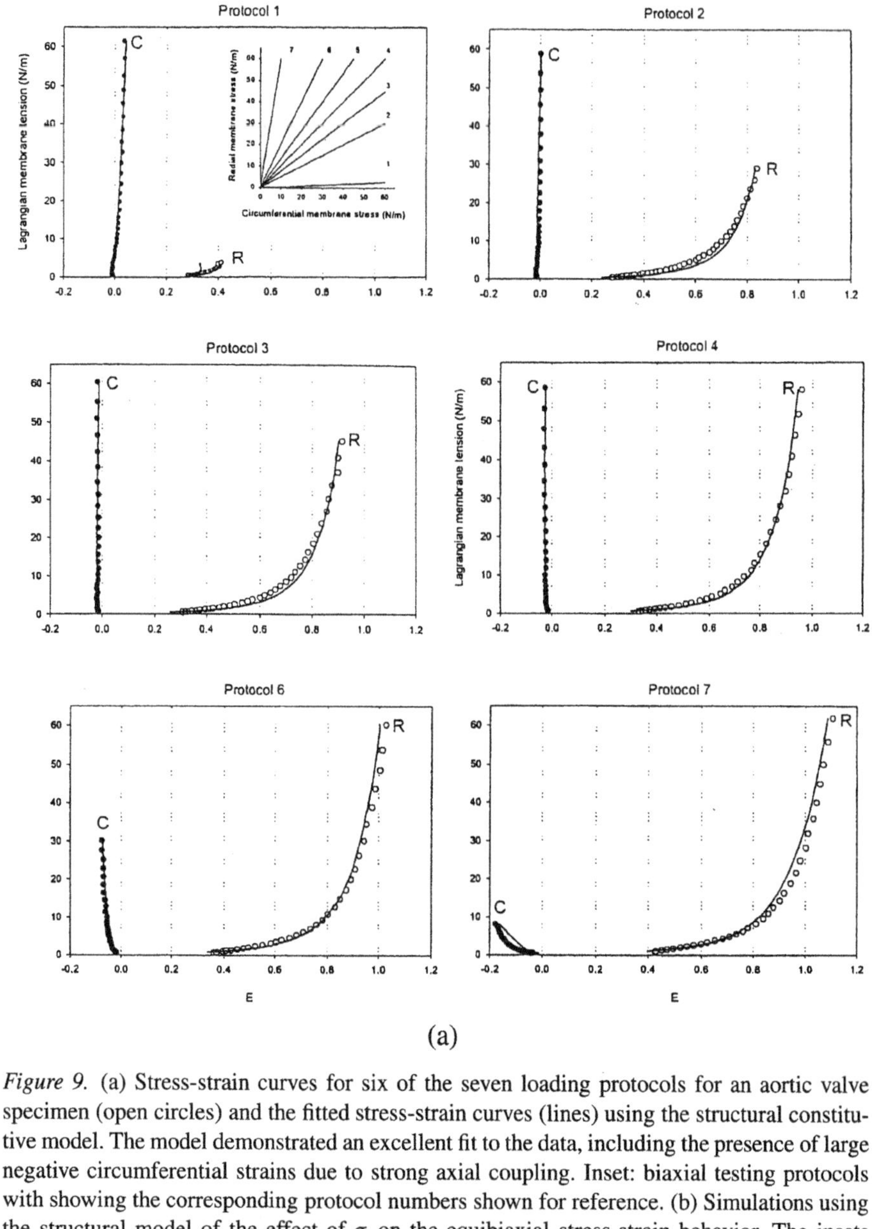

(a)

Figure 9. (a) Stress-strain curves for six of the seven loading protocols for an aortic valve specimen (open circles) and the fitted stress-strain curves (lines) using the structural constitutive model. The model demonstrated an excellent fit to the data, including the presence of large negative circumferential strains due to strong axial coupling. Inset: biaxial testing protocols with showing the corresponding protocol numbers shown for reference. (b) Simulations using the structural model of the effect of σ on the equibiaxial stress-strain behavior. The insets provide a graphical representation of the fiber probability density distribution for each σ value: (a) $\sigma = 90°$ approximately isotropic, (b) $\sigma = 35°$ response qualitatively similar to bovine pericardium, (c) $\sigma = 20°$ the circumferential strains are negative at low equibiaxial tensions, and (d) $\sigma = 10°$ the material behavior is highly anisotropic. The dotted lines indicating zero strain are included to highlight ability of the model to simulate the cross-over to negative strain observed in the pressure fixed cusps subjected to equibiaxial tension.

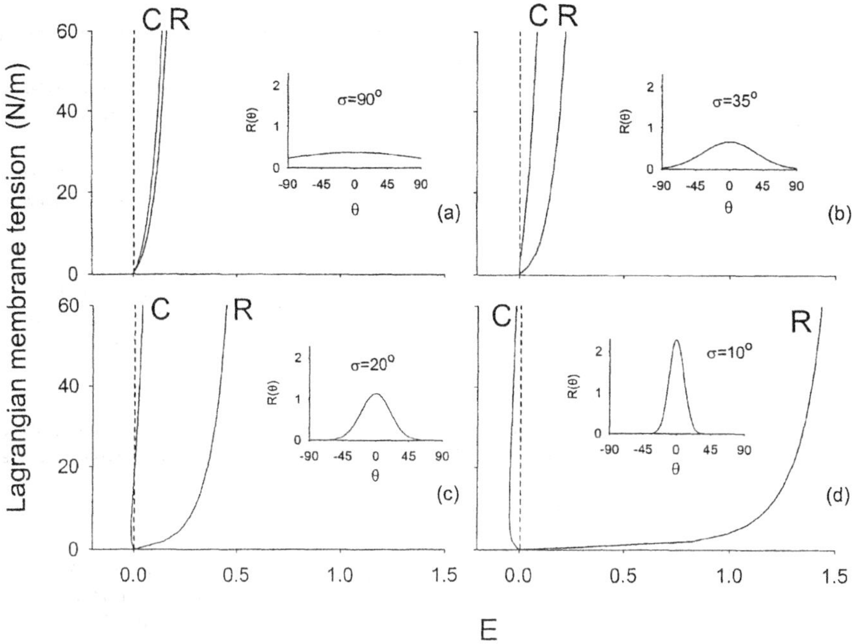

(b)

Figure 9. (Continued.)

Not surprisingly, the resulting equations could not fit the data well due, in part, to the lack of shear and coupling terms. Especially evident was the inability to simulate negative strains. Expanding the number of terms in Q would obviously improve the fit, but would also increase the complexity of the nonlinear optimization and may lead to over-parameterization. Alternatively, since the tests were run under tension control, use of a complementary strain energy density function would also appear to be attractive approach. However, regardless of the particular approach used it is unlikely that a simple analytical function can be found that reproduces the complex behavior with only three parameters. In addition to minimizing the number of parameters, the strength of the current structural approach is underscored by the insight given into the mechanics of the aortic valve cusp (e.g., fiber rotations allowing for the large radial strains).

Another important aspect of the structural approach is that the two distinguishing aspects of the aortic valve cusp biaxial behavior, namely, the extreme mechanical anisotropy and the strong mechanical coupling between the axes, can be explained by the angular distribution of fibers. To more clearly demonstrate this effect, we generated simulations under equibiaxial loading for a given set of A^* and B values by letting σ vary (Figure 9(b)). These simulations indicate that the value of σ is the primary determinant of the biaxial stress-strain response, as shown for (a) nearly random ($\sigma = 90°$), (b) moderately anisotropic ($\sigma = 35°$), (c) highly anisotropic, including contraction along one axis ($\sigma = 20°$), and (d) extremely anisotropic ($\sigma = 10°$). Although we assumed a simplified tissue structure in the formulation of the model, the structural approach highlighted the importance of the angular orientation of the fibers in determining the complex anisotropic tissue mechanical behavior.

4.6. MODIFICATION OF THE STRUCTURAL APPROACH FOR CHEMICALLY TREATED TISSUES

The therapy for aortic valve disease is typically complete heart valve replacement. For patients greater than 60 years in age, treatment with bioprosthetic heart valves is preferred. In the fabrication of bioprosthetic heart valves, native porcine aortic valves or bovine pericardium are chemically treated to reduce immunological reactivity and improve tissue durability. The treated tissue is then mounted into a flexible wire frame (stent) and covered with Dacron cloth to assist in surgical implantation. Although current bioprosthetic heart valves can last up to 15 years, there continues to be a need to improve their durability, which is limited by tissue mineralization and fatigue damage [52]. Clearly, the understanding of biomechanical events that occur during the fatigue process is critical to the improving bioprosthetic heart valve design. This will clearly require a constitutive model for normal and fatigued tissue.

When soft tissues are chemically treated changes in the mechanical properties in both the fibers and ground matrix will occur. Chemical treatments can form exogenous cross-links between collagen fibers and with the protein core of the proteoglycans. This will tend to stiffen the ground matrix, so that the ground matrix can now contribute significant mechanical loads. There is no available data on the mechanical properties of chemically treated matrix, or the effect of chemical treatment on pure collagen. However, we have previously shown that while chemical treatment substantially alters bovine pericardium mechanical properties, the degree of mechanical anisotropy is unchanged [48]. This result suggests that chemically treated ground substance acts as an isotropic, hyperelastic material. Further, while chemical treatment quantitatively alters the stress-strain curve, it remains qualitatively similar.

Based on these considerations, we developed a constitutive model for chemically treated tissues [45] using data from our previous study [48]. Briefly, we assume that the total strain energy of the chemically treated tissue is the sum of the fiber and matrix components, i.e., $W = c_f W_f + (1 - c_f) W_m$, where W_f and W_m are the fiber and matrix strain energies, and c_f is the fiber volume fraction. Thus, the total stresses along the PD and XD axes (Figure 2) are given by

$$\begin{aligned} \mathsf{S}_{PD} &= c_f \mathsf{S}^f_{PD} + (1 - c_f)\mathsf{S}^m_{PD}, \\ \mathsf{S}_{XD} &= c_f \mathsf{S}^f_{XD} + (1 - c_f)\mathsf{S}^m_{XD}, \end{aligned} \tag{32}$$

where S^f and S^m are stresses due to the fiber and matrix components, respectively. The following isotropic strain energy function was found to fit the matrix stress-strain data very well:

$$W_m = \frac{c_0}{2} \exp\left[c_1\left(E_{PD}^2 + E_{XD}^2 + 2E_{PD}E_{XD}\right)\right], \tag{33}$$

where c_0 and c_1 are constants.

The formulation for the structural model follows that for the aortic valve [45] (Section 4.5). However, unlike the aortic valve study we utilized SALS to directly determine $R(\theta)$. This advance is important in that unlike man-made composite systems, the angular orientation of the collagen fibers in soft tissues is not known *a-priori*. To determine this critical structural feature, we used SALS to directly determine the distribution of fiber angles. This is possible due to the property that in SALS the distribution of scattered light, $I(\theta)$, is directly proportional to the angular distribution of fibers [50], where θ is the fiber angle. In our previous study [48], a 2.54-mm rectilinear scanning grid was used resulting in ~625 tests per specimen, with $I(\theta)$ measured using 1° increments. For each specimen, the mean $I(\theta)$ was computed for each specimen by averaging the values of $I(\theta)$ from all test locations. For the constitutive model, the normalized fiber orientation distribution function, $R(\theta)$ is required (see below). $R(\theta)$ is defined such that $R(\theta)\,\mathrm{d}\theta$ is the fraction of fibers oriented between θ and $\theta + \mathrm{d}\theta$, with the normalization constraint $\int_{-\pi/2}^{\pi/2} R(\theta)\,\mathrm{d}\theta = 1$. In this study, $R(\theta)$ computed from the $I(\theta)$–θ curve using

$$R(\theta) = \frac{I(\theta)}{\sum_{\theta=-\pi/2}^{\theta=\pi/2} I(\theta)\Delta\theta}, \tag{34}$$

where, since $I(\theta)$ is measured in discrete 1° increments [50], $\Delta\theta = 1°$.

Using (30)–(32), and assuming $W = c_f W_f + (1 - c_f) W_m$, the final equations for the stresses for chemically treated tissues are given by:

$$\begin{aligned} S_{PD} &= c_f \sum_{-90°}^{90°} \left[R(\theta) S_f(E_f) \cos^2(\theta)\Delta\theta\right] \\ &\quad + (1 - c_f)c_0c_1\,(E_{PD} + E_{XD}) \exp\left[c_1\left(E_{PD}^2 + E_{XD}^2 + 2E_{PD}E_{XD}\right)\right], \\ S_{XD} &= c_f \sum_{-90°}^{90°} \left[R(\theta) S_f(E_f) \sin^2(\theta)\Delta\theta\right] \\ &\quad + (1 - c_f)c_0c_1(E_{XD} + E_{PD}) \exp\left[c_1\left(E_{PD}^2 + E_{XD}^2 + 2E_{PD}E_{XD}\right)\right]. \end{aligned} \tag{35}$$

Since $R(\theta)$ was determined by SALS, only the constants A, B, c_0 and c_1 need be determined. Equation (35) modeled the biaxial response for the chemically treated tissues quite well (see [45] for details). In additional to fitting the total tissue response well, the structural model allowed examination of the isolated contributions of the fibers and the matrix. For example, the fiber stress-strain responses indicated that the chemically treated tissue had a more gradual increase in stiffness. Further, on average the matrix stresses were approximately one half of the fiber stresses at any given strain level.

5. Future Areas

From this review, one can easily see that our understanding the biaxial mechanics of soft tissues is incomplete, and remains a challenging scientific area. There exists many areas where advances in constitutive modeling are required, including the development of experimental techniques and theoretical frameworks for the fundamental characterization of tissue response. The underlying motivations for these studies lie not only in the understanding of natural tissue function, but also in new biomedical applications. One such application is tissue engineering, a new therapeutic approach for the functional restoration of diseased or damaged organs that utilizes cells and related biological factors to generate living tissue replacements. In mechanically demanding applications such as heart valves, the design and development of these tissues will invariably require a detailed understanding of the biomechanical phenomena associated with the growth, development, and engineering of soft tissue replacements. In the following, ongoing issues in several key areas in the biaxial mechanical evaluation of tissues are identified, including those that have had a long-standing need for theoretical contributions.

5.1. UNIQUENESS OF PHENOMENOLOGICAL FORMULATIONS

As demonstrated in Sections 4.5 and 4.6, structural approaches can provide insights into the mechanisms underlying the tissue mechanical behavior. However, there are still many problems to overcome with structural approaches (see below). Phenomenological approaches continue to represent an important approach in the characterization of tissue mechanical behavior. They are inherently simpler in that they do not require an in-depth knowledge of the mechanical behavior of tissue components and mechanical interactions, and are also more amenable to computational applications. For example, in order for novel biologically-derived soft tissues to be utilized in artificial heart valve designs, robust constitutive models are required for computational simulations of valve function. The theoretical approaches required to develop constitutive models for heart valve biomaterials must be able to describe a wide range of combined states of normal and shear stresses.

Some of the issues associated with accomplishing this task are illustrated in our research on heart valve biomaterials. In our original technique to apply in-plane shear in a biaxial test [44], control of the normal Green's strain components alone was used, while the shear strain component E_{12} was allowed to vary freely (Figure 8). While this approach allowed for the application of in-plane shear, the range of stresses and strains were limited. The limitation restricts the ability to accurately simulate heart valve stresses, since they exhibit strains outside those used in [44].

Recently, we modified our biaxial testing device to allow for Lagrangian stress control to allow for a wider range of stress and strain. Here, chemically treated bovine pericardial specimens were prepared as before [44], with the preferred collagen fiber direction oriented 45° to the specimen axes (Figure 6). Seven Lagrange stress (T_{ij}) controlled biaxial test protocols were conducted using a maximum value of 1.8 MPa, with axial stress ratios of $T_{11} : T_{22} = 1 : 0.1$, $1 : 0.5$, $1 : 0.75$, $1 : 1$, $0.75 : 1$, $0.5 : 1$, $0.1 : 1$ were used (Figure 10(a)), which produced wide range of normal and shear stresses (Figure 10(b)). The resulting strains spanned the strain field well, and indicated a complex tissue response (Figure 10(c)).

Although able to accurately simulate data from our initial study [44], (27) did not adequately describe the data from the new stress-controlled protocol (Figures 11(a)–(c)). In particular, it was unable to simulate the complex shear stress response (Figure 11(b)). To aid in developing an alternative model, we developed a method to estimate the strain energy response functions with respect to each strain component. First, the data was fit using an exponential function fit each stress component as a function of the three strain components individually ($r^2 \geqslant 0.99$). Next, the stress component as a function of two strain components while the other was kept at zero. This allowed use to simulate the response of S'_{ij} against various combinations of E'_{ij} and guide the choice of the functional form of Q. All together nine response functions of S'_{ij} vs. E'_{ij} in 3D were generated, which provided useful insights in forming a new strain-energy function. Based on these results, we developed the following alternative Fung model Q:

$$\begin{aligned} Q = {} & \mathrm{A}_1 E_{11}^{\prime 2} + \mathrm{A}_2 E_{22}^{\prime 2} + 2\mathrm{A}_3 E'_{11} E'_{22} + \mathrm{A}_4 E_{12}^{\prime 2} + 2\mathrm{A}_5 E'_{11} E'_{12} + 2\mathrm{A}_6 E'_{12} E'_{22} \\ & + \mathrm{A}_7 E_{12}^{\prime 2} E_{22}^{\prime 2}. \end{aligned} \tag{36}$$

Equation (36) was able describe the data of the 5 "inner" test protocols ($T_{11} : T_{22}$ ratios $0.5 : 1$ through $1 : 0.5$, as shown in Figure 10(a)) with an r^2 of 0.982 (Figures 11(d)–(f)). The predictive capability of (36) was evaluated by extrapolating to the outermost protocols ($T_{11} : T_{22}$ ratios $1 : 0.1$ and $0.1 : 1$, Figure 10(a)), which fit the data reasonably well, especially the shear term (Figures 11(d)–(f)). Further, our experimental approach show a wide range of shear stresses and strains, with peak shear stresses of ±600 kPa and peak shear strains of ±0.2, which is comparable to the magnitudes of the corresponding normal components.

Although (36) was clearly able to "fit" the experimental data well, and that we utilized response functions to guide choosing the correct functional form (*à la* Rivlin [42]), this approach still represents a somewhat ad-hoc methodology. Clearly, the inability for a given constitutive relation to fit a broader range of experimental data for the same biomaterial represents a major pitfall in current constitutive models. Part of this problem may be statistical in nature, particularly with the amount of covariance that exists in current models. This is especially the case in the Fung form (18) where the 2nd Piola–Kirchhoff stresses will always contain an e^Q term, which contains all strain measures. Recently, Criscione et al. [11] proposed a new invariant basis, using natural (logarithmic) strain which yields orthogonal stress

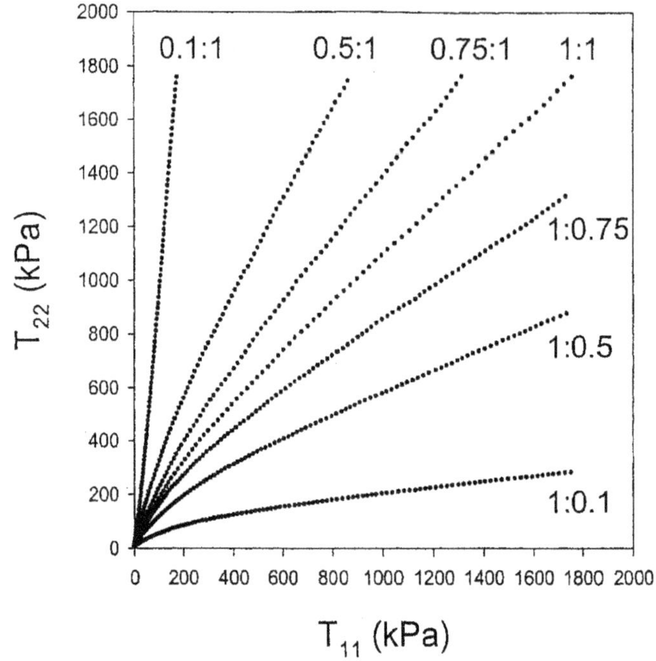

(a)

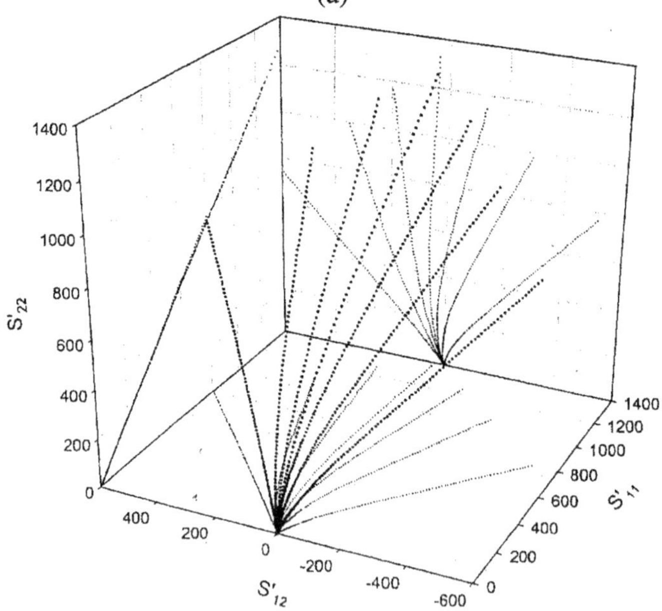

(b)

Figure 10. (a) The Lagrangian stress-controlled biaxial test protocol, which when applied to tissue specimens with their material axes oriented at 45° to the specimen axes produces a wide range of (b) 2nd Piola–Kirchhoff stresses and (c) Green's strains in the material coordinate system (Figure 6(a)), including a large range of shear stresses and strains.

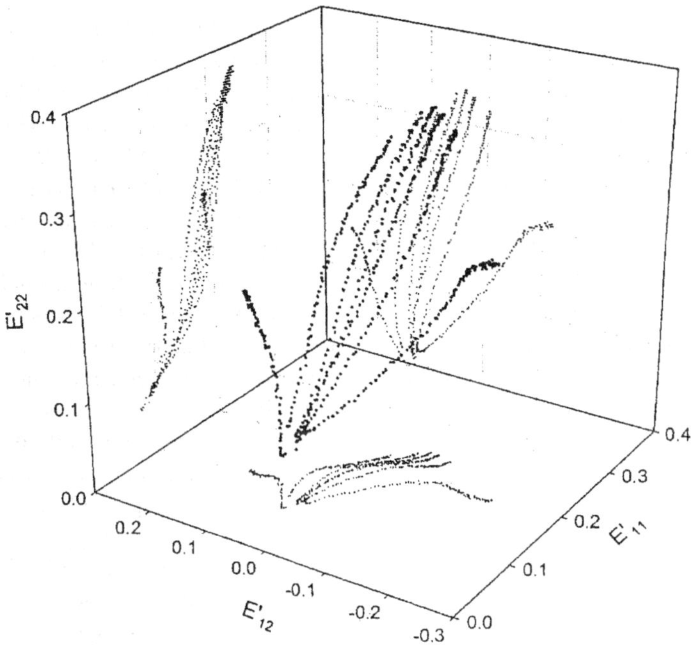

Figure 10. (Continued.)

response terms in isotropic hyperelasticity. In this approach, covariance between terms vanishes since orthogonal bases are used. In addition, the use of natural strain minimizes the dependence on accurate reference states, which are problematic in soft tissues. Thorough testing of this approach is limited to date, but may prove to significantly advance phenomological approaches.

5.2. PRECONDITIONING AND RELATED TIME DEPENDENT PHENOMENA

All experimental and modeling presented herein dealt only with quasi-static behavior. Time-dependent behavior, such as visco-elasticity and poro-elasticity, has not been examined in a biaxial stress-state, largely due to experimental technical limitations. Only one previous time-dependent biaxial experimental study known to the author is by Lee et al. [34]. Lee et al. conducted biaxial-stress relaxation experiments, although the loading times were very slow ($\sim$1 sec) due to experimental limitations. Of any of the time-dependent phenomena, preconditioning is perhaps the least studied and understood, yet paradoxically the most widely observed. As originally stated by Fung [16], preconditioning involves cyclically loading a tissue to the same stress-strain state until its response becomes stabile. Although originally developed for uniaxial loading, preconditioning has been extended to biaxial testing. An example of this is shown in Figure 12 for the urinary bladder wall.

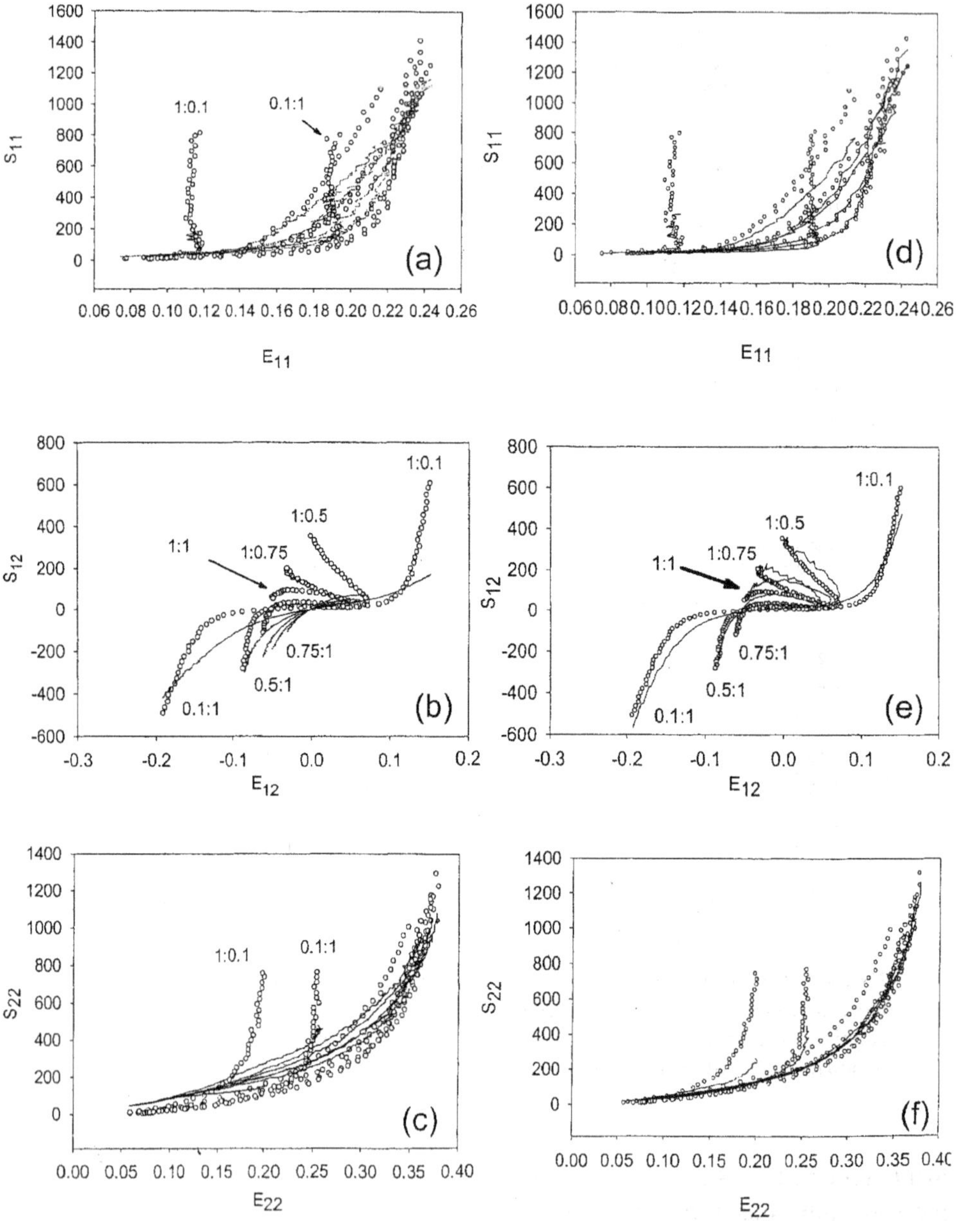

Figure 11. (a)–(c) Biaxial mechanical data from Figure 10 showing the fit of (27), indicating a relatively poor fit. (d)–(f) show the fit using (36), using only the middle five test protocols (Figure 10), and demonstrating much better fit, indicating that (36) is more a robust model. Ratios listed next to each curve indicate the corresponding $T_{11} : T_{22}$ ratio as shown in Figure 10(a).

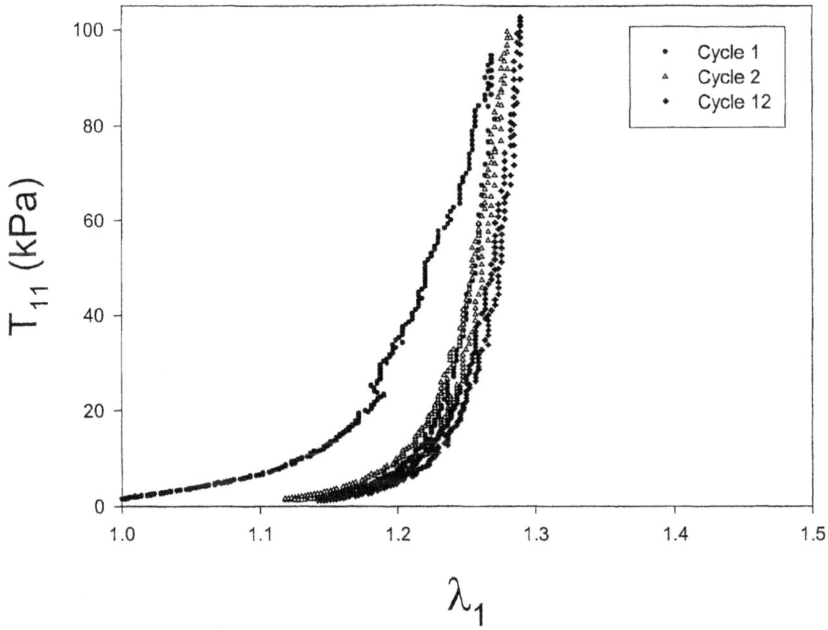

(a)

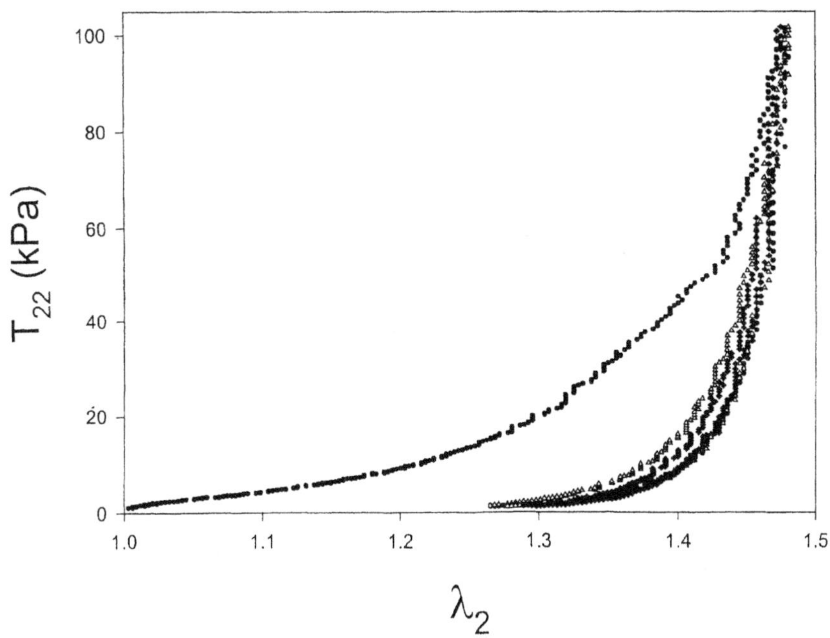

(b)

Figure 12. Demonstration of the preconditioning phenomena for biaxial loading, a specimen extracted from the rat bladder wall. The first cycle indicated a pronounced toe region (i.e., low stress), which disappeared in subsequent cycles. Further, note that cycles 2 and 12 are nearly identical, indicating that the mechanical response stabilized after 12 cycles.

Here, the first cycle is quite different from the second. Virtually no differences occur between the 2nd and 12th cycles, indicating at least short-term stability. However, the mechanisms of the preconditioning phenomena, methods to describe it mathematically, and most importantly the relation between preconditioned tissue response to actual in-vivo properties is unknown.

Our lab has recently completed a set of equibiaxial strain history and stress-relaxation tests on glutaraldehyde-treated bovine pericardium. Testing began with first preconditioning for 15 cycles of equibiaxial stretch ($E_{\text{PD}} : E_{\text{XD}} = 1 : 1$) to the maximum strain level at a strain rate of 1%-sec^{-1}. Next, one of two tests was performed. First, to examine the effects of increasing the maximum strain level, an equibiaxial test was performed with the maximum strain level increased by 5% strain. Following this test, the maximum strain level was reduced back to the previous strain level and another equibiaxial test performed. The specimen was then unloaded and allowed to mechanically recover in the stress-free state for 24 hours, and then a final equibiaxial test performed at the original maximum strain level.

The second test type was a three-hour equibiaxial stress relaxation experiment performed to the maximum strain level using a loading time of 100 msec (a loading rate ~100 times faster than cyclic tests). During the first 10 seconds of the stress relaxation test, video marker tracking was disabled (i.e. no strain data acquired) to allow for high speed data acquisition (~3.3 msec/data point) of load and time. Because specimen strain could not be measured during loading, the applied strain level was prescribed by using the same number of stepper motor steps used in the equibiaxial tests for each axis, but at a pulse rate sufficient for a 100 msec loading time. Afterward, normal data acquisition of the time, video marker tracking, and load signals was re-enabled, allowing for strain level accuracy check and monitoring of the strain throughout the test.

Results of these studies suggest a complex, highly visco-elastic response. When the loading time was decreased to 100 msec, a noticeable increase in peak stress and stiffness was observed in the stress-time response when compared to the normal loading rate data (Figure 13(a)). Another example of visco-elastic nature can be seen in an example of the normalized equibiaxial stress relaxation behavior (Figure 13(b)). There were no differences between PD and XD responses, indicating that the stress-relaxation behavior for glutaraldehyde treated bovine pericardium isotropic. When compared to earlier pilot stress-relaxation data that used a 1.0 second rise time (and run only for one hour), results from the current study (which used a 100 msec rise time) demonstrated substantially greater relaxation (Figure 13(b)). Overall, while biaxially-loaded tissues exhibit much of the same behavior uniaxial ones, the potential for complexity is much greater. For example, we utilized only equibiaxial strain states; non-equibiaxial tests could produce rather complex behavior if the stress-relaxation response is strain-level dependent.

Changes in maximum strain level indicated strain history effects (Figure 14). When the maximum strain level was increased, stress-strain curve did not contin-

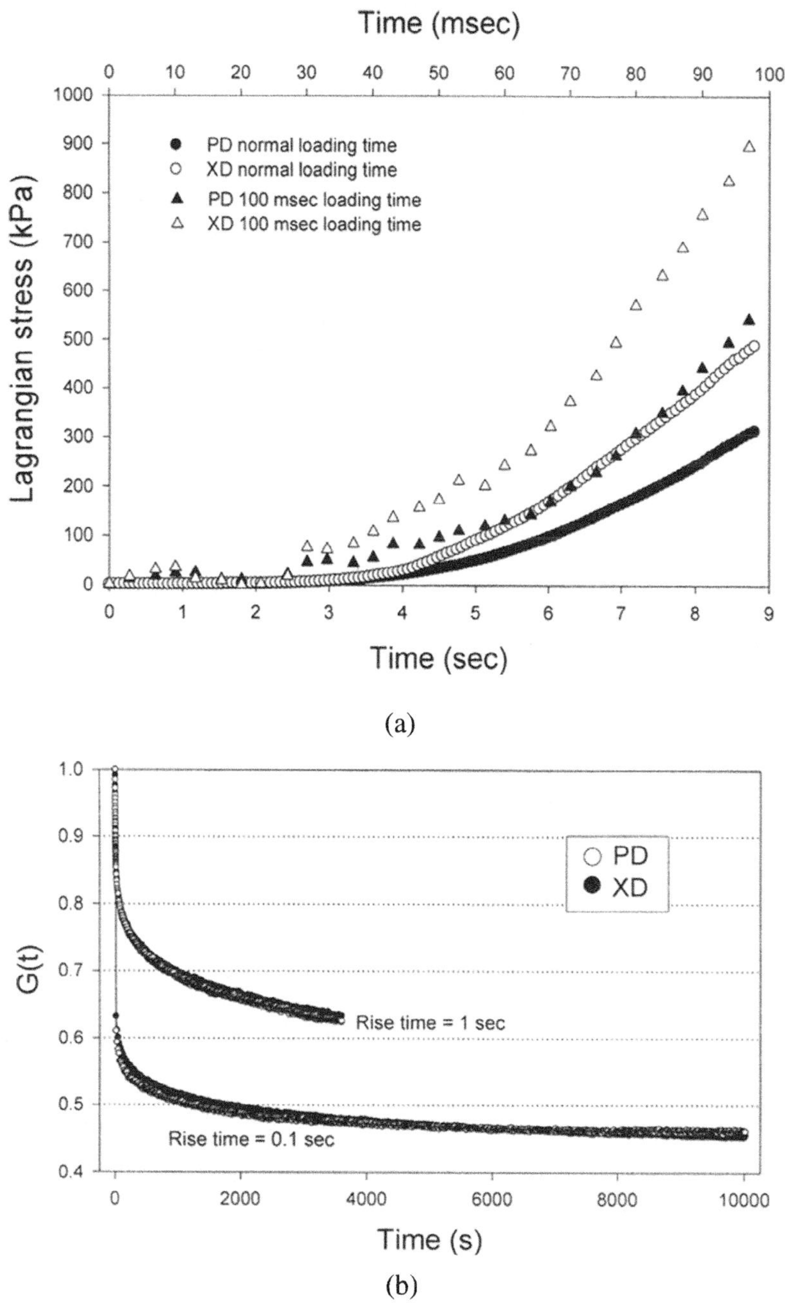

Figure 13. Effects of strain rate on the equibiaxial ($E_{11} = E_{22}$) mechanical properties of chemically treated pericardium. In (a) a change in loading time from 1 sec to 100 msec resulted a near doubling of stress levels. In (b), the same change in loading time resulted in a more pronounced degree of stress-relaxation. Note too that there was no difference between stretch axes (PD, XD), suggesting the viscoelastic response might be isotropic even though the tissue is orthotropic.

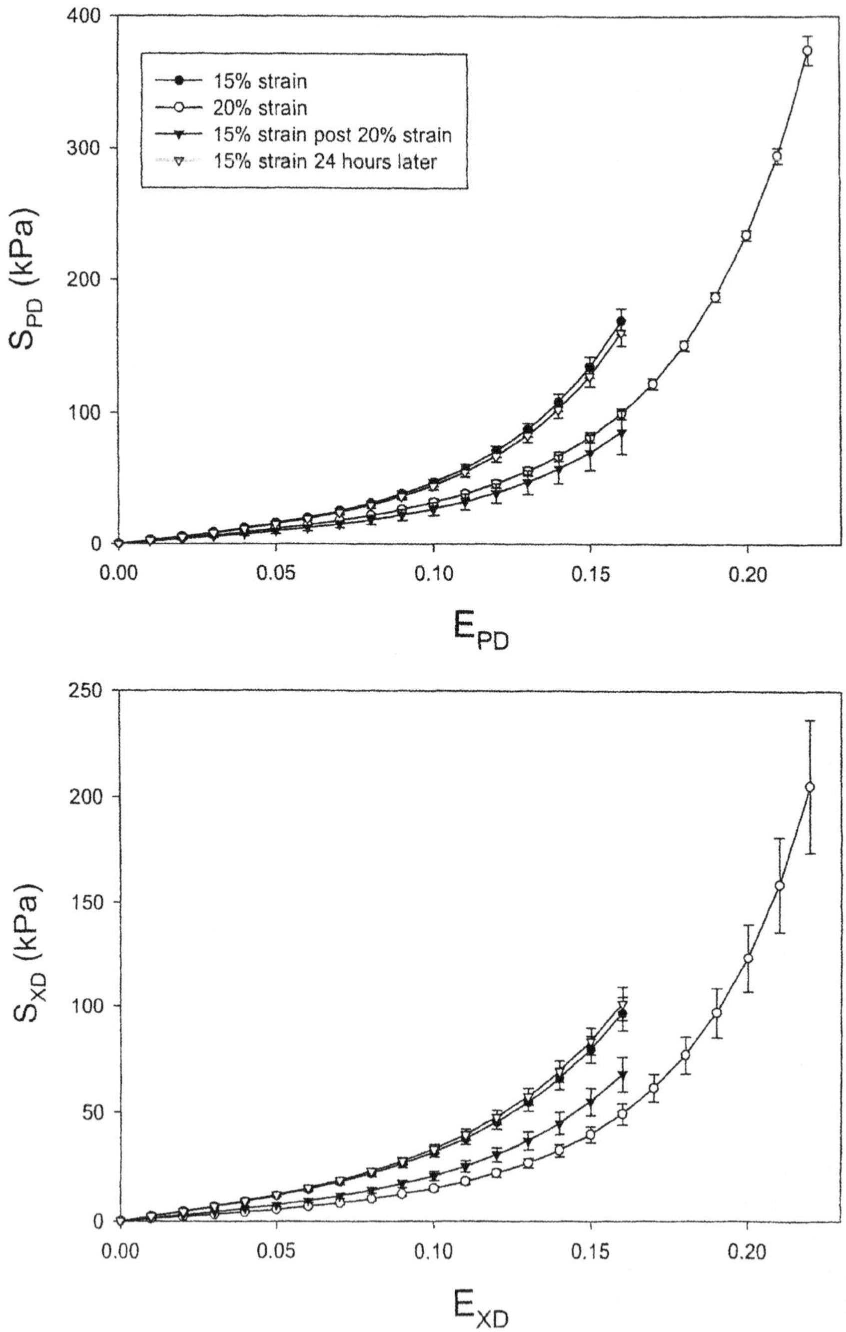

Figure 14. Effects of strain history on chemically treated pericardium first subjected to equibiaxial strain ($E_{11} = E_{22}$) first up to a maximum of 15% strain, followed by 20% peak strain, then a 15% peak strain test for the (a) PD and (b) XD axes. Although both axes demonstrated substantial effects, after 24 hours the original 15% peak strain response was recovered.

ued directly from the 15% strain level data. Instead, the 20% maximum strain level curve formed a new path, with stress levels below that of the 15% maximum strain level data. When the maximum strain level was reduced back to 15%, it generally followed the 20% maximum strain level curve (Figure 14). However, after the 24 hour rest period, the original 15% maximum strain level curve was recovered. This strain-level dependent preconditioned behavior was observed in both the control and chemically treated specimens.

The above results suggested that the exact form of the stress-strain curve is dependent on the preconditioning maximum strain level. To investigate this phenomenon further, we performed another set of biaxial tests where a single maximum strain level was used, 5% larger than the original level (i.e. 20%). When compared to the original test group, these tests demonstrated overlapping identical stress-strain curves (Figure 14). Thus, only changes in maximum strain level from the initial preconditioned value appear to alter the stress-strain curve.

5.3. THE FUTURE OF STRUCTURAL APPROACHES

In addition to the above issues, fundamental questions remain as to the optimal approaches for constitutive modeling of tissues. The structural approach taken by us represents a particularly attractive approach in that it can be used to help explain material behavior and minimize the number of parameters. However, critical to the formulation of any structurally guided model is knowledge of how the fibers deform in response to global tissue strains. Prior models (including ours) assume an affine deformation, where the fibers are assumed to rotate and stretch in the same way as the bulk tissue. When a tissue is chemically treated, and especially when it is bonded together to form biocomposites, fiber mobility behavior may change.

As example of the level of complexity that the fiber rotations can have within even relatively simply planar tissues, we utilized our integrated SALS/biaxial device [3] to examine how the collagen fibers rotate under biaxial strains for small intestinal submucosa (SIS), a thin (<100 μm) collagenous tissue. Three native SIS samples were prepared as per our established methods [49]. Each specimen was first SALS scanned before deformation and at three stretch levels of $\lambda_1 : \lambda_2 = 1.12 : 1.12, 1.30 : 1$, and $1 : 1.30$. The fiber distribution after an equibiaxial stretch of $1.12 : 1.12$ and $1.30 : 1$ were both very similar to the corresponding fiber distribution computed from affine deformation theory assumption (Figure 15(a)). However, in the sample stretched to $1 : 1.30$ the single fiber distribution split into a dual fiber distribution with peaks at ~50° from the center of the single distribution (Figure 15(b)).

The basis for this unusual behavior can be explained as follows. It has been shown that SIS consists of two fiber populations oriented at ±30 degrees from the preferred fiber direction [40, 41, 49]. We have also demonstrated that the two fiber populations usually overlap sufficiently so that only a single fiber population is detectable [49]. The current results indicate that under large circumferential strains,

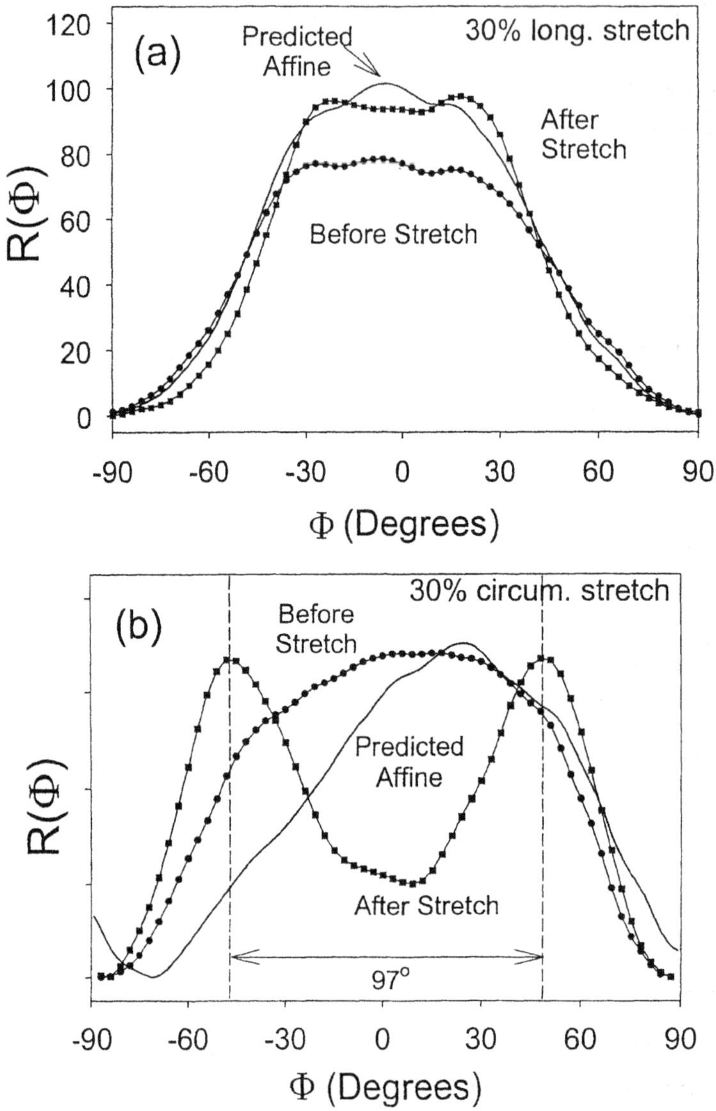

Figure 15. The angular distribution of collagen fibers of a SIS specimen determined before (circles) and after (squares) 30% stretch in the (a) preferred and (b) cross-preferred fiber directions. The separation of the single fiber population into dual populations after stretch was poorly predicted by the affine deformation model (lines).

the two populations rotate with respect to each other and become distinguishable. These results suggest non-affine fiber kinematics for native SIS. Clearly, a more in-depth understanding of the tissue fiber kinematics is required before realistic structural models can be developed for native tissues.

5.4. CLOSURE

Besides the questions raised in this section, many other ones need consideration. For example, theories that encompass growth of tissues under multi-axial stress states will be paramount in tissue engineering applications. Another challenging area is the constitutive modeling of actively contracting tissues. The little experimental data available and reviewed herein on active biaxial mechanical properties clearly indicate that one cannot simply model the active state as a change in reference configuration. Instead, complex changes in the constitutive behavior of the material as well need to be accounted for. Finally, the basic tenets of biomechanics as applied to biaxial testing, such as pseudo-elasticity, need to be put on firmer ground by establishing both underlying causes and better approaches to modeling and simulation.

Acknowledgments

The author would like to thank Dr. Steven C. Cowin for his kind invitation to write this review. In addition, the author would like to acknowledge the National Science Foundation, the Whitaker Foundation, and the American Heart Association for their generous financial support of the author's research presented in this review.

References

1. R. Abeyaratne, C. Chu and R.D. James, Kinetics of materials with wiggly energies: theory and application to the evolution of twinning microstructures in a Cu-Al-Ni shape memory alloy. *Philos. Mag. A* **73**(2) (1996) 457–497.
2. K.J. Bathe, *Finite Elements Proceedures in Engineering Analysis*. Prentice-Hall, Englewood Cliffs, NJ (1982).
3. K.L. Billiar and M.S. Sacks, A method to quantify the fiber kinematics of planar tissues under biaxial stretch. *J. Biomech.* **30**(7) (1997) 753–756.
4. K.L. Billiar and M.S. Sacks, Biaxial mechanical properties of fresh and glutaraldehyde treated porcine aortic valve cusps: Part II – A structurally guided constitutive model. *J. Biomech. Engrg.* **122**(4) (2000) 327–335.
5. K.L. Billiar and M.S. Sacks, Biaxial mechanical properties of the natural and glutaraldehyde treated aortic valve cusp – Part I: Experimental results. *J. Biomech. Engrg.* **122**(1) (2000) 23–30.
6. M.M. Black, I.C. Howard, X.C. Huang and E.A. Patterson, A three-dimensional analysis of a bioprosthetic heart valve. *J. Biomech.* **24** (1991) 793–801.
7. L.J.Brossollet and R.P. Vito, A new approach to mechanical testing and modeling of biological tissues, with application to blood vessels. *J. Biomech. Engrg.* **118**(November) (1996) 433–439.
8. P.H. Chew, F.C. Yin and S.L. Zeger, Biaxial stress-strain properties of canine pericardium. *J. Mol. Cell Cardiol.* **18**(6) (1986) 567–578.
9. H.S. Choi and R.P. Vito, Two-dimensional stress-strain relationship for canine pericardium. *J. Biomech. Engrg.* **112**(2) (1990) 153-159.
10. C. Chu and R.D. James, Biaxial loading experiments on Cu-Al-Ni single crystals. In: K.S. Kim (ed.), *Experiments in Smart Materials and Structures*, ASME–AMD (1993) pp. 61–69.

11. J.C. Criscione, J.D. Humphrey, A.D. Douglas and W.C. Hunter, An invariant basis for natural strain which yields orthogonal stress response terms in isotropic hyperelasticity. *J. Mech. Phys. Solids* **48** (2000) 2445–2465.
12. L.L. Demer and F.C.P. Yin, Passive biaxial mechanical properties of isolated canine myocardium. *J. Physiology* **339** (1983) 615–630.
13. H. Demiray, A note on the elasticity of soft biological tissues. *J. Biomech.* **5** (1972) 308–311.
14. J. Downs, H. Halperin, J. Humphrey and F. Yin, An improved video-based computer tracking system for soft biomaterials testing. *IEEE Trans. Biomed. Engrg.* **37** (1990) 903–907.
15. D. Flynn, G. Peura, P. Grigg and A. Hoffman, A finite element based method to determine the properties of planar soft tissue. *J. Biomech. Engrg.* **120** (1998) 202–210.
16. Y.C. Fung, *Biomechanics: Mechanical Properties of Living Tissues*, 2nd ed. Springer, New York (1993).
17. M.S. Hamid, H.N. Sabbah and P.D. Stein, Influence of stent height upon stresses on the cusps of closed bioprosthetic valves. *J. Biomech.* **19** (1986) 759–769.
18. A.H. Hoffman and P. Grigg, A method for measuring strains in soft tissue. *J. Biomech.* **10** (1984) 795–800.
19. J. Humphrey, D. Vawter and R. Vito, Quantification of strains in biaxially tested soft tissues. *J. Biomech.* **20**(1) (1987) 59–65.
20. J.D. Humphrey, R.K. Strumpf and F.C. Yin, Determination of a constitutive relation for passive myocardium: I. A new functional form. *J. Biomech. Engrg.* **112**(3) (1990) 333–339.
21. J.D. Humphrey, R.K. Strumpf and F.C. Yin, Determination of a constitutive relation for passive myocardium: II. Parameter estimation. *J. Biomech. Engrg.* **112**(3) (1990) 340–346.
22. J.D. Humphrey, R.K. Strumpf and F.C.P. Yin, A constitutive theory for biomembranes: application to epicardial mechanics. *J. Biomech. Engrg.* **114** (1992) 461–466.
23. J.D. Humphrey, D.L. Vawter and R.P. Vito, Mechanical behavior of excised canine visceral pleura. *Ann. Biomed. Engrg.* **14** (1986) 451–466.
24. J.D. Humphrey, D.L. Vawter and R.P. Vito, Quantification of strains in biaxially tested soft tissues. *J. Biomech.* **20** (1987) 59–65.
25. J.D. Humphrey and F.C. Yin, Biaxial mechanical behavior of excised epicardium (published erratum appears in *J. Biomech. Engrg.* **111**(3) (1989) 227). *J. Biomech. Engrg.* **110**(4) (1988) 349–351.
26. P. Khalsa, A. Hoffman and P. Grigg, Mechanical states encoded by stretch-sensitive neurons in feline joint capsule. *J. Neurophysiology* **76**(1) (1996) 175–187.
27. S. Krucinski, I. Vesely, M.A. Dokainish and G. Campbell, Numerical simulation of leaflet flexure in bioprosthetic valves mounted on rigid and expansile stents. *J. Biomech.* **26** (1993) 929–943.
28. Y. Lanir, A structural theory for the homogeneous biaxial stress-strain relationships in flat collageneous tissues. *J. Biomech.* **12** (1979) 423–436.
29. Y. Lanir, Constitutive equations for fibrous connective tissues. *J. Biomech.* **16** (1983) 1–12.
30. Y. Lanir and Y.C. Fung, Two-dimensional mechanical properties of rabbit skin. I. Experimental system. *J. Biomech.* **7**(1) (1974) 29–34.
31. Y. Lanir and Y.C. Fung, Two-dimensional mechanical properties of rabbit skin. II. Experimental results. *J. Biomech.* **7**(2) (1974) 171–182.
32. Y. Lanir, O. Lichtenstein and O. Imanuel, Optimal design of biaxial tests for structural material characterization of flat tissues. *J. Biomech. Engrg.* **118** (1996) 41–47.
33. J.M. Lee, D.W. Courtman and D.R. Boughner, The glutaraldehyde-stablized porcine aortic valve xenograft. I. Tensile viscoelastic properties of the fresh leaflet material. *J. Biomedical Mater. Res.* **18** (1984) 61–77.
34. J.M. Lee, M.C. LeWinter, G. Freeman, R. Shabetai and Y.C. Fung, Biaxial mechanical properties of the pericardium in normal and volume overload dogs. *Amer. J. Physiology* **249** (1985) H222–H230.

35. D.H. Lin and F.C. Yin, A multiaxial constitutive law for mammalian left ventricular myocardium in steady-state barium contracture or tetanus. *J. Biomech. Engrg.* **120**(4) (1998) 504–517.
36. K. May-Newman and F.C. Yin, Biaxial mechanical behavior of excised porcine mitral valve leaflets. *Amer. J. Physiology* **269**(4 Pt 2) (1995) H1319–H1327.
37. K. May-Newman and F.C. Yin, A constitutive law for mitral valve tissue. *J. Biomech. Engrg.* **120**(1) (1998) 38-47.
38. P.M.F. Nielsen, P.J. Hunter and B.H. Smaill, Biaxial testing of membrane biomaterials: Testing equipment and procedures. *J. Biomech. Engrg.* **113** (1991) 295–300.
39. V.P. Novak, F.C. Yin and J.D. Humphrey, Regional mechanical properties of passive myocardium. *J. Biomech.* **27**(4) (1994) 403–412.
40. J. Orberg, E. Baer and A. Hiltner, Organization of collagen fibers in the intestine. *Connective Tissue Res.* **11** (1983) 285–297.
41. J.W. Orberg, L. Klein and A. Hiltner, Scanning electron microscopy of collagen fibers in intestine. *Connective Tissue Res.* **9** (1982) 187–193.
42. R.S. Rivlin, Large elastic deformations of isotropic materials IV. Further developments of the general theory. *Philos. Trans. Roy. Soc. London Ser. A* **241** (1948) 379–397.
43. R.S. Rivlin and D.W. Saunders, Large elastic deformations of isotropic materials, VII. Experiments on the deformation of rubber. *Philos. Trans. Roy. Soc. London Ser. A* **243** (1951) 251–288.
44. M.S. Sacks, A method for planar biaxial testing that includes in-plane shear. *J. Biomech. Engrg.* **121** (1999) 551–555.
45. M.S. Sacks, A structural constitutive model for chemically treated planar connective tissues under biaxial loading. *Comput. Mech.* **26**(3) (2000) 243–249.
46. M.S. Sacks and C.J. Chuong, Biaxial mechanical properties of passive right ventricular free wall myocardium. *J. Biomech. Engrg.* **115** (May 1993) 202–205.
47. M.S. Sacks and C.J. Chuong, A constitutive relation for passive right-ventricular free wall myocardium. *J. Biomech.* **26**(11) (1993) 1341–1345.
48. M.S. Sacks and C.J. Chuong, Orthotropic mechanical properties of chemically treated bovine pericardium. *Ann. Biomed. Engrg.* **26**(5) (1998) 892–902.
49. M.S. Sacks and D.C. Gloeckner, Quantification of the fiber architecture and biaxial mechanical behavior of porcine intestinal submucosa. *J. Biomedical Mater. Res.* **46** (1999) 1–10.
50. M.S. Sacks, D.B. Smith and E.D. Hiester, A small angle light scattering device for planar connective tissue microstructural analysis. *Ann. Biomed. Engrg.* **25**(4) (1997) 678–689.
51. M.S. Sacks, D.B. Smith and E.D. Hiester, The aortic valve microstructure: Effects of transvalvular pressure. *J. Biomedical Mater. Res.* **41** (1998) 131–141.
52. F. Schoen and R. Levy, Tissue heart valves: Current challenges and future research perspectives. *J. Biomedical Mater. Res.* **47** (1999) 439–465.
53. A. Shacklock, Biaxial testing of cardiac tissue. In: M.S. University of Auckland, Auckland, New Zealand (1987).
54. R.K. Strumpf, L. Demer and F.C.P. Yin, Quantification of passive biaxial myocardial stress-strain properties. *Circulation* **68** (1983) 372.
55. R.K. Strumpf, J.D. Humphrey and F.C. Yin, Biaxial mechanical properties of passive and tetanized canine diaphragm. *Amer. J. Physiology* **265**(2 Pt 2) (1993) H469–H475.
56. P. Tong and Y.C. Fung, The stress-strain relationship for the skin. *J. Biomech.* **9**(10) (1976) 649–657.
57. L.R.G. Treloar, *The Physics of Rubber Elasticity*. Oxford Univ. Press, London (1975).
58. R.P. Vito, The mechanical properties of soft tissues – I. A mechanical system for biaxial testing. *J. Biomech.* **13** (1980) 947–950.
59. S.D. Waldman and J.M. Lee, Boundary conditions during biaxial testing of planar connective tissues: Part 1: Dynamic behavior. *J. Biomech.* (2000) submitted.

60. S.D. Waldman, M.S. Sacks and J.M. Lee, Imposed state of deformation determines local collagen fibre orientation but not apparent mechanical properties. *Biomed. Sci. Instrum.* **35** (1999) 51–56.
61. A.S. Wineman, Large axisymmetric inflation of a nonlinear viscoelastic membrane by laterial pressure. *Trans. Soc. Rheology* **20**(2) (1976) 203–225.
62. W.H. Yang and C.H. Lu, General deformations of neo-hookean membranes. *J. Appl. Mech.* (1973 March) 1–12.
63. F.C. Yin, R.K. Strumpf, P.H. Chew and S.L. Zeger, Quantification of the mechanical properties of noncontracting canine myocardium under simultaneous biaxial loading. *J. Biomech.* **20**(6) (1987) 577–589.
64. F.C. Yin, W.R. Tompkins, K.L. Peterson and M. Intaglietta, A video-dimension analyzer. *IEEE Trans. Biomed. Engrg.* **19**(5) (1972) 376–381.
65. F.C.P. Yin, P.H. Chew and S.L. Zeger, An approach to quantification of biaxial tissue stress-strain data. *J. Biomech.* **19** (1986) 27–37.
66. P. Zioupos and J.C. Barbenel, Mechanics of native bovine pericardium: I. The multiangular behavior of strength and stiffness of the tissue. *Biomaterials* **15** (1994) 366–373.

GPSR Compliance

The European Union's (EU) General Product Safety Regulation (GPSR) is a set of rules that requires consumer products to be safe and our obligations to ensure this.

If you have any concerns about our products, you can contact us on ProductSafety@springernature.com

In case Publisher is established outside the EU, the EU authorized representative is:

Springer Nature Customer Service Center GmbH
Europaplatz 3
69115 Heidelberg, Germany

Batch number: 10614940

Printed by Printforce, the Netherlands